Non-synaptic Interactions Between Neurons: Modulation of Neurochemical Transmission

Non-synaptic Interactions Between Neurons:
Modulation of Neurochemical Transmission

Pharmacological and Clinical Aspects

E. Sylvester Vizi
Institute of Experimental Medicine
Hungarian Academy of Science,
Postgraduate Medical School
Budapest

A Wiley–Interscience Publication

JOHN WILEY & SONS
Chichester · New York · Brisbane · Toronto · Singapore

Copyright © 1984 by John Wiley & Sons Ltd.

Library of Congress Cataloging in Publication Data:

Vizi, E. S.
  Non-synaptic interactions between neurons.

  'A Wiley–Interscience publication.'
  Includes bibliographical references and index.
   1. Neural transmission.   2. Neurotransmitters.
I. Title.
QP364.5.V59 1984   599'.0188   83–21605
ISBN 0 471 90378 7

British Library Cataloguing Data:

Vizi, E.
  Non-synaptic interactions between neurons.
  1. Neurotransmittors
  I. Title
  511'.0188   QP3647
ISBN 0 471 90378 7

Phototypeset by Input Typesetting Ltd., London SW19 8DR
Printed by St. Edmundsbury Press Ltd., Suffolk.

For Vera

Contents

Acknowledgements

This book is dedicated to all who have stimulated, encouraged, and helped the author in his work. Among his scientific colleagues and friends the author would like to mention, first, Sir William Paton and the group of colleagues associated with him. The years spent in the Pharmacological Department of Oxford University in the 1960s, at that time a centre with many brilliant scientists, had a decisive impact on the author's scientific education.

But it is also fitting to mention the author's *alma mater*, the Department of Pharmacology of Semmelweis University, Budapest (Director: Joseph Knoll), where he spent a very stimulating period of his life learning the importance of bearing in mind the interrelationship between the part and the whole in biological systems before any conclusion can be drawn from experimental data.

E. SYLVESTER VIZI
Budapest, 1983

Preface

The elucidation of neuronal circuitry has been a major goal in neuroscience, and our understanding of how neuronal networks function largely depends on our knowledge of the ways in which neurons are related to each other.

About a hundred years ago, neuroanatomists, including Ramón y Cajal, For, and His, realized that nerve cells were independent units, and it was proposed that neurons must function by making connections with each other. Light-microscopic techniques, which had established the character of neuronal systems before the beginning of the twentieth century, also revealed the existence of knobs, baskets, or boutons in neuronal systems, and that there are areas of contact between neurons.

The first reference to the concept of the synapse was given by Sherrington:

> . . . It seems therefore likely that the nexus between neurone and neurone . . ., involves a surface of separation between neurone and neurone: and this as a transverse membrane across the conductor must be an important element in intercellular conduction.
> . . . In view, therefore, of the probable importance physiologically of this mode of nexus between neurone and neurone, it is convenient to have a term for it. The term introduced has been synapse. (Charles S. Sherrington, *The Integrative Action of the Nervous System*, 1906)

Since Sherrington's classic work it has become a statement of neurophysiology that the synapse, the 'surface of separation' between neurons, is the primary site of neuronal information-processing. It is now accepted that intercellular communication in the nervous system generally involves the release of chemical transmitters from such synapses. The synapse is the most common and generally accepted structural basis for the interaction between neurons, and it provides a one-way communication system between them.

However, there is another possibility for interneuronal communication; that one neuron communicates with many others without making synaptic contact. In the past few years neurochemical, morphological, and pharmacological evidence has shown that some neurotransmitters, or modulators, may be released from both synaptic and non-synaptic sites for diffusion to target cells more distant than those observed in conventional synaptic transmission. This would be a transitional form of communication between discrete classical neurotransmission and the relatively non-specific neuroendocrine secretion. This type of communication between neurons is different from that given by our traditional views on neurochemical transmission: for example, a chemical substance released from varicosities with no synaptic contact would reach its target cell only by diffusion over a distance of some microns. It appears therefore that chemical interaction between nerve cells does not only take place across the gap between pre- and postsynaptic membranes in well-defined synapses but may also occur in the absence of such specialized contacts (i.e. non-synaptically).

In the past few years neurochemical, anatomical, pharmacological, and neurophysiological observations have suggested that the amount of transmitter released at nerve terminals may be influenced by modulators released from other axonal varicosities or from dendrites or axons. In addition, recent developments have shown that some chemical signals released from nerve terminals do not act as phasic transmitters but rather as tonic modulators (neuromodulators, neurohormones). It is likely, therefore, that in the complex arrangements of the various regions of the central nervous system, a variety of combinations of neurons can act upon each without making synaptic contact. The modulator may be released locally (not necessarily from the axon terminal), and diffuse within a large amount of nervous tissue, and the effects of this reaction depend upon the properties of the nearby cells that may or may not have receptors sensitive to the modulator, thus allowing discrimination. Anatomical observations support this possible modulatory action. For instance, a large number of dopaminergic, noradrenergic, and serotonergic neurons projecting to various brain structures have failed to show the membrane specializations which characterize synaptic junctions.

During recent years, considerable evidence has been discovered in favour of interactions between neurons through pre-synaptic receptors. It has been shown that there are pharmacological differences in sensitivity of pre- and post-synaptic receptors to agonists and antagonists. For example, selective activation or inhibition of pre-synaptic modulatory receptors can be achieved with drugs which have a negligible effect on the post-synaptic receptors. In clinical practice there are drugs which have been developed to utilize the pre-synaptic modulation of neurochemical transmission.

At a time when research in neuroscience is rapidly expanding and is being documented in the literature it is important to summarize new results and concepts. Therefore this book will not deal with classical neurotransmission or with neurohumoral interactions. Both are adequately discussed in the literature. Instead we will seek to summarize data and concepts related to non-synaptic interactions between neurons and the pre-synaptic modulation of transmitter release, taking into account morphological, electrophysiological, neurochemical, and pharmacological evidence.

E. Sylvester Vizi
Budapest, 1983

Foreword

It is a pleasure to be asked by my old colleague, Sylvester Vizi, to write a foreword to this book. Since we worked together 15 years ago, there has been an enormous growth in knowledge of the factors modifying transmitter release. For a time, the term 'modulation' left me uneasy. A cynic might say that it was simply a means of dignifying by an impressive name an effect on transmitter output whose physiological significance was not understood. But today, so numerous and sometimes so substantial are these effects, that a term of this sort is necessary; and with the availability of receptor-binding techniques, of a much wider range of drugs with new specificies, and of great advances in knowledge of ultrastructure, it is beginning to be possible to give both functional and structural meaning to the word in its different contexts.

This book is concerned particularly with one of these new insights—that modulation of transmitter release by the same or another transmitter may include that by 'action at a distance', not involving a direct synaptic contact yet not so diffuse in effect as a hormone acting systematically. One can only be grateful that Professor Vizi has brought together both his own distinguished work and that of others, to allow us to see the implications of such a mechanism. One welcomes, too, his proposal to clarify discussion by distinguishing three types of process: neurotransmission, neuromodulation, and neurohormonal action. Perhaps 'neuromodulation' is of especial interest to the pharmacologist. One of his great problems is that while, with his drugs, he usually wishes to hit sharply specific targets, yet in administering them it is hard to avoid peppering the whole body with 'magic bullets'. But one might hope that, with the more diffused target of neuromodulation, then at least the distortions introduced by our indiscriminate aim may be reduced. It will be very interesting to see both the physiological and the therapeutic outcome of this approach, with all the potentialities it offers.

W. D. M. Paton

Chapter 1

The history and meaning of chemical neurotransmission

Early in the nineteenth century Du Bois-Reymond showed that a flow of electric current was used in the transmission of impulses between nerve and muscle (Du Bois-Reymond, 1848). Although the eighteenth-century idea of 'animal electricity' had a lasting influence on nineteenth-century scientific thought, as early as 1877 Du Bois-Reymond suggested that there was a chemical (NH_4 or lactic acid) transmission across neuromuscular junctions. It is interesting to note that the philosopher, mathematician, and physiologist Descartes (1596–1680) regarded the peripheral nerves as tubes and believed that the nerve inflated the muscle, like a balloon, with a 'spiritous gas'. He was, in fact, the first to suggest the possibility of transmission between nerve and muscle (Descartes, 1644).

The idea of chemicals being involved in neuronal communication is usually credited to T. R. Elliott (1904), who suggested that stimulation of peripheral autonomic nerves might release amounts of a chemical substance that produced effects on a target cell. Elliott put forward the proposal that 'adrenaline might be the chemical stimulant liberated on each occasion when the impulse arrives at the periphery'. This suggestion was not received with any great enthusiasm and his discovery remained for others to prove. The concept of chemical neurotransmission was extended to parasympathetic nerves by Dixon (1906), who noted the similarities between acetylcholine (ACh) and parasympathetic neuronal activity. Otto Loewi (1921) discovered that the vagal inhibition of the cardiac muscle was mediated by a chemical substance that was later identified as ACh. Since chemical transmission to the heart was found to be slow, it was suggested that the shorter transmission process was not chemical but mediated by electrical currents. Until an ultrastructural analysis of the heart had been made it was not known that a vagal innervation of the heart resulted in a much more remote release of ACh compared with the very close contact of the neuromuscular synapse. Since ACh is released from axon terminals of the vagal nerve which do not make synaptic contact it takes longer for the ACh to reach the heart muscle cells. Dale (1934) suggested that the differences in responses (fast and slow) by the effector cells were due to the difference in the morphological arrangement of the innervation: fast responses by the effector cells are due to a very close proximity of the releasing axon terminals and the slow responses to a remote release.

Nevertheless, the notion of chemical transmission was not accepted. The scientific community was not yet prepared for such a revolutionary idea as chemical transmission. Subsequently, Henry Dale and his co-workers (Brown, Feldberg, Gaddum, and MacIntosh) showed that ACh is not only the transmitter at neuroeffector junctions but that it also transmits impulses between neurons.

In the 1940s Eccles and his colleagues (Eccles *et al.*, 1942) studied the endplate potential of a curarized muscle and the effect of cholinesterase inhibitors. They showed that even the initial fast transmission, as well as the slow response, was due to ACh. Eccles, who pioneered the understanding of electrical transmission, '. . . went on record in complete support of ACh transmission' (Eccles, 1982).

The next significant step in neurophysiology was the discovery of noradrenaline (NA) as a transmitter substance for noradrenergic nerves (von Euler, 1946). With this information, scientists began to understand the chemical transmission of nerve impulses. Through extensive work pioneered by Eccles and his colleagues (cf. Eccles, 1982), much became known about the mechanism of synaptic transmission.

Because synaptic communication between cells has been the most comprehensively studied form of intercellular communication, it has been generally accepted that it is the only form of interaction, and that any changes in neuronal excitability are the result of electrical or synaptic transmission.

1.1 CHEMICAL INTERACTION BETWEEN NERVE CELLS: TRANSMISSION

The hypothesis that synaptic communication might utilize specific chemical agents, as first suggested by Elliott (1904), has now been generally accepted, and it seems likely that chemical interaction between nerve cells not only occurs across the gap between closely opposed pre- and post-synaptic membranes in well-defined synapses but may also occur in the absence of such specialized contacts (i.e. non-synaptically).

In contrast to skeletal muscle, there is a considerable diversity of innervation of smooth muscle cells (Burnstock, 1970). The typical axon innervating smooth muscle is studded with varicosities which represent the *en-passage* release sites. The neuro-effector transmission sites of the vegetative nervous system are examples of chemical transmission without synaptic specialization. Similar non-synaptic boutons were found in the central nervous system. Therefore transmitters could be grouped into synaptic (conventional) and non-synaptic (unconventional).

Recent developments have stressed that some chemical signals released from nerve terminals do not act as phasic transmitters but rather as tonic modulators (neuromodulators, neurohormones) (cf. Iversen, 1978; Eccles and McGeer, 1979; Rotsztejn, 1980; Vizi, 1979, 1980a).

When the effector cell is closely coupled in time, space, and function with the releasing cell we speak of a conventional, synaptic interaction. However, when there is a loose interaction between the target cell and secreting cell, this is an unconventional non-synaptic communication (cf. Vizi, 1980a, 1981a, b, 1982a, 1983, cf. Cuello, 1983).

1.1.1 Neurotransmitters

The difficulties in defining a neurotransmitter increased expontentially as an increasing number of new substances were proposed for carrying out the transmitter function. Many scientists used this term to describe new substances before sufficient data was available to prove that they were transmitters (cf. Cooper *et al.*, 1982). It must be emphasized that not all of the substances proposed in the literature had an established neurotransmitter function.

When a signal is transmitted from one neuron to another, a transmitter, released from the nerve terminal, interacts with specific post-synaptic receptor sites, producing an alteration in electrochemical properties and excitability of the post-synaptic membrane. In the central nervous system most excitatory synapses cannot alone generate an action potential, however, they shift the resting membrane potential, resulting in an excitatory postsynaptic potential (EPSP), and the summation of a number of EPSPs initiates a post-synaptic spike.

According to Paton (1958), the following list would appear to characterize the pre-synaptic neuron. It should

(1) Synthetize the transmitter, and
(2) Release it in a pharmacologically identifiable form
(3) reproduce the specific events of normal transmission and
(4) be antagonized by competitive blocking agents
(5) that there should exist an enzyme destroying the transmitter is also important.

According to Burnstock (1976) a neurotransmitter (1) must be synthetized and stored in nerves (being released during nerve activity) and (2) its interaction with specific receptors in the post-synaptic membrane, should lead to changes in post-synaptic activity.

Criteria for establishing the identity of a transmitter has been also specified by Barchas *et al.* (1978):

(1) The neurotransmitter must be present in pre-synaptic sites of neuronal tissue, possibly in an uneven distribution throughout the brain.
(2) Precursors and synthetic enzymes must be present in the neuron, usually in close proximity to the site of potential activity.
(3) Stimulation of afferents should cause release of the substance in physiologically significant amounts.

(4) Direct application of the substance to the synapse should produce responses which are identical to those of stimulating afferents.

(5) There should be specific receptors present which interact with the substance, and these should be in close proximity to pre-synaptic structures.

(6) Interaction of the substance with its receptors should induce changes in post-synaptic membrane permeability, leading to excitatory or inhibitory post-synaptic potential.

(7) Specific inactivating mechanisms should exist and an interaction of the substance with its receptor within a physiologically reasonable period of time.

(8) Intervention at post-synaptic sites or through an inactivation mechanism afferents or direct application of the substance should be equally responsive.

While synaptic, phasic transmission is often mediated by the opening of specific ionic gates by using excitatory transmitters such as acetylcholine, glutamate, and aspartate, which increases g_{Na} and g_K (cf. Krnjevic, 1980; Nicoll and Alger, 1979) or by using inhibitory transmitters, such as γ-aminobutyric acid or glycine, increase g_{Cl} (cf. Dudel and Kuffler, 1961), the effect is short-lasting. Therefore is there reason to believe that non-synaptic tonic modulation is mediated by the triggering of chemical reactions (production of cyclic AMP activation of a protein kinase, stimulation of membrane ATPase, etc) which alter the release of transmitter from other terminals. This metabotropic action is long-lasting (Eccles and McGeer, 1979). Thus, any substance released from nerves that exerts a tonic modulation of chemical neurotransmission will be termed a *modulator*.

1.1.2 Neuromodulators

Current work in neuroscience suggests that interneuronal communication is not limited to classical neurotransmission. Transmitters are not the only substances that can influence the excitability of cells, just as the post-synaptic membrane is not the only part of the cell that accepts signals from another neuron. There are substances called modulators that are released from neurons that are able to affect transmitter release and cell excitability in a rather unconventional way.

According to Florey (1967), electrical synaptic transmission is genuine synaptic transmission, that is, the transmission of a signal from one cell to another. One pre-synaptic action potential triggers one post synaptic action potential, and this propagates over the entire effector cell membrane. Chemical transmission, on the other hand, creates local potentials at the post-synaptic effector cell that affect the excitability of the target cell. Nerve cells respond to pre-synaptic nerve impulses with graded post-synaptic potentials. The action of the transmitter is restricted to the sub-synaptic membrane. However, there could be a substance released from neurons which might affect the membrane region outside the specific synaptic areas, and which is not released from the synaptic region. Florey (1967) had already suggested the term 'modulator' substance, claiming that there was good reason to believe that excitable cells are controlled not only by transmitter substances but also by modulator substances that reach excitable cells by humoral pathways, and which originate in nerve cells, gland cells, neurosecretory cells, or ependima (glia) cells.

The concept of neuromodulation is therefore not new, but is, as yet, incompletely developed. It arises from the discovery of a number of compounds which are important in general communication between nerve cells but which do not act trans-synaptically but in a hormonal-like manner (Vizi, 1974a, 1979; Barchas *et al.*, 1978). The unconventional action of modulators suggests that they may be released by non-synaptic mechanisms in an equally unconventional fashion. In the past few years neurochemical (Paton and Vizi,1969; Vizi, 1979) and morphological (cf. Beaudet and Descarries, 1978) evidence has suggested that some neurotransmitters are released from non-synaptic sites, and diffuse to target cells distant from the release sites to modulate the release of transmitters. Therefore it was suggested that they be termed neuromodulators (Vizi, 1979) or non-synaptic neurohumours (Dismukes, 1979).

A large number of articles have been published on the meanings of the terms neurotransmitter, neuromodulator, and neurohormones. Dismukes (1979) suggested a flexible generic term for substances released from neurons by stating that a neurohumour is any substance that is released from a neuron (Fig. 1). Although the proposed classification of Dismukes (1979) seems very useful, it takes into account only the effect of transmitters on target cells;

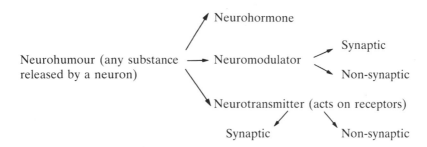

Fig. 1. Classification of substances released from a neuron (according to Dismukes, 1979, and Vizi, 1979)

their mode of action and their function is not considered. According to these criteria, adrenaline released from the adrenal medulla that has a receptor-mediated action on the Auerbach plexus of the gut cannot be classified: its effect is mediated via receptors, but it reaches the target cells via the bloodstream.

The question as to which criteria establish a substance as a modulator must be considered: it has been suggested (Vizi, 1981a,b) that more emphasis should be placed on the *site* of action of a substance and the *duration* of this action. Transmitters act on sub-synaptic areas, and this action is very short and phasic. Modulators affect membrane excitability, resulting in changes in transmitter (modulator) release, and their action is long-lasting and tonic (Vizi, 1981a).

Barchas *et al.* (1978) formulated the neuromodulator properties of a substance:

(1) The neuromodulator is not acting as a neurotransmitter in that it does not act trans-synaptically.
(2) The substance must be present in physiological fluids and must have access to the site of potential modulation in physiologically significant concentrations.
(3) Alterations in endogenous concentrations of the substance should affect neuronal activity consistently and predictably.
(4) Direct application of the substance should mimic the effect of increasing its endogenous concentrations.
(5) The substance should have one or more specific sites of action through which it can alter neuronal activity. Inactivating mechanisms should exist which account for the time course of effects of endogenously or exogenously induced changes in concentration of the substance.
(6) Innervations which alter the effects on neuronal activity of increasing endogenous concentrations of the substance should act identically when concentrations are increased by exogenous administration.

A neuromodulator is intermediate between classical transmitter function and hormonal remote regulation (Table 1), and the modulation, depending on the site of action, could be pre-and post-synaptic. This is a new type of communication system among neurons.

The neuromodulator communication which does not require direct synaptic contact between the nerve cell releasing the chemical substance and the target cells can be compared to a radio broadcast system, where there is public transmission of a signal that can be picked up by any properly equipped receiver, within range (i.e. the cells are properly equipped with receptors).

In addition to neurotransmitter substances acting at close range in chemical synaptic neurotransmission, there are chemical interactions between neurons without any close synaptic contact, i.e. interneuronal modulation of transmission which operates over some distance (Vizi, 1979, 1980a). This would be a transitional form between classical neurotransmission and the broadcasting of neuroendocrine secretion.

Although it is generally accepted that intercellular communication between neurons usually involves the release of chemical transmitter substances from nerve terminals, our knowledge of the modulation of neurochemical transmission is still incomplete. In the past few years, however, several observations

Table 1. The criteria upon which a substance may be classified as a neurotransmitter, neuromodulator, and neurohormone

	Neurotransmitter	Neuromodulator	Neurohormone
Origin	Axon terminals	Dendrites Axons Axon terminals	Axon terminals
Target	Sub-synaptic area Target cell	Pre- and/or post-synaptic elements Free nerve endings Varicosities, axon hillock, dendrites	
Acts via	Synaptic gap	Synaptic and/or non-synaptic route	Non-synaptically, via blood-stream
Aim and mechanism[a]	Transmits signals enhancing or reducing the permeability of the membrane to Na, K, etc.	Reduces/enhances transmitter (modulator) release (presynaptic) or reduces/enhances the sensitivity of the effector cell to the transmitter (post-synaptic)	
Duration of action	Short and phasic (few ms)	Long and tonic (> s)	Long and tonic (> min)

[a]The major effect of transmitters is to increase or decrease the peak magnitude of specific ionic conductances involved in generating the action potential. Conversely, a modulator can act differently (1) the time course of its action is quite slow, lasting in the order of seconds and minutes, compared to the brief depolarization (few ms), (2) it does not interact directly with action potential channels. Its action is mediated by messengers including Ca^{2+} and cyclic AMP or via NaK-activated ATPase. The effects of modulator are associated with a variety of changes in basic cellular electrical properties, including alterations in the threshold for the action potential, changes in the amplitude and/or duration of the action potential, reduction of the probability of varicose axon terminal to be invaded.

have been reported which suggest how the amount of transmitter released at nerve terminals may be modulated, either by a transmitter/modulator released from another nerve terminal (cf. Vizi and Knoll, 1971; Vizi, 1974a; Muscholl, 1973b) or by an auto-inhibition mechanism where the transmitter released into the synaptic cleft inhibits its own release (cf. Langer, 1973, 1974, 1981a; Stjäne, 1975a; Starke et al. 1977a; Starke, 1977, 1981; Westfall, 1977).

The modulation of synaptic transmitter release seems to play an important role in information-processing in the nervous system, and may be involved in the cellular mechanisms of learning and behaviour and in the regulation of the vegetative nervous system. Several mechanisms have been reported to modulate synaptic transmission, including repetitive activity in the pre-synaptic terminals which can produce facilitation, depression, or post-synaptic potentiation of signal transmission (cf. Schmidt, 1971; Kuffler and Nicholls, 1976). Nevertheless, no modulatory chemicals are involved in this type of modulation.

1.1.3 Neurohormones

According to Guillemin (1977), a hormone is a '. . . substance released by a cell, carried by the blood or extracellular liquid and which affects another cell near or far'. The term 'neurohormone' was first used by the Scharrers in the early 1950s to describe the release of hormones from neurosecretory cells in response to conventional neuronal signals relayed through synapses. This definition also applies to various monoamines. In this context dopamine (DA) can act as a neurohormone in some cases. Since neurons also release hormones (such as the neurohypophysial hormones and the hypothalamic releasing factors) that act on targets that are not close to the releasing cell, the proximity of the cell to the one that releases the transmitter has been used as a factor of differentiation. It is, of course, possible that the same substance may act in some regions as a hormone and in others as a trans-mitter. It is also possible that the same cell may release a substance that acts as a hormone from some of its terminals and that in other terminals the same substance is released and acts locally as a modulator.

The author believes that focusing on function, rather than on pre- or post-synaptic sites, provides a better system for classifying substances with neuroactive action. Therefore a very simple terminology of neuroactive substances is suggested and will be used throughout this book:
(1) Neurotransmitters;
(2) Neuromodulators; and
(3) Neurohormones.

1.2 MODULATION OF CHEMICAL NEUROTRANSMISSION

The nerve impulse is an all-or-none event, subject to little change during normal physiological activity. The transmission of excitation at chemical synapses, on the other hand, may be modulated by different physiological mechanisms (Fig. 2). Such modulation of synaptic transmission may be an important part of information-processing in the nervous system, and may therefore be of physiological importance. The signals can be classified into two groups:

(1) Passive or localized potentials that depend on the cable-like properties of the cell; and
(2) The impulses or action potentials that travel rapidly at a speed of over 400 km/h.

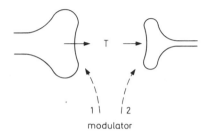

Fig. 2. Classification of modulation:
(1) pre-synaptic, (2) post-synaptic.
T, transmitter

The main characteristics of passive or localized potentials are that they are continuously graded and cannot diffuse over distances greater than 1 mm without being seriously attenuated. They occur at specialized regions of nerve cells, where responses are initiated or where they are inhibited. At synapses they are called excitatory post-synaptic potentials (EPSPs) if they tend to give rise to impulses and inhibitory post-synaptic potentials (IPSPs) if they suppress excitation. The impulses start in a special region of the axon called the initial segment or axon hillock, next to the cell body. This part of the neuronal membrane has a lower threshold for the initiation of a regenerative impulse than the cell body, whose dendrite may not conduct impulses at all. In contrast to these passive, localized potentials the nerve impulse is not attenuated with distance, and is an all-or-none explosive event which has much the same amplitude in all nerve cells of the body, and varies mainly in the velocity with which it propagates. Transmission of information in the nervous system can, however, be modulated at synaptic junctions, pre- and post-synaptically on a varicose axon terminal, or on the axon hillock where action potential generation takes place. Two types of modulation can therefore be differentiated: (1) pre-synaptic and (2) post-synaptic.

1.2.1 Pre-synaptic modulation

The term *pre-synaptic modulation* means an inhibition of the excitation of a transmission signal which results from a change of transmitter release rather than from a decrease in the post-synaptic sensitivity of effector cells to transmitters. The significance of pre-synaptic inhibition is that one input to the post-synaptic cell can be selectively inhibited without changing the excitability of the effector cell as a whole. In this case the sensitivity of the post-synaptic membranes of effector cells to the transmitter is not changed. The first evidence for the existence of pre-synaptic inhibition came from electrophysiological studies (see Nicoll and Alger, 1979). In the early 1950s Fatt and Katz (1953) observed that the depression of the excitatory junction potential due to stimulation of the inhibitory input was not fully accounted for by changes post-synaptic permeability (Frank and Fuortes, 1957). In 1961 Eccles and his colleagues (Eccles *et al.*, 1961) in mammals, and Dudel and Kuffler (1961) in crustaceans, provided clear evidence that the action of the inhibitory nerve can also be exerted on excitatory axon terminals, reducing the release of the neurotransmitter. Crustacean claw opener and closer muscles are innervated by excitatory and inhibitory axons. Dudel and Kuffler (1961) observed that the inhibitory junction potential (i.j.p.) reversal potential was more positive than the amplitude of an excitory junction potential (e.j.p.). A single i.j.p. released simultaneously with an e.j.p. could only enhance the size of the e.j.p., since of Dudel and Kuffler found a reduction in amplitude of e.j.p. Quantal analysis showed that the effect involved a reduction in quantal content with a change in quantal size, i.e. that the inhibition observed was pre-synaptic in nature.

The pre-synaptic effect is relatively brief, its action lasting a few milliseconds (6 to 7 ms). In mammalian neurons (cf. Eccles, 1964) the duration of the pre-synaptic inhibitory action is much longer (100 to 200 ms) than in crustaceans. The inhibitory effect is mediated via receptors that give rise to increases in either chloride or potassium conductances or in both (Dudel and Kuffler, 1961; Dudel 1965). There is some evidence that γ-aminobutyric acid (GABA) is the inhibitory substance. (We will not discuss in detail the electrophysiological evidence of pre-synaptic inhibition and therefore the reader should read the excellent review of Nicoll and Alger (1979.)

Two different types of pre-synaptic inhibition seem to occur (Table 2):

Table 2. Differences in the synaptic and non-synaptic inhibitory interactions between neurons

| | Inhibitory interaction between neurons | |
	Synaptic	Non-synaptic
Communication	One-to-one	One to many
Delay	Short (100–200 ms)	Long (>1 s)
Action	Short and phasic	Long and tonic
Discrimination		Depends on the presence of receptors
Chemical agent has to	Cross the gap (5–20 nm)	Diffuse distances of some micrometres
Morphological characteristics	Axo-axonic synapse (post-junctional specializations)	Free nerve-ending, axonal varicosities (no post-junctional specializations)

1. *Synaptic* (conventional). The inhibitory neuron forms a synapse on the initial segment of the axon, or on its terminals, depressing or inhibiting transmitter release from the neuron (Dudel and Kuffler, 1961).

2. *Non-synaptic* (unconventional). Here the inhibitory substance (modulator) is released from axon terminals at a distance from effector cells that are devoid of synaptic morphological characteristics. The modulator diffuses over some micrometres to exert its inhibitory effect on a neighbouring neuron. The target may be the initial segment the axon terminals, or the varicosity, and thereby only transmitter release is affected, and not the sensitivity of the effector cell (Fig. 3).

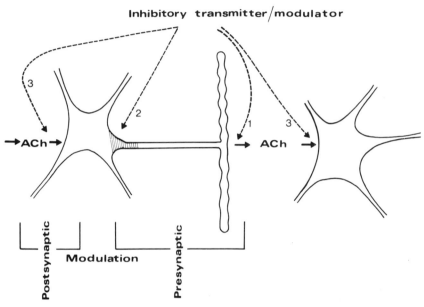

Fig. 3. Two types of pre-synaptic modulation. Pre-synaptic inhibition describes a type of modulation where the amount of transmitter released is reduced or inhibited without changing the sensitivity of receptors located on the cell body. Pre-synaptic inhibition of transmitter release: (1) action on the nerve terminal: (2) effect on the initial segment of the axon where action-potential generation takes place. Post-synaptic inhibition: (3) when the incoming input is ineffective, because the membrane sensitivity to transmitter is reduced. (Figure is redrawn from Vizi, 1979). (Reproduced by permission of Pergamon Press Ltd)

The morphology (the origin of the modulator) of non-synaptic modulation can be classified as follows:

Interneuronal modulation (for review, see Vizi, 1979), when the modulator released from one neuron affects the release of substances secreted from another (Fig. 4a),

Trans-synaptic modulation (for review, see Vizi, 1979; Fredholm, 1981), when a neurotransmitter acting on the target cell causing depolarization releases substances which are able to influence the release of the transmitter trans-synaptically (Fig. 4b),

12

Negative feedback modulation (for reviews, see Langer, 1977, 1981; Starke, 1977, 1981; Vizi, 1979; Erulkar, 1983) when the transmitter/modulator inhibits its own release (Fig. 4c).

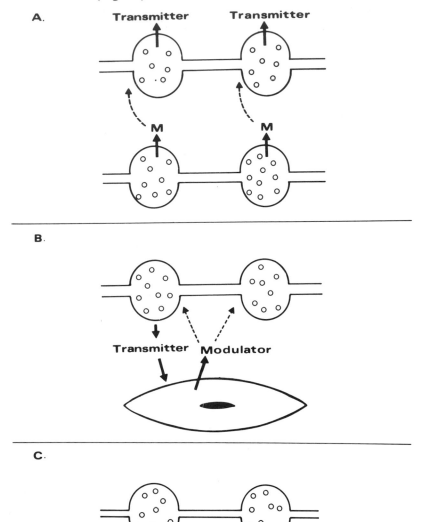

Fig. 4. Different types of modulation: (a) 'interneuronal' (Vizi, 1979, 1980, 1981a,b), when the modulator (M) released from a neuron influences the release of a principal transmitter or other modulator from another neuron; (b) 'trans-synaptic' (Vizi, 1979; Fredholm, 1981; Hedqvist, 1981); when the modulator acts retrogradely, it is released from the effector cell in response to depolarization; (c) 'negative feedback' (cf. Langer, 1977; Starke, 1977, 1981) when the transmitter/modulator released inhibits its own release

An interesting possibility is that auto-receptors might be stimulated by the modulator/transmitter released not from axon terminals supplied with receptors, but from varicosities originating from another neuron. So far, this type of modulation has not been experimentally verified. Although it would operate with the same transmitter, because it comes from another neuron it can be looked upon as a form of inter-neuronal modulation (Vizi, 1981).

1.2.2 Axon hillock modulation

The best strategic locus for an inhibitory synapse would be on the axon hillock (Fig. 3). Inhibitory synapses with this type of localization can be found on Purkinje cells and Mauthner cells (Hámori and Szentágothai, 1965). The initial segment region is the main site of action-potential generation of the conducted spike (Coombs *et al.*, 1957). It has been found electrophysiologically (Andersen *et al.*, 1963) and morpholoigcally (Hámori and Szentágothai, 1965) that transmission from basket cell to Purkinje cell is inhibitory Somogyi and Hámori (1976) carried out a quantitative electron-microscopic study of the Purkinje cell axon initial segment and found that inhibitory axo-axonic synapses on the axon hillock are not accidental but regular structures. It is very probable that the modulator either released nearby from the axo-axonic synapse or from a more distant nerve terminal might control neurochemical transmission very effectively, inhibiting the generation of action potential.

Inhibition of generation of action potentials by an action at the axonal hillock can be regarded as a mode of truly pre-synaptic inhibition since the excitability of the soma or dendrites is not affected. While somatic (post-synaptic) inhibition may be selective, depressing somatic excitability to some of its inputs without affecting others, pre-synaptic inhibition by a modulator acting on the axonal hillock is non-selective. The same holds true for pre-synaptic inhibition of the release of a transmitter from the terminals.

It must be emphasized that although axo-axonic synapses have been described, and are assumed to form the structural basis of pre-synaptic inhibition, this inhibition also seems to occur in many cases, e.g. in the absence of close synaptic contact between the inhibitory neuron and its targets, as stated above. Anatomical, electrophysiological, neurochemical, and pharmacological evidence indicates that both amines (e.g. noradrenaline, dopamine, and serotonin) and peptides (e.g. enkephalins) may mediate such non-synaptic modulatory control of neurotransmission, both in the brain and in the periphery.

1.2.3. Axon terminal modulation

A modulator released in the vicinity of axon terminals might inhibit the release of a transmitter or modulator (Fig. 5) by different mechanisms. The site of action might be on the axon terminal, the varicosity, or the axon hillock.

14

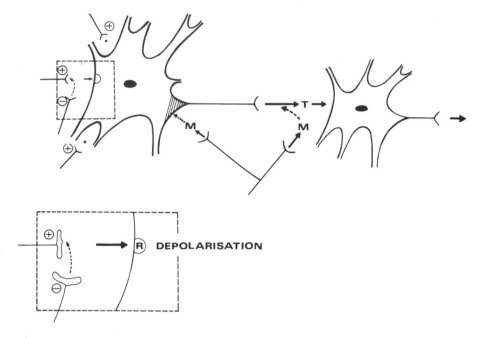

Fig. 5. Scheme of pre-synaptic inhibition. The site of action of the modulator (M) is either on the axon hillock or on the axon terminal, or on both, but the ability of transmitter (T) to affect the post-synaptic membrane is not influenced. The significance of pre-synaptic inhibition is that one input to the post-synaptic cell can be selectively inhibited without changing the excitability of effector cells as a whole

Certain obvious questions, arise when one considers that, for example, noradrenaline (NA) fails to inhibit transmitter release at the neuromuscular junction but does reduce release from other cholinergic neurons. How is it possible to explain this? There is an important difference between the two nerve terminals as autonomic nerve terminals are equipped with varicosities while at the neuromuscular junction there is a specialized nerve terminal. It seems very likely that pre-synaptic modulation on axon terminals mainly occurs in those places where varicosities are present (Vizi, 1979). Since a large body of evidence is available showing that the noradrenergic, dopaminergic, and serotonergic neurons in the central nervous system are varicose, it seems very likely that this type of modulation is of physiological and pharmacological importance.

1.2.4 Post-synaptic modulation

The term *post-synaptic modulation* means an inhibitory effect on the post-synaptic (post-junctional) membrane which inhibits the local effect of an excitatory transmitter on the post-synaptic membrane (Fig. 6) simply by preventing the action of the transmitter on, for example, sodium influx. The post-synaptic inhibitory effect was, in fact, first described by Fatt and Katz

(1953), when they found a depression of the excitatory junctional potentials and an increased Cl⁻ ion conductance of muscle fibre in response to stimulation of inhibitory fibres. Different types of post-synaptic inhibitions have different consequences for convergent synaptic potentials. An increase in post-synaptic conductance produced by the inhibitory transmitter results in a reduction of amplitude of convergent EPSPs, whereas a decrease of conductance would potentiate them. By augmenting rather than diminishing a synaptic input, conductance decreases result in long-lasting potentiation of synaptic transmission. Since convergent synaptic excitation must stimulate action potentials in the post-synaptic neuron to transfer information to the next neuron of effector cells, it seems likely that any inhibition in a synaptic pathway may play an important role in information-processing in the nervous system.

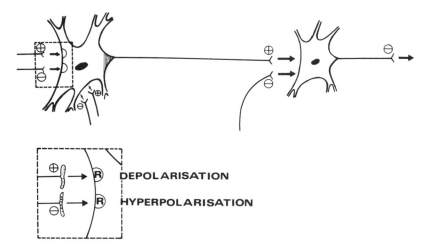

Fig. 6. Scheme of a post-synaptic inhibition. Note that the effect of the transmitter on the post-synaptic, effector cell is prevented or reduced (depolarization–hyperpolarization)

The brain is often compared to computers, a comparison which is not absolutely valid. One of the critical differences between them is that although both use electrical signals for communication, nerve cells also use chemical substances whose release from axon terminals can be modulated.

1.2.5 Pre- and post-synaptic receptors

The term pre- and post-synaptic receptors is used to explain the site of action and to designate the drug used. This type of terminology, however, provides no more than an anatomical guide. This does not necessarily mean that these receptors are located at the two sites of a classical synapse; they may not even be within the synapse. In this book these terms will, however, be used, since they describe more comprehensively those receptors which are located on target cells and which are accessible to the modulator released from a nearby neuron or to a circulating drug (agonist or antagonist).

The term pre-synaptic receptors will be used for those receptors that are acted on by other neurotransmitters/modulators, or by the axon terminals' own neurotransmitter/modulator. Hence, the term pre-synaptic autoreceptors will be used for those receptors that are located in the axon terminals and stimulated by the neurotransmitter/modulator released from the same bouton.

Chapter 2

Conventional and non-conventional release of transmitters/modulators

The concept of synaptic transmission generally accepted in the first half of the twentieth century suggests that neurons consists of receiving and transmitting zones, summate their excitatory and inhibitory synaptic inputs, and transmit information in all-or-none impulses along axons, whose terminals release a single type of chemical neurotransmitter (Fig. 7). The classical view of neurotransmission is that it takes place at the synapse, where the nerve-ending contains the transmitter substance and comes into close contact with the effector cell containing the receptors.

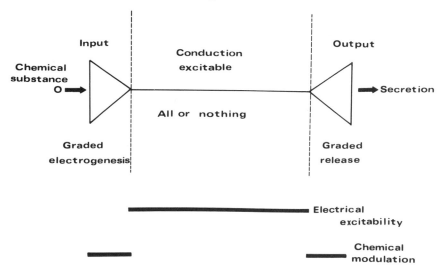

Fig. 7. Possible sites of chemical modulation in a neuron. Since the nerve impulse conduction is an all-or-none event, the site of modulation is either on the cell body (soma, dendrites) or on the axon hillock (where the action-potential generation takes place), or on the axon terminal. The input on the cell body and the output on the nerve terminals are caused by a graded mechanism. The input membrane is electrically inexcitable: it receives chemical information which is transformed at the initial segment of the axon into an electrical signal

Recently, neurochemical and morphological evidence has shown that some substances may be non-synaptically released from neurons for diffusion to target cells more distant than those found in conventional synaptic transmission (cf. Dismukes 1979; Vizi, 1979; 1980a). In both noradrenergic and

17

18

serotonergic pathways the transmitter/modulator is not only contained in nerve terminals but is also concentrated in varicosities closely spaced (1–3μm) along tiny unmyelinated axons. A low incidence of synaptic contact by varicosities has been established for catecholamines (Ajika and Hökfelt, 1973) and indolamines (Calas et al., 1974) in the median eminence, for serotonin in cerebral ventricles (Richards et al., 1973; Chan-Palay, 1976), presumptive dopamine in the neostriatum (Tennyson et al., 1974), and for serotonin (Descarries et al., 1975) and noradrenaline (Descarries et al., 1977) in the neocortex. Examples of transmitter release from regions of the nerve cell other than axon terminals (Fig. 8) have been already shown, and an ACh-like substance release from cholinergic neurones has been demonstrated (Lissák, 1939; McIntosh, 1941). Direct evidence was provided (Vizi et al., 1983a) that ACh is released from cholinergic axon. Esquerro et al. (1980a,b) have also shown that noradrenaline can be released from the ligated cat hypogastric nerve. Johnson and Pilar (1980) provided evidence for somatic release of ACh and dendritic release of dopamine from dendrites of pars compacta in the substantial nigra (Korf et al., 1976; Geffen et al., 1976; Nieoullon et al., 1977). It was even suggested that dendritically released DA could act as receptors located either on dendrites, where it is released from autoreceptors (Aghajanian and Bunney, 1977), or at receptors which are located on striato-nigral afferent fibres.

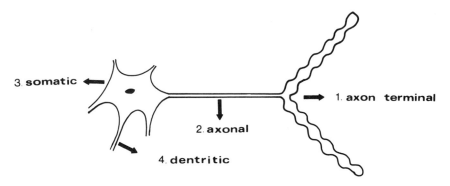

Fig. 8. Possible sites of transmitter release, conventional: (1) from the axon terminal; non-conventional, (2) from the axon; (3) from the soma; and (4) from dendrites

In addition, a number of papers have been published on the possibility that more than one transmitter may be released from the same neuron, although there is no direct evidence that substances co-released have post-synaptic effects characteristic of transmitter molecules. Nevertheless, there is a great need for revision of textbook statements about nervous system function, particularly in regard to the modulation of chemical neurotransmission.

2.1 CONVENTIONAL SYNAPTIC RELEASE FROM AXON TERMINALS

A nerve impulse travelling toward its target cell releases a chemical substance in order to transmit information, and this substance crosses the synaptic cleft and acts on the post-synaptic membrane. In general, the impulse reaches an axon terminal that is in contact with another structure. The site of contact is called a synapse or synaptic junction. The pre- and post-synaptic elements are separated by a distance of 15–100 nm, and the transmission in chemical synapses is unidirectional.

The arrival of the action potential at the nerve terminal greatly increases the rate of quantal liberation, several hundred quanta are released within a millisecond, and this 10^5-fold increase in the probability of quantal liberation produces a large post-synaptic depolarization. In addition, there is an non-quantal form of transmitter release. It is generally accepted that the following parameters are able to affect this:

(1) Pre-synaptic depolarization;
(2) Extraneuronal Ca; and
(3) Intraneuronal Ca.

2.1.1 Pre-synaptic depolarization

Under physiological conditions, the nerve terminal is activated by an action potential, resulting from an inward sodium current which depolarizes the membrane. The currents produce voltage changes according to the membrane resistance and capacitance. This means that a decrease in membrane resistance makes the depolarization less effective and less transmitter is released. This could happen in the case of pre-synaptic inhibition (Dudel and Kuffler, 1961), where the inhibitory transmitter reduces the membrane resistance of the pre-synaptic nerve terminal, and hence less excitatory transmitter is liberated when an action potential takes place.

By contrast, some drugs are known to inhibit the voltage-dependent potassium current and thus cause a broadening of the action potential increase. Tetraethylammonium (Katz and Miledi, 1969) and aminopyridines (Lev-Tov, 1978; Pelhate and Pichon, 1974) cause an increase in inward calcium current resulting in an increase in Ca_i (Katz and Miledi, 1969) and an increase in transmitter release.

2.1.2 Extraneuronal Ca

It is generally accepted that extracellular calcium (Ca_o) ions are essential for transmitter release (cf. Katz and Miledi, 1969). The relationship between Ca_o and quantal release (m) is described by

$$m = K \frac{Ca/K_1}{1 + (Ca/K_1)}$$

where K and K_1 are constants (Dodge and Rahamimoff, 1967). This equation shows that small changes in Ca_o results in a very large change in transmitter release (Rahamimoff *et al.*, 1980).

2.1.3 Intracellular Ca

It is now generally accepted that the concentration of free calcium ions inside the nerve terminal $(Ca)_{in}$ is a key factor in transmitter release (Miledi, 1973). The concentration of free calcium inside the nerve terminals is approximately $10^{-7}M$. Therefore release of calcium from intra-cellular binding sites (mitochondria, vesicles, endoplasmic reticulum, cytosolic molecules) can raise the level of free $(Ca)_{in}$; conversely, an increased uptake by intracellular stores can lower $(Ca)_{in}$. Under physiological conditions the increase in $(Ca)_{in}$ is mainly dependent on the entry of Ca_o during the action potential. The following factors are capable of influencing Ca_{in} (Rahamimoff *et al.*, 1980).

Voltage-dependent calcium channels

In the case of Ca there is a very large inward chemical gradient. In addition, the electrical potential can produce an additional inward-driving force for calcium ions. Two types of voltage-dependent calcium channels have been described (Baker *et al.*, 1971):

(1) The tetrodotoxin-sensitive sodium channel and
(2) the channel that is blocked by a variety of ions such as Mn^{2+}, Co^{2+} and Mg^{2+} and which is responsible for transmitter release.

Na^+-Ca^{2+} *exchange*. The extrusion of calcium ions is coupled to the entry of sodium ions. In this exchange the nerve terminals utilize the sodium electrochemical gradient to transport calcium ions against its own very large gradient.

ATP-driven calcium extrusion

Calcium can also be extruded from nerve cells by an ATP-driven process (Di Polo, 1978).

An intracellular buffer (binding) system

Mitochondria, endoplasmic reticulum, and vesicles are able to bind Ca, thereby affecting Ca_{in} concentration.

2.2 NON-CONVENTIONAL (NON-SYNAPTIC) RELEASE

The generally accepted form of neurotransmission is that the transmitter (e.g. acetylcholine) is liberated from nerve terminals into the synaptic cleft in quantal packages. However, where the gap is large, where there is no synaptic specialization, or where the transmitter must cross distances of micrometres to reach the target cell, the transmitter released from the cytoplasm also plays a critical role in signal transmission (Vizi *et al.*, 1982). The idea that transmitters can be released from non-synaptic areas was first suggested for transmitters in the autonomic nervous system, where the axon terminals very rarely make synaptic contact with the target cells (Fig. 9). Since then, a large amount of evidence has shown that transmitters/modulators can be released from regions other than the nerve-ending. Dismukes (1979) stated that 'non-synaptic release would fit rather nicely with the merging concepts of the electrophysiological action of amine and peptides'.

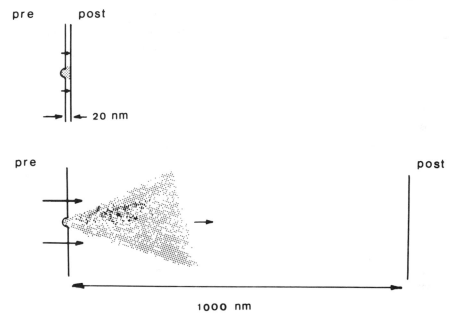

Fig. 9. A proportional diagram of two synaptic clefts: a typical synapse with a gap of 20 nm and a free axon terminal with its remote target cell (e.g. the vegetative nervous system). Note the difference in volume. Let us assume that both transmission sites are cholinergic. The average diameter of a vesicle containing ACh is 50 nm, therefore its volume is about 65 000 nm^3 and its ACh concentration is about 0.1 M (cf. Marchbanks, 1979). If we assume that its content is completely discharged into a small synaptic cleft whose volume is about 200 000 nm^3 (20 × 100 × 100), the final concentration of ACh in the cleft is 30mM (0.1/(200 000/65 000)), which is an extremely high concentration. Let us suppose that, for example, in Auerbach's plexus, where the target smooth muscle cell is far from the varicose axon terminals (100–1000 nm), only one vesicle is released. The volume in which the ACh is released is 10^9 nm^3 (1000 × 1000 × 1000), i.e. a volume 10^4 times larger than that of a vesicle. Therefore the ACh released from the vesicle is diluted by a factor of 10 000. Provided that the cholinesterase (ChE) is not active, the final concentration of ACh which reaches the muscarinic receptors of the smooth muscle is about 10μM (0.1M/10 000). The cytoplasmic release has not been taken into account. Note the concentration of ACh in the cleft

2.2.1 Non-synaptic release from axon terminals

Electron-microscope and histochemical studies of the relationship of nerves to smooth muscle cells have shown that the essential feature of a transmitter is that it is *en passage* from large numbers of terminal varicosities. The minimum width of the cleft between nerve varicosities and effector cells varies considerably in different tissues. The nerve-muscle separation in the regions of closest proximity in the vas deferens and sphincter pupillae is about 15–20 nm. In blood vessels, the closest proximity between varicosities in the perivascular plexus at the adventitial–medial border and smooth muscle cells varies from about 50 to 80 nm in small muscular arteries and large arterioles as to much as 1–2 μm in large elastic arteries (see Burnstock *et al.*, 1970a; Bevan and Su, 1974; Burnstock, 1975b, 1978a). No post-junctional specializations have been found with any consistency for wider neuromuscular junctions (Burnstock, 1979a). Release from axon terminals devoid of synaptic membrane specialization has recently been suggested to be the function of the central monoamine terminals (Beaudet and Descarries, 1978). Beaudet and Descarries (1978) claimed that the release of biogenic amines solely from varicosities making synaptic contact could hardly account for the total amount released from axon terminals. A transmitter or modulator released non-synaptically could affect very large neuronal assemblies.

There is an interesting difference in localizations of receptors on effector cells where the transmitter (ACh, NA) is released into a small synaptic gap compared with those where it is released into a large extraneuronal space. In the former case there is a small area of the cell where the ACh receptors are concentrated. The extrasynaptic area is relatively insensitive to the transmitter. However, when the transmitter release site and the target cells (e.g. Auerbach's plexus varicose axon terminals and smooth muscle cells' are widely separated from each other (100–1000 nm) there is no specific sub-synaptic arrangement, and the receptors are evenly distributed along the whole surface of the smooth muscle cell. This morphological arrangement fits any diffusion type of transmission where the advantage of quantal release cannot be utilized (Fig. 9). There seems to be no evidence that non-quantal release can occur solely at diffuse synapses.

2.2.2 Axonal release

The concept of synaptic chemical transmission suggests that neurons transmit information in all-or-none impulses along axons, whose terminals release a single type of transmitter into the synaptic cleft. The synaptic release of the transmitter is one of the basic principles of classical neuro-transmission. Nevertheless, some evidence shows that a neuron is able to release substances from parts of it other than the axon terminal (Vizi, 1983a).

Acetylcholine release

It has been shown that both the axon and the axon terminal of the cholinergic neuron are able to synthesize ACh. This may not be confined to the endings but may also occur along the whole length of a cholinergic nerve (Brown and Feldberg, 1936; Feldberg, 1943). An ACh-like substance was shown to be released from cholinergic neurons (Lissák, 1939; McIntosh, 1941). Whereas basic features of the transport of acetylcholinesterase (Kása, 1968; Lubinska and Niemierko, 1971; Ranish and Ochs, 1972; Kása et al., 1973; Tucek, 1975), choline acetyltransferase (Fonnum et al., 1973; Kása et al., 1973; Tucek, 1975), and acetylcholine (Häggendal et al., 1973; Dahlström et al., 1974) in cholinergic axons are well understood, neurochemical evidence has shown that the axon resembles the axon terminal in that it is capable of releasing ACh (Vizi et al., 1983a). In response to potassium (KCl, 49.4 mM) depolarization or electrical stimulation, ACh was released from isolated pre-ganglionic axons of the truncus sympathicus (the cervical pre-ganglionic sympathetic branch) of the cat. This release was shown to be dependent on extracellular Ca^{2+}. The question arose as to whether this release from the axon was a direct efflux of ACh or a leakage from the periaxonal Schwann cells (Dennis and Miledi, 1974). There was no increase in the release when the axons were ligated, and no synapse was seen electron microscopically (ruling out a pre-terminal origin of ACh), thereby excluding the Schwann cell origin of ACh. It was suggested that the axonal membrane was the release site.

The distribution of ACh in the pre-ganglionic sympathetic nerve differs markedly from the distribution in the axon terminal, where most ACh is found in vesicles. No vesicle or vesicular clusters were seen in the axon when it was serially sectioned. Therefore most of the ACh was assumed to be in the cytoplasm. Thus when ACh was released, it was from the cytoplasm. It was concluded, that regardless of the exact mechanism of ACh release from the axon, it is obvious that ACh is released when the tissue is depolarized and Ca_o is present. The release process from isolated axons resembles the axon terminal release in many respects, except that membrane ATPse is not involved: inhibition by ouabain of Na, K-ATPase fails to enhance the release of ACh as in other tissues (cf. Vizi, 1978). As with axon terminals, the release from the axon is Ca-dependent, but it cannot be inhibited by noradrenaline, a phenomenon established for enteric cholinergic neurons (cf. Vizi, 1979) and for cholinergic neurons in the cerebral cortex (Vizi, 1980b; Vizi and Pásztor, 1981). These findings also suggest that the modulation is confined to distal parts of the axon (for details, see Section 4.1).

Noradrenaline release

It has been shown that the proximal segment of a ligated sympathetic nerve takes up and retains much more ^{3}H-NA than the distal portion of the same nerve (Kirpekar et al., 1968; Dahlström and Waldeck, 1968). Electrical stimulation in vitro of cat hypogastric and splenic nerves released ^{3}H-NA into the incubation medium, to a level of five to six times the background release (Kirpekar et al., 1968). Veratridine also produced a release of endogenous noradrenaline from a ligated cat hypogastric nerve (Esquerro *et al.*, 1980a). In addition, there is a significant increase of NA in the presence of the ionophore X537A or high K^+ concentrations (Esquerro *et al.*, 1980b).

2.2.3 Dendritic release

Dopamine release

Using histochemical techniques Björklund and Lindwall (1975), observed a dopamine containing varicose dendritic network in the pars reticulata of the rat substantia nigra. Geffen *et al.* (1976) and Cuello and Iversen (1978) provided evidence that endogenous DA is released from these dendrites. Björklund and Lindvall (1975) have demonstrated histochemically that the dendrites of the nigrostriatal neurons of the rat brain also contain dopamine. They suggest that DA may function not only within the neostriatum but also in the nigra itself, not least in the pars reticulata, where most of the striatonigral fibres have been reported to terminate. In 1976 Korf *et al.*, provided indirect evidence for a release of DA from dendrites although its function remained unclear. Later it was shown that ^{3}H-DA is released from substantia nigra slices depolarized by K^+ (Geffen *et al.*, 1976) or incubated in d-amphetamine (Paden *et al.*, 1976). Nieoullon *et al.* (1977) demonstrated therefore ^{3}H-DA into superfusates from cat substantia nigra and showed that this release was greatly enhanced by the addition of K^+ and d-amphetamine to the medium.

To extent to which dendrites resemble terminals with respect to transmitter release is an important question. Nieoullon *et al.* (1977) reported different reactions of dendrites and terminals, whereas superfusion with textrodotoxin inhibited release of ^{3}H-DA in caudate-putamen, the same substance increased its release in substantia nigra. The finding that textrodotoxin failed to reduce the spontaneous release of ^{3}H-DA from dendrites indicates that fast sodium channels are not involved in the release process. In fact, dendritic spikes, resistant to the neurotoxin, have already been described (Fukuda and Kameyama, 1980; Llinas and Hess, 1976).

Endogenous dopamine can be released by a high-concentration of K^+. The release is Ca^{2+} dependent and can be antagonized by Mg^{++2+} (Cuello and Iversen, 1978). The fact that an ever more efficient release of dopamine takes place from the pars reticulata under the same conditions as the previously mentioned release strongly suggests that dendrites are the main sites of release of dopamine in the substantia nigra.

A stimulation of ^{3}H-DA release has been induced by glycine, acetylcholine, and serotonin, but substance P induces the opposite effect (Michelot *et al.*, 1979; Cheramy *et al.*, 1979). Although GABA by itself is ineffective,the GABA agonist muscimol enhances the dendritic release of ^{3}H-DA (Chéramy *et al.*, 1979), whereas the reverse is shown with diazepam.

Bunney and Aghajanian (1976) have shown the presence of dopaminergic receptors on dopaminergic cell bodies or dendrites in the rat substantia nigra.

Microiontophoretic application of DA or of apomorphine reduces the firing rate of the dopaminergic cells and this effect is inhibited by neuroleptics (Aghajanian and Bunney, 1977). Groves *et al.* (1975) have reported that amphetamine, a drug which releases DA, inhibits the firing rate of pars compacta neurons after its local application, and this effect is prevented by the peripheral injection of α-methyl-paratyrosine. In 1976. Groves *et al.* suggested that DA could be released from dendrites and was involved in an auto- or lateral inhibition of the nigral dopaminergic neurones (Groves *et al.*, 1975).

Any changes in the dopaminergic transmission in the caudate nucleus which may in turn affect the activity of the nigral dopaminergic cells through the influence of striato-nigral fibres involved in the activation of the inhibition of the dopaminergic cells, has yet to be conclusively determined, although substance P neurons may be responsible for activation (Michelot *et al.*, 1979; Cheramy *et al.*, 1979; Cheramy *et al.*, 1977, 1978) and GABA neurons for inhibition (Bunney and Aghajanian, 1976; Dray, 1979).

After the serotonergic input to the substantia nigra was lesioned the release of exogenously applied ^{3}H-dopamine from the partially denervated substantia nigra was found to be very similar to that observed from slices of control substantia nigra (Tagerud and Cuello, 1979). These results show that the release of exogenously applied ^{3}H-dopamine at the level of the substantia nigra occurs mainly from dopaminergic dendrites rather than from terminals of 5-hydroxytryptamine-containing neurons. Veratridine-induced release of ^{3}H-dopamine from the pars reticulata of the substantia nigra is also textrodotoxin-sensitive (Tagerud and Cuello, 1979).

The active dopamine uptake process in dendrites has already been suggested in biochemical, histochemical, and autoradiographic studies (cf. Cheramy *et al.*, 1981). Exogenous dopamine can be taken up in slices of the rat substantia nigra and can be visualized in dendrites. Labelled dopaminergic dendrites were seen after an intraventricular or local injection of ^{3}H-catecholamines. The dopaminergic dendrites contain tyrosine hydroxylase, as indicated by a recent immunohistochemical study (Cheramy *et al.*, 1981).

Dopamine released into the substantia nigra seems to have a physiological role. Aghajanian and Bunney (1977) postulated that the amine inhibits the activity of dopaminergic cells by acting on dopaminergic autoreceptors. Dopamine (10^{-7} M) administered into the substantia nigra inhibited the release of newly synthesized ^{3}H-dopamine in the ipsilateral caudate nucleus (Cheramy *et al.*, 1977). Since dendrodendritic contacts have been observed in the substantia nigra (Cuello and Iverson, 1978; Hajdu *et al.*, 1973) it is possible to speculate that dopamine released from dendrites may affect the activity of neighbouring dopaminergic neurons (lateral inhibition).

Lateral inhibition is a process which has been extensively described by neurophysiologists. Groves *et al.* (1976) have suggested that dopamine released *in vitro* from dopaminergic dendrites may modulate nigral afferents at a pre-synaptic level (Fig. 10).

26

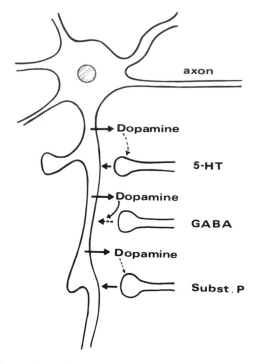

Fig. 10. The substantia nigra (dopaminergic neuron). Serotonergic, GABA-ergic, and substance P-ergic inputs are influenced by dopamine. Note the non-synaptic interactions between inputs and dendrites. Dopamine released from the dendrites inhibits (dotted arrows) the release of serotonin (5-HT) and substance P (Subst. P). For details, see the text. (A. C. Cuello and L. C. Iversen in *Interactions Between Putative Neurotransmitters in the Brain.* © 1978. Reproduced by permission of Raven Press, New York)

Indirect evidence suggests that some dopamine released from nigral dendrites may diffuse to some targets non-synaptically. GABA fibres from the striatum terminate in the substantia nigra and cause inhibition of cell firing. (Groves *et al.*, 1975; Iversen, 1979). These GABA terminals may be the site of pre-synaptic dopamine receptors coupled to adenylate cyclase as observed in the substantia nigra (Spano *et al.*, 1975; Gale *et al.*, 1977). Dopamine at low concentrations has been found to specifically stimulate the release of ^{3}H-GABA from nigral terminals (Reubi *et al.*, 1977). Iversen (1979) has proposed that dopamine released from dendrites in the substantia nigra may stimulate the release of GABA from terminals pre-synaptic to the dopaminergic cell bodies, thus providing a negative feedback loop. The pre-synaptic dopamine dendrites observed by Wilson *et al.* (1977) all appeared to be in contact with other dendrites. (Fig. 11) and whenever the post-synaptic element could be identified, it appeared to be dopaminergic. No evidence was found for synaptic contacts between pre-synaptic dopaminergic dendrites and axon terminals. Thus if the model drawn by Iversen (1979) is correct, dopamine released from dendrites may diffuse to GABA terminals that are not in direct synaptic contact. Similarly, Groves *et al* (1975) have proposed a model in which dopamine released from dendrites produces self-inhibition of firing (Fig. 11). The stimulation of dopamine autoreceptors results in a decrease of DA synthesis (Argiolas *et al.*, 1982). This inhibition might be produced either by dendrodentritic synapses between adjacent neurons, by non-synaptic diffusion of dopamine to autoreceptors, or by a combination of both.

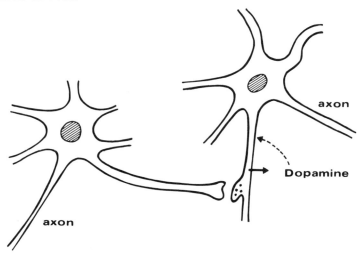

Fig. 11. Self-inhibition by dopamine of firing of substantia nigra dendrites (Groves, 1979). Note that there is dendro-dendritic contact (Cuello and Iversen, 1978)

There is no evidence of synaptic vesicles or dendrodentritic contacts in nigral cell bodies and dendrites labelled by either ^{3}H-dopamine or 5-hydroxy-dopamine. Nevertheless, the observations of Hefti and Lichtensteiger (1978) indicate that dendrites of the DA neurons in substantia nigra can form particles which behave like synaptosomes in density gradient centrifugation: these may be termed dendrosomes.

Cheramy *et al.* (1981) asked the question as to 'how DA reaches the dopaminergic autoreceptors located on parent or other dopaminergic neurones in the SN. Are there functional dendrodendritic synapses (Mercer *et al.*, 1979) or appositions (Cuello and Iversen, 1978), as suggested by some authors, or are the dendrites always separated by glial elements connected by multiple gap functions, as claimed by other workers (Reubi and Sandri, 1979)? The latter features could observe a modulatory role of DA in the substantia nigra which would explain the long-term changes in activity of the dopaminergic neurones seen in several of our *in vivo* experiments.'

Dendrites in substantia nigra are always in a post-synaptic position to other neurones (Grofova and Rinvik, 1970; Gulley and Wood, 1971; Hajdu *et al.*, 1973). They have been reported to contain vesicles which, however, are not connected to pre-synaptic structures (Hajdu *et al.*, 1973). These observations make it unlikely that DA released from dendrites exerts its effect on typical synapses, but has rather a more general role, affecting a larger area. Interestingly, like the dendrites of the DA neurons in the substantia nigra, the axons of the dopaminergic tubero-infundibular system in the median eminence do not form classical synapses (Ajika and Hökfelt, 1973) (for details, see Section 3).

In addition to dopamine, other neurotransmitters such as 5-hydroxtryptamine (Palkovits *et al.*, 1974), GABA (Fahn and Cote, 1968; Kanazawa and Jessel, 1976; Kim *et al.*, 1971) and substance P (Brownstein *et al.*, 1976, Kanazawa and Jessel, 1976) are present in the substantia nigra. The GABA terminals and the substance P-containing neurons (Kanazawa *et al.*, 1977) originate from the caudoputamen-globus pallidus region. There is evidence that the serotoninergic terminals in substantia nigra originate from the raphé nuclei (Dray *et al.*, 1976; Kuhar *et al.*, 1972a). Therefore it is very likely that their release is also modulated in the substantia nigra by dopamine released from dendrites.

One important question is to determine whether all central neurons share the capacity to release their transmitters from dendrites. It has already been suggested that GABA could be released from dendrites in the olfactory bulb and Cuello *et al.* (1982) have shown that in *in vitro* turtle olfactory bulb preparation there is a depolarizing potential due to direct feedback of dendritically released excitatory transmitter into the same and neighbouring neurons.

According to Cuello's hypothesis, therefore, it is highly probable that DA released from dendrites of substantia nigra acts pre-synaptically (Fig. 10), modulating the release of other neurotransmitters/modulators. In addition, DA released from dendrites may exert a direct feedback inhibition of the neuron from which it is released (Fig. 11).

2.3 CO-RELEASE OF MODULATORS/TRANSMITTERS

It is generally accepted that, according to Dale's Principle, each nerve cell has only one transmitter; therefore it synthesizes and releases only one transmitter (Dale, 1935). However, in the last few years Dale's Principle has been challenged by different experimental data and a general review of this concept has been published by Burnstock (1976) and Osborne (1981).

A neuron possesses specific mechanisms for synthesis, storage, axonal transport, release and uptake of a neurotransmitter. If more than one transmitter is released from a neuron, then it should synthesize, store, and transport more than one substance. Evidence for the co-existence of transmitter/modulator substances in the same neuron mainly derives from immunohistochemical studies. Therefore until there is direct neurochemical evidence of the co-release of transmitter/modulator substances from a single neuron care is needed in the interpretation of any new data or proposals concerning co-release mechanisms. However, even when two substances are stored in and released from a neuron, it does not necessarily mean that both substances have a transmitter/modulator function. This function has yet to be demonstrated. For example, Osborne (1977, cf. 1981) found an extremely small amount of serotonin in a cholinergic cell of the snail: there was a 10 000-fold difference in their contents. It is very unlikely that serotonin could serve as a transmitter or modulator and it is a possibility that the presence of serotonin is only of phylo- and/or ontogenetic significance, with no physiological importance. Each neuron, equipped with a set of genes, has the potential to build up the complete enzymatic machinery for all transmitter/modulator substances. For example, it is known that isolated sympathetic neurons can have either noradrenergic or cholinergic characteristics (Reichardt and Patterson, 1977), depending on the circumstances. Therefore conclusions drawn from seeking different putative transmitter/modulator substances in one neuron should be treated with caution.

2.3.1 Synthesis

It has been shown (Eisenstadt et al., 1973) that several identified cells of Aplysia californica convert choline into acetylcholine (ACh), and that other cells synthesize serotonin from triptophan when the precursors are directly injected into the cell bodies. Synthesis of both transmitters does not occur in the same neuron. However, Goldman and Schwartz (1974) showed that when 5-hydroxytryptophan was injected into the cell body of the cholinergic neurons, a significant amount of serotonin was synthesized by the cholinergic neuron but it was neither stored in vesicles nor transported along the axon. In contrast, newly synthesized serotonin was stored and transported by the serotoninergic neuron.

2.3.2 Storage and axonal transport

A neuron possesses specific mechanisms for the storage and axonal transport of a transmitter which in fact characterize its function. It seems that these mechanisms are neuron-specific. Goldman and Schwartz, (1974) have shown that the R_2 cell of Aplysia californica is able to synthesize serotonin, but this alien transmitter is neither stored in vesicles nor transported along the axon of the cell. This cell does, however, transport acetylcholine from the cell body to the axon (Koike et al., 1972). Certain nerve cells in gastropod molluscs and flatworms have been reported to contain both serotonin and dopamine (Welsh and Williams, 1970) and Brownstein et al (1974) have evidence that ACh is present together with serotonin in certain cells of Aplysia californica, but its concentration is nearly two orders of magnitude less than that of serotonin. Storage of both serotonin and noradrenaline has been shown in one nerve terminal of the rat vas deferens (Thoa et al., 1969) and, an accumulation of serotonin has been observed in nerve vesicles containing noradrenaline in the rat vas deferens by Zieher and Jaim-Etcheverry, 1971. In addition, it has been shown (Jaim-Etcheverry and Zieher, 1975) that the sympathetic nerve-endings in the rat pineal gland store no fewer than three different amines (serotonin, NA, and octopamine), although, there is no evidence that all three amines are released.

Burn and Rand (1959, 1965) have suggested that ACh is present in noradrenergic fibres and forms an intermediate link between nerve impulses and the release of noradrenaline: when the nerve impulse reaches the noradrenergic nerve terminals there is a release of ACh followed by stimulation of nicotinic receptors which, in turn, enhances the influx of calcium resulting in a release of noradrenaline. In addition, Burn (1976) suggested that in the course of evolution the sympathetic fibre has changed from cholinergic to adrenergic. This hypothesis was supported by the misleading report that small agranular vesicles were present in small number in adrenergic and peptidergic neurons. It has since been demonstrated that these vesicles were empty and were an artefact due to insufficient fixation (Tranzer and Thoenen, 1967). It has been shown (cf. Muscholl, 1973a) that in low concentration, ACh reduces the release of NA, and only in high-concentrations does it enhance NA release, and therefore ACh should be present in a significant amount in the adrenergic neuron. To test this hypothesis, 6-OH-dopamine was used, since it has been shown that the effect of this substance in the central and peripheral nervous systems is specific to noradrenergic neurons without affecting nerves containing other transmitters. If a significant amount of ACh were stored in sympathetic neurons, the ACh tissue content should be decreased when the sympathetic nerves are destroyed. This was not the case (Vizi, 1977a); 6-OH-dopamine pretreatment failed to significantly change the ACh content and Table 3 shows that chemical sympathectomy failed to affect the ACh content of rat vas deferens (Vizi, 1977a). Consolo et al (1972) have also observed that chemical sympathectomy, like surgical sympathectomy (Ehinger et al., 1970), does not affect the content of ACh in the cat iris.

Table 3. The failure of chemical sympathectomy to decrease acetylcholine content in the central and peripheral nervous systems

	Controls	6-OHD[a]	
Cortex (rat)	4.82±0.55 (4)	5.01±0.23 (4)	ns
Diencephalon (rat)	8.62±1.0 (4)	12.4±3.63 (4)	ns
Vas deferens[b] (rat)	7.14±2.56 (3)	6.51±0.39 (4)	ns
Auerbach's plexus (guinea-pig) ileum	78.1±6.6 (8)	70.1±8.1 (5)	ns

[a]6-OH-dopamine (6-OHDA) (100 µg) was administered intraventricularly 7 days prior to experiments when rat cortex or diencephalon were studied. It was given intravenously (2×170 mg/kg) 16 and 6 n prior to experiment when guinea-pig Auerbach's plexus and rat vas deferens was studied.
[b] Data taken from the paper of Knoll et al. (1972). Number of experiments given in brackets.
ns: not significant.

Further evidence that different transmitters are not stored together is that following 6-OH-dopamine treatment the 5-HT content in different cat tissues is unaltered. (Votavova et al., 1971).

Immunohistochemical data (Hökfelt *et al.*, 1980a, Hökfelt *et al.*, 1980b) has indicated that a cholecystokinin-like peptide co-exists with dopamine (DA) in neurons of the ventral tegmental area and bilateral 6-OHDA lesions induce a parallel decrease in CCK-8 and DA levels in the nucleus accumbens area (Studler *et al.*, 1981). However, 6-OHDA or electrolytic lesions were without effect on CCK-8 levels in the anterior part of the nucleus accumbens, the rostral part of the striatum, or the parietal cortex (Emson *et al.*, 1980).

More recent evidence, however, indicates that opiate-like peptides may co-exist with noradrenaline in sympathetic C-fibres, possibly in the large, dense-cored noradrenergic vesicles of this nerve (Wilson *et al.*, 1980), and it is also possible that opioid peptides may be co-transmitters with noradrenaline in the adrenal medulla as well as in some sympathetic nerves. Nevertheless, there is as yet no evidence for the presence of endogenous opiate peptides in the peripheral tissues in which the presence of pre-synaptic inhibitory opiate receptors has been clearly established (mouse vas deferens, cat nictitating membrane).

2.3.3 Release

The first serious challenge to Dale's Principle was the finding that hypothalamico-neurohypophysial neurosecretory fibres which release oxytocin and vasopressin might contain and release ACh from the same nerve terminals (Abrahams *et al.*, 1957). Burn and Rand (1965) postulated that ACh was liberated from the adrenergic nerve terminals first, and was followed by a release of NA. In addition, Singh (1964) claimed that four transmitters, ACh, serotonin, histamine, and a polypeptide, are released from the same nerve terminal of the vagus nerve running to the stomach of the frog. The main criticism of these data is that there is no direct evidence that the transmitters released came from the same neuron; the possibility exists that they came from different axon terminals. When a nerve bundle is stimulated and either the release or the response of effector cells is measured, the transmitter released could come either from the same nerve terminals or from different axons containing different transmitters.

Chemical sympathectomy by 6-OHDA (Vizi, 1977a) failed to affect the rate of ACh release (Vizi, 1977a) at rest and in response to stimulation (Table 4). If the adrenergic nerve was able to release ACh in significant amounts the destruction of the nerve should have influenced the release of ACh. The fact that the selective destruction of the adrenergic nerves failed to affect this release indicates that there is no significant ACh release derived from adrenergic fibres.

Cottrell (1976) has presented evidence that the giant cerebral neuron of *Helix* may release ACh as well as 5-HT, but a similar phenomenon was not observed in mammalian tissue.

Table 4. The effect of 6-OH-dopamine on the acetylcholine release from the vas deferens of the rat

Conditions	Control	6-OH-dopamine pre-treatment[a]	
Resting	12.2±2.86 (5)	9.85±1.8 (5)	ns
0.1 Hz	8.9±1.2 (5)	8.86±3.2 (5)	ns
10 Hz	170.9±20.6 (4)	231.1±74.0 (4)	ns

[a]6-OH dopamine (2 × 70 mg/kg; i.v.) administered 16 and 6 h prior to experiment.
±: Standard error of the mean.
(): Number of experiments.
ns: Not significant.

2.3.4 Co-release or release of acetylcholinesterase and other peptides

An unexpected observation was made by Chubb and Smith (1975a,b), who showed that acetylcholinesterase (AChE) was released from the adrenal gland. This is the first case in which an enzyme responsible for the destruction of a transmitter has been shown to be released from a neuron. This observation raised the question of the possible physiological role of the secreted AChE. Since a release of ChE from an isolated ganglion and its appearance in rabbit cerebrospinal fluid (Chubb *et al.*, 1976; Greenfield and Smith, 1976) and in perfusates from cat cerebral ventricles (Fuenmayer *et al.*, 1976) has been observed, it is very likely that neuronal AChE is secreted from the nerve terminals, and that it plays a physiological role in the modulation of neurochemical transmission. Once it is released, it should protect the nerve terminals from ACh released from the cholinergic neurons which otherwise might reduce the release of NA by stimulation of the muscarinic receptors (cf. Muscholl, 1972). Alternatively it may only affect the actual concentration of ACh in the synaptic cleft.

The question arises as to how widespread the ability of neurons to secrete AChE is. AChE is secreted from CNS neurons into the cerebrospinal fluid, and this may be only one of several forms in which it is secreted (Chubb *et al.*, 1976). Cytochemical evidence presented by Kreutzberg and Schubert (1975) has shown that motor nerves probably also secrete AChE from their dendrites and there is some evidence that protein is secreted from the cholinergic nerve-endings of the mouse hemidiaphragm (Musik and Hubbard, 1972). The protein released by nerve stimulation is dependent upon calcium, and is independent of whether nicotinic receptors of the muscle are paralysed (d-tubocurarine). There is reasonable evidence therefore that motor nerves can release some protein from their terminals (Musik and Hubbard, 1972), from their axons (Skangiel-Kramska and Niemierko, 1975), and from their dendrites (Kreutzberg and Schubert, 1975).

The most important question that arises is whether the proteins which are secreted from nerves have any extracellular functions. The secretion of proteins from sympathetic nerves (e.g. chromogranin) and cholinergic motor nerves has been reviewed by Smith (1971) and by Chubb (1977) with the conclusion that the chromogranins released have no function outside the cell.

In 1905, Scott proposed that '. . . nerves act upon one another and on other cells by the passage of a chemical substance of the nature of a ferment or proferment', i.e. the secretion of an enzyme or protein was suggested. More than sixty years after Scott's proposal Smith and his colleagues demonstrated that nerves secrete proteins from their terminals (De Potter *et al.*, 1969a,b; Smith, 1971). Whether these molecules have any role in the change of sensitivity of neurons or influence the termination of released transmitter (e.g. whether AChE affects the biophase concentration of ACh) is still a matter of speculation.

2.3.5 ATP as co-transmitter

There is a growing body of evidence that nerve activity is accompanied by release of purines. Holton and Holton (1953) were the first to describe the release of adenine nucleotides at sensory nerve-endings and Burnstock *et al.* (1970b) later showed that ATP or related nucleotides are released from purinergic nerves present in Auerbach's plexus and in other tissue (cf. Burnstock, 1972). ATP has been shown to be stored together with NA and ACh and it is also thereby co-released with transmitters, during exocytosis (see section 6.1).

2.4 NON-CONVENTIONAL SIGNAL TRANSMISSION BY PEPTIDES

The suggestion that peptides may also act as neurotransmitters/neuromodulators is a relatively recent one in neurobiology, only becoming respectable within the past five years. At present about thirty peptides, are under consideration as transmitters/modulators (cf. Emson, 1980).

A number of biologically active peptides, many traditionally considered as hormones of the gastrointestinal tract or as constituents of skin are also stored in peripheral (Costa and Furness, 1982) and central neurons (Hökfelt *et al.*, 1980c, Pearse, 1976, 1977) and may be transmitters, modulators, or both.

Both skin and nervous tissue arise from ectoderm and therefore, there may be a possible link between peptides in brain, skin, and enteropancreatic organs (brain-skin-gut axis) through their common ectodermal origins (Teitelman *et al.*, 1981). Since many of them exert endocrine, hormonal action as well as a paracrine action, they are able to influence neuronal activity and they might be non-synaptic transmitters/modulators.

2.4.1 Transmitter-like peptides

This group of peptides, are those which are only able to excite other neurons, mainly on the dendrites and/or soma. Whether they are released far from the target cell or whether there is some contact between the release site and effector cell has not yet been studied in detail. Nevertheless, in the case of cholecystokinin (CCK) it seems that there is a remote interaction between two neurons, however, substance P appears to work as a classical transmitter, released into a small synaptic cleft, transmitting sensory impulses.

Substance P

Substance P is a peptide consisting of eleven amino acid residues. In 1931 von Euler and Gaddum were looking for acetylcholine in extracts of brain and intestine when they found a substance that caused intestinal contractions and lowered blood pressure. This substance was not acetylcholine, so they simply designated it as substance P because it was sorted in powder form in a certain preparation of their tissue extracts.

The suggestion that the peptide might be a neurotransmitter dates back to the mid 1950s, when Fred Lembeck (1953), of the University of Graz in Austria, found that the dorsal root, a nerve tract carrying sensory impulses from the periphery to the spinal cord, contains ten times more substance P than the ventral root. He suggested that the substance P in the dorsal root might be a neurotransmitter.

The gastrointestinal tract

Hökfelt et al. (1980d) identified immunoreactive substance P in nerves of the autonomic nervous system. In addition, there is histochemical evidence that substance P is present in the gut plexus (Costa *et al.*, 1980; Schultzberg *et al.*, 1980). In isolated intestinal preparations both tetrodotoxin and atropine failed to alter the contractile response induced by substance P (Rosell *et al.*, 1977). However, Holzer and Lembeck (1980) noted that after the initial contraction elicited by substance P had reached its peak, the recovery phase to the pre-stimulated baseline could be modified by tropicamide, physostigmine, morphine, or tetrodotoxin. It has been recently observed (Yau and Youther, 1982) that substance P releases ^{3}H-acetylcholine in a dose-related fashion. Both tetrodotoxin and (D-Pro^2, D-Phe^7-D-Trp^9) – substance P, a synthetic antagonist of substance P, completely blocked this release and provided evidence for a cholinergically mediated mechanism of substance P on enteric neurons. This finding strongly supports a neural mechanism of substance P through the release of acetylcholine from enteric cholinergic components. The failure in previous studies (Rosell *et al.*, 1977; Yau, 1978) to detect this cholinergically mediated action in isolated gut is related to the method used. Only the initial peaks of mechanical contraction were measured, and these were perhaps not related to ACh release.

The central nervous system

There is growing evidence that substance P is a neuromodulator or transmitter in the mammalian central nervous system. It has been shown that it may be a transmitter of brain signals carried by the sensory nerves into the spinal cord and then relayed to the brain. This concept is supported by the finding that the endorphins and enkephalins, (the brain's built-in opiates,) may produce their analgesic effects by reducing substance P release in the spinal cord. Nevertheless, substance P has powerful excitant effects on neurons (Konishi and Otsuka, 1975; Barker, 1976) and it is unevenly distributed in the central nerous system. The peptide is released from new-born rat spinal cord during dorsal-root stimulation (Otsuka and Konishi, 1976) and from hypothalamic slices after high-potassium stimulation.

Lembeck has suggested that substance P is the transmitter by which the incoming sensory fibres communicate with neurons in the dorsal horn of the spinal cord. His finding was confirmed by Otsuka (Otsuka and Konishi, 1976). Substance P has fulfilled about half the criteria (see section 1.1) in support of its postulated role as a transmitter of pain impulses. Several lines of evidence suggest that substance P is present in nerve fibres that carry pain signals. Hökfelt has shown that many of the small-diameter nerve fibres of the dorsal horn, which have traditionally been considered to be pain fibres, contain substance P.

Moreover, experiments with capsaicin, a chemical substance isolated from Hungarian red peppers, suggest that substance P neurons are pain nerves. Administration of capsaicin causes an immediate sensation of intense pain in humans and experimental animals: after prolonged administration, however, the animals become insensitive to painful stimuli. It has been shown (Jessel et al., 1978; Nagy et al., 1980) that capsaicin causes the release of substance P from the dorsal horn, which could account for the initial pain sensations. Eventually the peptide becomes depleted, a circumstance that may explain the decrease in pain sensitivity caused by prolonged capsaicin treatment.

Using dispersed cell culture Mudge et al. (1979) provided convincing evidence that high K^+ concentration (30–120 mM) causes the release of a substance with P-like immuno-reactivity by a Ca^{2+}–dependent mechanism, and that an enkephalin analogue at 10^{-5} M inhibits this evoked release. Hunt et al. (1980) failed to show a clear axo-axonic relationship between primary afferent terminals and enkephalin-positive terminals in the superficial layers of the spinal cord.

In contrast to the above-mentioned effect of substance P, where it acts either on the cell body or on the axon hillock producing depolarization, a pre-synaptic effect of substance P is also described (Starr, 1978). Substance P (1 μM) facilitates the potassium induced release of ^{3}H-dopamine from dopaminergic nerve terminals in the rat striatum but not from the cell bodies in the substantia nigra (Starr, 1978).

Cholecystokinin

Since the identification of the carboxyl (C)-terminal octapeptide of cholecystokinin (CCK-8) in brain extracts, a new proposal for a function of gastrointestinal CCK has emerged (cf. Dockray, 1982). It has been found that numerous CCK-8-containing interneurons are present in the cerebral cortex and other brain structures such as the hippocampus or the amydala (Hökfelt *et al.*, 1980a,b). The ventral part of the pre-frontal cortex contained the highest levels of CCK-8 and DA. DA levels are negligible in the parietal cortex but substantial amounts of the peptide were found in this part of the anterior cortex. These data do not support a localization of CCK-8 immunoreactivity in cortical DA terminals since 6-OHDA lesions, which induced a near-complete degeneration of the mesocortico-pre-frontal DA pathway, were without significant effect on CCK-8 levels in the ventral part of the pre-frontal cortex. In Dockray's experiments (1982) the highest level of CCK was found in the amygdala (Fig. 12), but in the cerebellum, however, its content was negligible. Considerable progress has been made to prove

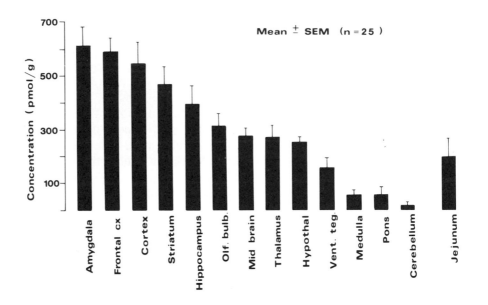

Fig. 12. Distribution of CCK in rat brain (redrawn from Dockray, 1982). (Reproduced by permission of Longman Group Limited)

the transmitter role of CCK-8: it is synthesized in neurons, packaged into granules or vesicles, and transported to nerve-endings (Dockray *et al.*, 1981). CCK-8 is released from synaptosomes and brain slices by depolarization in the presence of Ca (Dodd *et al.*, 1980; Emson *et al.*, 1980). Electrophysiological evidence has suggested (Dodd and Kelly, 1981; Jeftinija *et al.*, 1981) that CCK-8 has an excitatory depolarizing action on neurons (pyramidal cells of the rat hippocampal slices, spinal cord). In addition, neurochemical evidence has shown that CCK is able to stimulate cell bodies of the enteric plexus, resulting in a release of ACh (Vizi *et al.*, 1972, 1973a; Vizi, 1973). It was suggested that CCK is a long-distance stimulant of a population of intrinsic enteric cholinergic neurons (Vizi *et al.*, 1973a,b; cf. Bertaccini, 1982). This idea was based on pharmacological and neurochemical evidence (Vizi *et al.*, 1972, 1973a,b; Vizi, 1973) that cholecystokinin and related polypeptides (gastrin, coerulein) release ACh from the Auerbach's plexus of the intestine (Fig. 13). Evidence has been presented that CCK and related peptides stimulate the cell body and/or dendrites, enhancing neuronal firing in the Auerbach's plexus (Fig. 14). Octapeptide amide of CCK-PZ, gastrin I, coerulein, and pentagastrin are all able to release acetylcholine from the nerve elements of longitudinal muscle strip of guinea pig ileum. The threshold concentration was as low as 9×10^{-10}M. Noradrenaline and tetrodotoxin inhibited the effect of polypeptides on acetylcholine release. In contrast, hexamethonium failed to alter the acetylcholine-releasing effect of the polypeptides studied. The contractile effect of gastrointestinal hormones and related polypeptides on the intestine can be regarded as being due to acetylcholine release only, but some other mechanism may also be involved.

Physostigmine, potentiated, whereas atropine and tetrodotoxin completely abolished the contractions produced by both electrical stimulation and peptides.

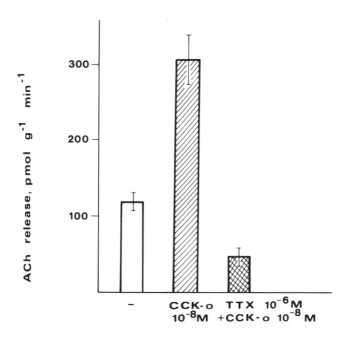

Fig. 13. The release of acetylcholine by cholecystokinin-octapeptide (CCK-o). The data of Vizi *et al.* (1973a) are redrawn. Longitudinal muscle strip of guinea-pig ileum, TTX, tetrodotoxin. Note that tetrodotoxin inhibits the ACh-releasing effect of CCK-o.

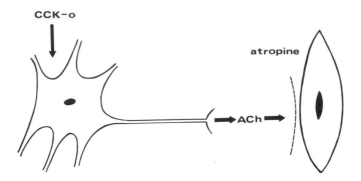

Fig. 14. Possible site of action (soma or dendrite) of cholecysto-kinin-octapeptide (CCK-o) released from remote CCK-containing cells in Auerbach plexus's of the gut. Tetrodotoxin prevents its action, therefore its site of action is proximal to the axon. Atropine also blocks the effect of cholecystokinin by preventing the effect of ACh released by CCK-o) on the smooth muscle cell

It has also been found that the response by isolated rabbit jejunum to polypeptides is blocked by sympathetic stimulation under conditions in which a possible action of noradrenaline on effector cells is excluded. The inhibitory action of sympathetic stimulation was completely prevented by a receptor blocking agent. These data provide evidence for the possibility of sympathetic control of acetylcholine release by gastrointestinal hormones under physiological conditions.

The question now arises as to whether, under physiological conditions, the gastrointestinal hormones can act as factors influencing gastrointestinal motility through acetycholine release (Vizi *et al.*, 1973a).

2.4.2 Pre-synaptic modulators

Angiotensin

Pre-synaptic stimulation of transmitter release is also a way of modulating synaptic transmission. Pre-synaptic receptors may be the sites of action of modulators, such as angiotensin, which originate from a remote part of the organism and are transported by the bloodstream. Angiotensin is one candidate, as pre-synaptically effective concentrations of angiotensin occur in plasma, at least when renin secretion is high.

The peripheral nervous system

Angiotensin II facilitates the release of noradrenaline caused by nerve stimulation from several post-ganglionic autonomic nerves, probably through the activation of specific pre-synaptic angiotensin receptors. This phenomenon has been demonstrated *in vitro* (Zimmerman and Whitmore, 1967; Zimmerman, 1978; Zimmerman and Kraft, 1979; Starke, 1971; Garcia-Sevilla *et al.*, 1981).

The central nervous system

Angiotensin II does not modify the release of ^{3}H-noradrenaline when it is evoked from slices of cerebral cortex or hypthalamus of the rat (Starke *et al.*, 1975; Taube *et al.*, 1977), but, when using the rabbit hypothalamus as an experimental model, angiotensin II facilitates the potassium-induced release of ^{3}H-noradrenaline (Garcia-Sevilla *et al.*, 1979). This facilitatory effect is reduced by the receptor antagonist, saralasin (Garcia-Sevilla *et al.*, 1979). Captopril, which inhibits the angitotensin-converting enzyme, may reduce sympathetic tone through the decrease in the circulating levels and the local formation of angiotensin II.

The physiological role of angiotensin in the central nervous system is currently a subject of research. Evidence has been obtained that it may play some role in control of thirst, neurogenic pressor action, and antidiuretic hormone release (Ganten *et al.*, 1971; Fischer-Ferraro *et al.*, 1971). There is increasing evidence of an endogenous brain renin–angiotensin system (cf. Ramsey, 1979), particularly from immunhistochemical findings, which suggests that angitotensin is synthesized locally in the brain. Fuxe and his colleagues (Fuxe *et al.*, 1976) have reported immunoprecipitates in nerve terminals but not in nerve cell bodies, and it has been shown that the brain possesses an angitotensin receptor binding capacity higher than in any other body tissue. Nevertheless the release of angiotensin from brain slices or synaptosomes has not yet been demonstrated, although Haas and his colleagues (1982) suggested a release of angiotensin from neurons. Angitotensin-sensitive neurons were described in the hypothalamus, the septum, and particularly in the subfornical organ, and these are believed to be the sites for the dipsogenic action of angiotensin II. Saralasin often decreased spontaneous activity of neurons, which supports the argument that endogenous brain angitotensin II maintains the spontaneous neuronal firing frequency. Angiotensin may have a dual function: as a true hormone circulating in the blood and as a modulator which is released from angiotensinergic neurons that might act on neighbouring cells.

2.4.3 Post-synaptic modulators

Vasoactive intestinal polypeptide (VIP)

More than hundred years ago it was found that salivary secretion was accompanied by a marked increase in glandular blood flow (Bernard, 1858). Later, it was shown that atropine completely blocked salivary secretion produced by para-sympathetic nerve stimulation, whereas the accompanying vasodi-

40

lation was practically unaffected, and it was therefore postulated that a special set of vasodilatory nerves existed (Heidenhain, 1872). However, this is not the case, since botulinum toxin, an inhibitor of ACh release, simultaneously paralyses the salivary secretion and the vasodilation upon post-ganglionic parasympathetic nerve stimulation (Hilton and Lewis, 1955).

In agreement with Bloom and Edwards (1980), Lundberg *et al.* (1981) found that parasympathetic nerve stimulation at 15 Hz, which gives maximum vasodilation and secretion, caused a marked increase in VIP release. He found that VIP output at 2 Hz was not significantly changed after atropine administration, but at 6 Hz there was a clear-cut increase in VIP overflow after the atropine was introduced. At 15 Hz the release was markedly increased especially during the first stimulation after atropine (8–10 fold). The enhancement of VIP output by atropine may imply that there is a close anatomical relationship between the sites from which ACh and VIP are released. It is suggested that, at the level of the neuro-effector junction, the transmission process is mediated and regulated by the two agents ACh and VIP (Fig. 15). VIP seems to potentiate ACh-induced secretion; local infusion of exogenous VIP causes marked atropine-resistant vasodilation but no salivary secretion and ACh infusion induces both vasodilation and salivary secretion. As a vasodilating agent. VIP is about a hundred times more potent than ACh. The salivary secretion induced by ACh is potentiated by addition of VIP and ACh and VIP seems to be involved in both the vasodilatory and secretory responses. For the secretory response, ACh may serve as a neurotransmitter in a classical sense, while VIP acts as a post-synaptic modulator. However, ACh may act as a modulator to regulate VIP release and both ACh and VIP may be transmitters for vasodilation (Lundberg, 1981).

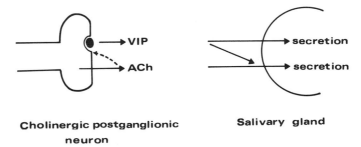

Cholinergic postganglionic neuron   **Salivary gland**

Fig. 15. VIP as post-synaptic (Lundberg, 1981) and acetylcholine as pre-synaptic modulators. Both VIP and acetylcholine (ACh) enhance salivary secretion, and VIP potentiates the effect of ACh. Acetylcholine inhibits the release of its co-transmitter (VIP)

2.4.4. Neurotensin

In 1967 Leeman and co-workers found a peptide with sialogogic activity in hypothalamic extracts which were later discovered to be substance P. With the isolation of substance P, Leeman and her colleagues searched for the ACTH-releasing factor from the hypothalamus. Instead they found a biologically active substance that lowered blood pressure in anaesthetized rats, which is a simple method of evaluating ACTH release (Carraway and Leeman, 1975). After purification, this substance, a tridecapeptide, was named neurotensin. Using immunocytochemical techniques, neurotensin was found to be most abundant in the anterior and basal hypothalamus, in the nucleus accumbens and septum, in the spinal cord and brainstem, in small interneurons of the substantia gelatinosa, and in the gut (cf. Rosell, 1980). Interestingly, 85% of rat neurotensin is stored in the mucosa of the ileum, in a specific type of glandular cell (N cell). Rosell (1980) suggested that it might have a paracrine or endocrine function, although it was also found that it is stored in the sympathetic pre-ganglionic neurons (Lundberg *et al.*, 1982).

Neurotensin is a potent hypotensive and hyperglycaemic peptide.

Effect on dopamine release

There is pharmacological evidence to suggest that there is a neurotensin–dopamine interaction in the central nervous system. Neurotensin stimulates the dopaminergic system (Andrade and Aghajanian, 1981) and *in vitro*, it enhances the potassium-induced ^{3}H-DA release from rat striatal slices (de Quidt and Emson, 1983).

Effect on acetylcholine release

Direct evidence has been obtained (Vizi, E. S. and Rosell, S., unpublished) that shows that neurotensin, like cholecystokinin, is able to excite cholinergic interneurons in the Auerbach plexus of the ileum and release ACh. Tetrodotoxin completely prevents this effect.

2.4.5 Luteinizing hormone-releasing factor (LHRH)

In frog sympathetic ganglia, both ACh and a LHRH-like peptide satisfy the criteria of transmitter (Kuffler, 1980, Jan and Jan, 1983a). A very slow synaptic potential mediated by a peptide was described (Kuffler, 1980). It was shown that LHRH-like peptide is present in the presynaptic terminals, its release is Ca^{2+}–dependent and the LHRH-induced depolarization mimicks the nerve-evoked late slow EPSP. It is highly probable that the peptide may be released together with ACh, and it acts to enhance the action of ACh, i.e. it serves a postsynaptic modulator action. In addition the LHRH-like peptide can be diffused and act on cells tens of micrometers away where the ACh has no action, i.e. it can act non-synaptically (Jan and Jan, 1983b).

2.5 PHYSIOLOGICAL IMPLICATIONS

This non-conventional non-synaptic release provides a possible mechanism for inter-neuronal communication and it seems likely that even peptides can be used for transmitting signals. Since Sherrington's classical work. *The Integrative Action of the Nervous System* it has become a statement of neurophysiology that the synapse, the surface of separation, is the primary locus of neuronal information-processing. It is now generally accepted that intercellular communication in the nervous system involves the release of chemical transmitters/modulators from an axon terminal. An important implication of recent findings is that this release may not be the exclusive property of axon terminal; it has been suggested (Vizi, 1979, 1981a,b, 1983, Cuello, 1983) that in addition to the classical synaptic form of interaction there is a non-synaptic form of communication between neurons when the transmitter or modulator released from axon terminals may affect remote target cells. Evidence has been obtained that a transmitter can also be released from the axon, although this release, unlike from axon terminals, cannot be modulated. It seems very likely that the released transmitter/modulator can reach surrounding target cells, neighbouring axons, or axon terminals, especially when they are not insulated, for example, in the case of the non-myelinated nerve.

Chapter 3

Morphological evidence of non-synaptic interactions between neurons

Axon terminals devoid of synaptic membrane specialization have long been the rule in the peripheral autonomic nervous system (Merrillees *et al.*, 1963; Taxi, 1965). It has been suggested that in the central nervous system the post-synaptic membrane specialization is not always apparent (Shepherd, 1974), and relative paucity or even a lack of synaptic junctions might be a feature common to all types of central monoamine terminals (Table 5). A similar situation has been documented for catecholamine- and indolamine-containing terminals in the median eminence (Ajika and Hökfelt, 1973; Calas *et al.*, 1974), 5-HT (Calas *et al.*, 1976), and presumptive dopamine-containing boutons in the neostriatum (Hökfelt, 1968; Tennyson *et al.*, 1974), 5-HT-containing nerve-endings in the cerebral ventricles (Richards *et al.*, 1973; Chan-Palay, 1976), one of the morphologically different types of varicosities labelled with ^{3}H–^{5}HT in the cerebellum (Chan-Palay, 1975), and 5-HT-containing nerve terminals in the locus coeruleus (Léger and Descarries, 1978) in organum vasculosum laminae terminalis and hypothalamus (Beaudet and Descarries, 1978).

Table 5. Relative paucity of synaptic junction, a feature common to central mono-amine axon terminals

Noradrenaline	*References*
Median eminence	Ajika and Hökfelt, 1973 Calas *et al.*, 1974
Frontoparietal neocortex	Descarries *et al.*, 1977
Dopamine	
Neostriatum	Hökfelt, 1968
	Tennyson *et al.*, 1974
Serotonin	
Median eminence	Calas *et al.*, 1976
Cerebellum	Chan-Palay, 1975
Locus coeruleus	Descarries and Leger, 1978
	Leger and Descarries, 1978
Cerebral ventricles	Richards *et al.*, 1973
Organum vasculosum laminae terminalis	Bosler, 1978
Hypothalamus	Beaudet and Descarries, 1978
Cerebral cortex	Beaudet and Descarries, 1976

The mode of action of central monoaminergic terminals lacking morphologically defined synaptic contacts was first envisaged in the case of presumptive dopamine-containing boutons in the neostriatum (Tennyson *et al.*, 1974) and 5-HT-containing varicosities in the cerebral cortex (Descarries *et al.*, 1975). The varicosities may have dynamic properties, and may be translocated and/ or reshaped along their parent fibres, allowing incessant fluctuation of their functional domains.

There is, as yet, no firm evidence for the non-synaptic release of the communicator (transmitter, modulator). A very critical question is, are serotonin and norepinephrine normally released from non-synaptic boutons? There is no direct evidence on this point: indeed, it would be methodologically difficult to study. However, there is good reason to believe it. As Descarries and co-workers pointed out (1977), non-synaptic varicosities appear to have all the apparatus normally associated with synaptic release. Direct evidence for non-synaptic release of a peptide has been obtained in *Aplysia* since neurosecretory bag cells of the abdominal ganglion (Coggershall, 1970) terminate in connective tissue, releasing products directly into the hemolymph for diffusion, affecting the activity of various types of target cell in different locations.

In the spinal cord, substance P has been detected immuno-cytochemically in extracellular spaces, in patterns suggesting that it can be released from neurons in packets that diffuse to neighbouring cells and capillaries (Chan-Palay and Palay, 1977).

There may be some doubt about the non-synaptic release of modulators in the central nervous system, but this concept is well established for the autonomic nervous system, where for example acetylcholine and noradrenaline are released from axon terminals devoid of synaptic contact (see section 3.1), which makes it a plausible concept for the central nervous system.

3.1 NON-SYNAPTIC INTERACTIONS IN THE AUTONOMIC NERVOUS SYSTEM

There are two autonomic systems; sympathetic and parasympathetic. The enteric nervous system is a part of the autonomic nervous system and in fact was originally classified by Langley (1921) as the third division of the autonomic nervous system (Furness and Costa, 1980; Gershon, 1981). All autonomic functions are mediated through the release of a transmitter substance. The autonomic nervous system controls functions that have many sites of action among the organs and most of the functions are visceral. Its postganglionic fibres are unmyelinated (C) fibres and their axon terminals are varicose.

3.1.1. Neuroeffector transmission

The anatomical arrangements at neuroeffector transmission sites are different from those at the neuromuscular junction. 'The mechanism by which each muscle fibre is influenced by many nerves is certainly not close contact between each muscle fibre and many nerve endings or axons' (Merrillees *et al.*, 1963). A transmitter-substance is released from areas of bare axon terminals, many of them within a micron of muscle membrane, which produces high concentrations of transmitter over discrete rgions of the smooth muscle cell surface, as well as a general increase in the concentration of transmitter in the extracellular space. This concept has some similarity with that developed by Rosenblueth (1950). Bare axons are also found in some of the larger extracellular spaces where drugs would have easy access to bare axons and smooth muscle membranes. 'There seems no reason why the release of transmitter should be limited to those regions of bare axon which make intimate contact with the smooth muscle membrane. It may well be that the transmitter can pass through the basement membrane surrounding those parts of the axon which are free of Schwann cells, but situated more than 200 Å from the muscle membrane' (Merrillees *et al.*, 1963).

The neuroeffector junctions of the peripheral nervous system has been reviewed by Burnstock and Costa (1975) and by Burnstock (1979a). Gaps between noradrenergic boutons and effector cells in the periphery vary considerably, and clefts range from 15–20 nm (e.g. vas deferens and iris) to 1000–2000 nm (e.g. large arteries).

It is interesting to note that the sensitivity of effector cells to transmitter is directly related to the distance between the varicosity, where the transmitter is released from, and the smooth muscle cell (Table 6).

Table 6. Distances between axon terminals and effector cells and their sensitivity to transmitter

Tissue	Transmitter	Distance[a] (nm)	Transmitter concentration (M) on target cell[b]
Longitudinal muscle—Auerbach's plexus	ACh	100–150	10^{-6}
Noradrenergic axon terminal			
cerebral cortex	NA	100–100	?
Pulmonary artery (rabbit)	NA	1000–2000	10^{-7}
ear artery (rabbit)	NA	500	5×10^{-6}
vas deferens (rat)	NA	10	10^{-4}
Neuromuscular junction	ACh	5	10^{-4}

[a]Data from Burnstock (1970).
[b]Determined by matching responses to neural stimulation and exogenous transmitter administration.

In neuroeffector junctions, whose membrane separation is greater than 20 nm, no post-junctional specialization is apparent, and in blood vessels, the closest proximity between varicosities and smooth muscle cells in about 80 nm. Generally, the greater the vessel diameter, the greater the separation of axon terminal and target muscle cell. Thus in arterioles and in small arteries and veins the separation is about 80–120 nm, while in large arteries the distance is 200–500 nm. However, in large elastic arteries where the innervation is sparser, the separations are as great as 1000–2000 nm (Burnstock, 1979a) and in the longitudinal muscle layer of the intestine and stomach, the distance between smooth muscle and autonomic or enteric nerves is rarely less than 100 nm. The essential features of this type of junction are that the terminal portions of autonomic nerve fibres are varicose, the transmitter being released *en passage* from varicosities (Fig. 16).

Auerbach's plexus receives a dense noradrenergic innervation, as seen through the light microscope using histochemical fluorescence techniques (Norberg, 1964; Jacobowitz, 1965; Furness, 1970; Gillespie and Maxwell, 1971; Costa and Gabella, 1971). It was found much earlier that many adrenergic fibres supply the myenteric plexus (Henle, 1879; Cajal, 1893, 1911). Langley and Anderson (1895) stated: 'It is hardly conceivable that some of the sympathetic nerve fibres should not become connected with the numerous nerve cells which are present in these plexuses, and the observations of Ramón y Cajal, von Gehuchten, Von Kölliker, and Dogiel show fairly conclusively that such a connexion exists.'

The myenteric plexus (Auerbach's plexus) lies between the longitudinal and circular muscle coats of the gut wall and adrenergic terminals ramify extensively amongst elements of the plexus and contribute to the internodal strands. At an ultrastructural level, adrenergic axons have been identified by their content of granulated vesicles, found from synapses in enteric neurones (Honjin *et al.*, 1963; Taxi and Droz, 1969; Gabella, 1971).

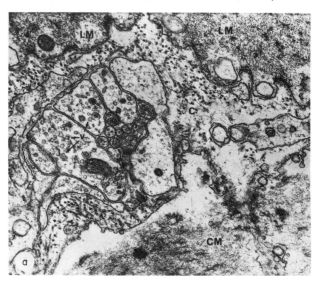

Fig. 16. Examples of varicosities without making synaptic contact. (a) Auerbach's plexus from the guinea-pig ileum; (b) arteria hepatica of the dog. Note the distance between varicosity and target cell

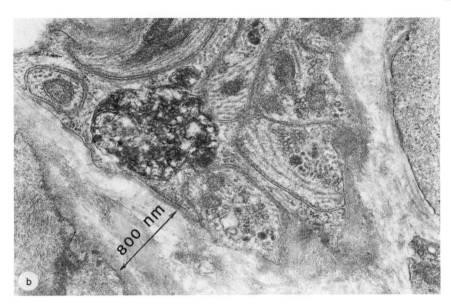

800 nm

b

It has been generally accepted that in the peripheral autonomic nervous system noradrenergic varicosities can be identified, using the electron-microscope, by the presence of synaptic vesicles containing small, electron-dense granules (Geffen and Livett, 1971). These techniques revealed that noradrenergic, sympathetic nerves were absent from the smooth muscle layers but were concentrated in the submucosal and myenteric (Auerbach's) nerve plexuses (Jacobowitz, 1965; Norberg, 1964). These findings prompted Norberg and Sjögvist (1966) to propose in their review article that the action of enteric nor-adrenergic neurones was to inhibit ganglionic transmission. Direct neurochemical evidence has since been obtained (Vizi and Knoll, 1971) showing that the stimulation of noradrenergic sympathetic nerves results in an inhibition of ACh release from the cholinergic interneurons of the Auerbach's plexus. In the light of these findings the intestinal relaxation that follows stimulation of sympathetic neurons may be attributed to a removal of an excitatory cholinergic tone. The observations made earlier by Norberg and Jacobowitz were misleading; these authors used freeze-dried material, and were unable to appreciate the three-dimensional arrangement of noradrenergic axons.

Ross and Gershon (1970), who used 6-hydroxydopamine to identify adrenergic nerves, reported that most adrenergic terminals in the myenteric plexus in the guinea-pig were remote from ganglion cells but were close to the smooth muscle. On the other hand, Gabella's (1972) description of this plexus indicates that none of the axons close to the smooth muscle have the ultrastructural characteristics of adrenergic nerves.

Recent morphological observations suggest that norad-renergic axon terminals are not on the ganglion cells. In fact, they have been found in electrophysiological studies (Holman *et al.* 1972) to be largely unresponsive to

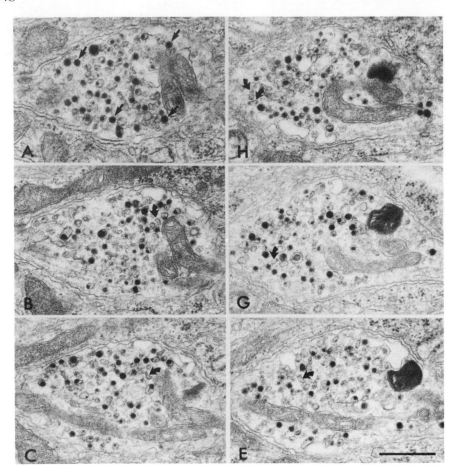

Fig. 17. Morphological evidence that there are two types of small granular vesticle-containing varicosities in Auerbach's plexus. The first type is noradrenergic and do not form synapses. The second is not noradrenergic, although it was formerly thought to be so. It is intrinsic to the gut and is resistant to the serotoninergic neurotoxin, 5, 6-dihidroxy tryptamine (Gordon-Weeks, 1982). Serial sections (17a) of a deep-lying, small granular vesicle-containing varicosity in Auerbach's plexus of the guinea-pig ileum pretreated by 5-hydroxydopamine, a marker of noradrenergic neurons. Note that many synaptic vesicles are filled with the marker (arrows in A). Low-power electron-micrograph (17b, on p. 49) of Auerbach's plexus in the guinea-pig ileum 24 h after extrinsic denervation. Note that there are degenerating varicosities (arrowheads). A small granular vesicle-containing varicosity (star) can be seen synapsing on the cell body of an intrinsic neuron. Bar marker: 1 μm. (Micrographs from Gordon-Weeks, 1981) (Reproduced by permission of Pergamon Press Ltd.)

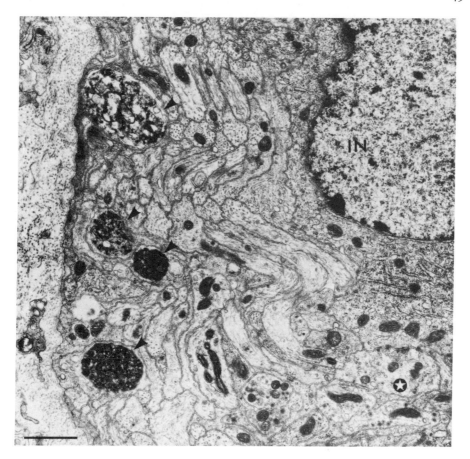

NA, but are around the cholinergic axon terminals. It has been shown that noradrenergic axon terminals do not make many axo-dendritic or axo-somatic synapses (Manber and Gershon, 1979). Small (50 nm) granular vesicles containing varicosities, some of which form axo-somatic and axo-dendritic synapses on intrinsic neurons, have been seen in Auerbach's plexus in the small (Cook and Burnstock, 1976; Gabella, 1971; Gabella, 1972; Llewellyn-Smith *et al.*, 1981) and large (Gordon-Weeks and Hobbs, 1979) intestine. These synapses were at one time considered to be noradrenergic (Gabella, 1971; Gordon-Weeks, 1981a), despite the fact that electrophysiological recording from intrinsic neurons in the small intestine has failed to demonstrate a direct synaptic input from noradrenergic axons (Hirst and McKirdy, 1974; Holman *et al.*, 1972, Nishi and North, 1973a,b); Takayanagi *et al.*, 1977). More recent indirect evidence suggests, however that noradrenergic varicosities do not form synapses in Auerbach's plexus (Furness and Costa, 1980; Gershon, 1981; Llewellyn-Smith *et al.*, 1981; Manber and Gershon, 1979; Gordon-Weeks, 1983). Gordon-Weeks (1981a,b) showed recently (Fig. 17) that there are two types of varicosities containing small granular vesicles in Auerbach's plexus of the guinea-pig ileum. The first type was located predominantly at the surface of the plexus and did not form synapses on intrinsic neurons. This type became labelled with 5-hydroxydopamine, a specific marker for noradrenergic axons, and was destroyed by 6-hydroxydopamine and extrinsic denervation, procedures which degenerate noradrenergic nerves in the gut. The second type formed axo-dendritic and axo-somatic synapses on intrinsic neurons, and the morphology of its synaptic vesicles differed subtly from that of the first type. It was unaffected by 5-hydroxydopamine, 6-hydroxydopamine, or extrinsic denervation.

Therefore it seems very likely that NA released from the varicose axon terminals reaches its target cells, the cholinergic varicosities, by diffusion, and inhibits the relase of ACh. That this kind of interaction between two axon terminals is of physiological importance is shown by the discovery that the stimulation of the sympathetic nerve results in an inhibition of ACh release from Auerbach's plexus (Vizi and Knoll, 1971; Manber and Gershon, 1979), although there is no synaptic contact between them.

Nevertheless, Manber and Gershon (1979) provided evidence that there is also some noradrenergic–cholinergic axo-axonic contacts in the mammalian gut, but these are negligible in number.

3.2 NON-SYNAPTIC INTERACTIONS IN THE CENTRAL NERVOUS SYSTEM

New data on the existence and distribution of dopamine (Berger et al., 1974, Thierry et al., 1973, Hökfelt et al., 1974) and serotonin-containing (Descarries et al., 1975) axon terminals in the cortex suggest that both dopamine and serotonin nerve-endings have a different distribution from the axonal boutons labelled with ^{3}H-NA (cf. Beaudet and Descarries, 1978). Within the same region, noradrenergic terminals predominate more clearly in the superficial layers (Fuxe et al., 1968; Lapierre et al., 1973), whereas dopaminergic terminals appear to be mainly confined to the deeper layers (Berger et al., 1974; Hökfelt et al., 1974; Lindvall et al., 1974). It has been suggested (Beaudet and Descarries, 1978) 'that noradrenaline- and 5-HT-containing neurons may be ideally designed to exert sustained influences in various regions of innervation and thus modulate several integrative and/or specific functions'.

3.2.1 Noradrenergic neurons

Ungerstedt discovered (Ungerstedt, 1971) that the NA neurons of the nucleus locus coeruleus (Corrodi et al., 1970; Dahlström and Fuxe, 1964; Fuxe et al., 1970a; Koslov et al., 1972; Kuhar et al., 1972b; Swanson, 1976; Ungerstedt, 1971) gives rise to the ascending dorsal NA bundle (Fuxe et al., 1970a; Ungerstedt, 1971).

The NA arborizations in the cortex are widely dispersed unmyelinated axons of very fine calibre, bearing small spherical enlargements spaced at short intervals (1–3 μm) (Descarries et al., 1977). The diameter of the intervaricose segments vary between 0.1 and 1 μm (average, 0.35 μm) and those of the varicosities range from 0.3 to 2 μm (average 1 μm). All varicosities seem to be deprived of glial ensheathment. In the varicosity two main categories of vesicles are present: small, round pleomorphic synaptic vesicles, 36–55 nm in diameter, and large, round or ovoid granular vesicles, 80–120 nm in diameter. Axonal varicosities contain agranular synaptic vesicles, often accompanied by one or more large granular vesicles and mitochondria. Within intervaricose segments, microtubules and smooth endoplasmic retiulum are visible, as well as scattered large granular vesicles and mitochondria (Descarries *et al.*, 1977).

When the tissue is exposed to tritiated noradrenaline, the ³H-NA is highly concentrated inside the varicosities and to a lesser degree within intervaricose segments. Only a very low proportion of cortical NA varicosities are found in a synaptic relationship (Descarries *et al.*, 1977). Among 1835 terminals, 341 of which were viewed in two or three adjacent sections, less than 5% exhibited typical junctional complexes, as opposed to 50% of unlabelled boutons similarly sampled in the neighbouring neuropil.

It is likely that endogenous cortical noradrenaline may be released from all axonal varicosities where it is concentrated, and not only from the small number which form typical synapses. Figure 18 illustrates the concept of

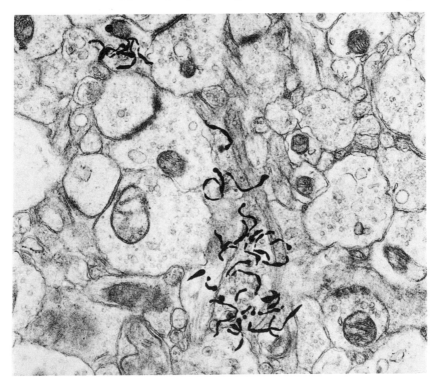

Fig. 18. Electron-microscope radioautograph of noradrenaline-containing varicosities labelled after topical application of ³H-NA (with the permission of Professor L. Descarries, unpublished). Adult rat frontoparietal cortex. The animal had been pre-treated with a monomine oxidase inhibitor. Note the clear identification of the labelled varicosities without making synaptic contact

an unconventional mode of innervation by noradrenaline afferents to the neocortex. The NA afferents may exert a diffuse, desynchronized, and tonic influence on large neuronal assemblies (Descarries *et al.*, 1977).

The recent investigation of axon terminal ultrastructure in the dentate gyrus of the hypocampus (Koda *et al.*, 1978) has shown that approximately 20% of boutons in the dentate gyrus displayed synaptic junctions irrespective of whether they possessed small granular vesicles. This study suggests a higher incidence of synaptic contact by noradrenergic varicosities than that found by Descarries and co-workers (cf. Beaudet and Descarries, 1978). The question arises as to why thsese differences exist. The procedures used for tagging aminergic boutons for electron-microscope identification can affect the apparent incidence of synaptic contact observed.

Whatever the distribution between synaptic and non-synaptic noradrenergic varicose axon terminals, the biogenic amine could diffuse over a short distance, possibly up to a micron or more, and have a relatively widespread influence on the neuropil (Fig. 19). This concept is analogous to that which occurs in the sympathetic nervous system, i.e. noradrenaline from sympathetic nerves is released into the connective tissue space before it reaches the smooth muscle or cholinergic varicosities. The observation that the chemical (by 6-OHDA) or anatomical (substantia nigra lesion) removal

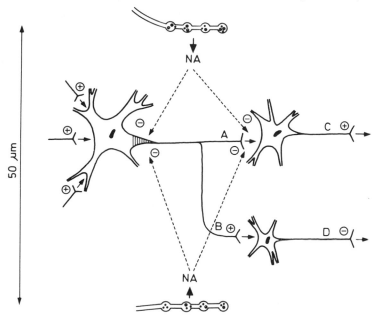

Fig. 19. Widespread inhibitory effect of noradrenaline (NA) released from remote varicose axon terminals on a neuron in the cortex. Dismukes (1979) calculated the longest distance between the noradrenergic axon terminal and the target neuron to be 15 μm. Note that only A axon terminal possesses receptors sensitive to NA; B, C, and D do not. This would be a way of discriminating them

of noradrenergic innervation in the cerebral cortex results in an increase of ACh release (Vizi, 1981b) suggests that there is a functional interaction between the two neurons.

3.2.2. Dopaminergic neurones

Hökfelt (1968) counted over 2500 boutons in the neostriatum and found that 16.4% contained tagged vesicles and that most varicosities did not make synaptic contacts. Tennyson *et al* (1974) studied dopamine-containing 'tagged' boutons in the rabbit neostriatum and found that out of 7800 vesicle-filled processes only 8.9% had tagged vesicles; about 2% of the tagged varicosities showing a distinct junctional contact. Most of these seemed to be symmetrical junctions, and only very few tagged boutons had asymmetrical junctions. This sparsity of synapses exhibiting asymmetrical post-synaptic thickenings is not consistent with the observations of others (Hattori *et al.*, 1973; Kemp and Powell, 1971). Kemp and Powell (1971) observed that after 6-OHDA administration the degenerated terminals formed assymetrical junctions, although these authors did not look for non-synaptic terminals. Hattori *et al.* (1973) injected ^{3}H-leucine into the substantia nigra, and reported that only 25% of the synapses labelled with silver grains ended in symmetrical junctions, whereas 75% ended in asymmetrical junctions. Since only the asymmetrical terminals degenerated after intraventricular injection of 6-hydroxydopamine, they suggested that dopaminergic terminals formed asymmetrical junctions (Hattori *et al.*, 1973). The discrepancy between the findings of Hattori *et al.* (1973) and that of Tennyson *et al.* (1974) is difficult to explain, but it does emphasize the fact that symmetrical junctions are difficult to find, and boutons lacking specializations may not have been considered in the analysis of the studies of Hattori *et al.* (1973).

It is possible that a tagged beaded axon containing an amine, such as dopamine, may have a dual-release mechanism, i.e. diffusion into the neuropil from the bare pre-terminal varicosities lacking junctions, as well as release limited to the cleft, at the synaptic endings.

It has been shown that the density of dopamine-containing axon terminals is very high in the neural lobe of the pituitary (Saavedra *et al.*, 1975) and that these axon terminals are in close proximity to the secretory axons and processes of pituicytes, but membrane-thickenings have not been demonstrated (Baumgarten *et al.*, 1972).

3.2.3 Serotonin neurons

It has been shown that when the number of labelled varicosities was expressed per nm³ of cortical layer, the serotonin nerve terminals appeared to be concentrated within the upper four layers, the overall density of cortical 5-HT innervation being estimated at almost 1×10^6 varicosities/nm³ of cortex (Beaudet and Descarries, 1976). This pattern of innervation differs markedly from that of other types of monoamine-containing afferents; the density of serotonin innervation is ten times higher than that estimated for noradrenergic afferents within the fronto-parietal region (Lapierre *et al.*, 1973), corresponding to an incidence of 0.07–0.12% of all cortical nerve-endings. This figure appears relatively low when compared with the proportion of nerve terminals labelled *in vitro* with tritiated DA (30% of all cortical synapses); (Bloom and Iversen 1971; Iversen and Bloom, 1972).

In the neocortex, endogenous 5-HT can probably be released from all axonal varicosities, not only from the small proportion making typical synaptic contacts (Descarries *et al*, 1975). It is thus likely that similar to noradrenergic fibres, serotonergic neurons do not exert effect solely upon restricted areas of post-synaptic membrane differentiation but can diffuse through tissue. The ^{3}H-5-HT-labelled serotonergic varicosities in Fig. 20.

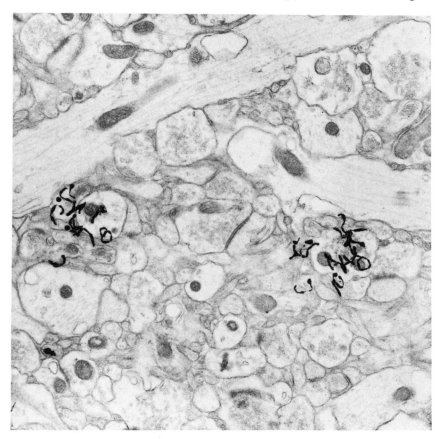

Fig. 20. Electron-microscope radioautograph of serotonin-containing varicosities in adult rat frontoparietal cortex, identified after topical application of ^{3}H-serotonin. Two adjacent varicosities, presumably belonging to the same axon, show no synaptic contact (with the permission of Professor L. Descarries, unpublished)

(Descarries, unpublished) appear to be deprived of a junctional complex, typical of the great majority of such terminals in the neo-cortex.

A recent immunocytochemical study by Pickel *et al.* (1977) has proved the existence of serotonin (5-HT)-containing nerve fibres in the locus coeruleus of the adult rat, thus accounting for the high 5-HT levels (Palkovits *et al.*, 1974). It is well established that most, if not all, of the 1400–1600 nerve cell bodies comprising the rat locus coeruleus (Descarries and Saucier, 1972; Swanson, 1976) are noradrenergic (Corrodi *et al.*, 1970; Dahlström and Fuxe, 1964; Fuxe *et al.*, 1970a,b; Swanson, 1976); 5-HT afferents are presumed to originate mainly from the mesencephalic nuclei raphé dorsalis and centralis superior.

Electron-microscope radioautographs show that all sites of radioactive accumulation in the locus coeruleus correspond to small axonal enlargements, measuring 0.5–1.3 μm and averaging 0.9 μm in diameter. These varicosities are sometimes exposed in the neuropil, but most often appeared unsheathed by glial processes.

Reactive varicosities in the locus coeruleus are not found in direct apposition or in synaptic contact with neuronal perikarya (Leger and Descarries, 1978): less than 10% of all reactive varicosities exhibit a junctional complex, as opposed to almost 50% of unreactive terminals selected at random in the surrounding neuropil. These are synapses made with dendritic processes rather than noradrenergic perikarya (Leger and Descarries, 1978) and it was therefore concluded that mainly non-synaptic interactions are involved in chemical transmission in the locus coeruleus. The 5-HT innervation of the locus coeruleus is relatively dense, and 5-HT varicosities seem to be approximately ten times more numerous (Descarries and Léger, 1978) in the locus coeruleus than in the fronto-parietal neocortex (Beaudet and Descarries, 1976). Thus in the locus coeruleus 5-HT afferents may have a wider influence on noradrenergic neurons independently of their synaptic connections.

Chan-Palay (1976) has examined the extensive plexuses of serotoninergic neurons originating in raphé nuclei which form supra- and subependymal systems in the walls of the cerebral ventricles. Electronmiscroscopy followed by autoradiography has shown that, these serotoninergic fibres contain varicosities with dense-core synaptic vesicles. No evidence was found for specialized synaptic contact with ependymal cells or axonal processes. Chan-Palay (1977) has suggested that these fibres release serotonin into the cerebrospinal fluid (CSF) providing a means of transporting the transmitter to widely distributed targets. Although there is no direct evidence for such a function it is not clear what purpose these fibres would serve if they did not release serotonin. Various authors have suggested that the ventricular system may be more than just a sewer!

Using high-resolution autoradiography asynaptic serotonin-containing axon terminals have been found in the rat neostriatum (Arlusion and De La Manche, 1980). Analysis of serial sections of monkey neostriatum showed that many varicosities do not have membrane specializations, and that their contents are similar to those making typical synaptic contacts (Pasik and Pasik, 1982). These authors demonstrated that a relatively low proportion of labelled synapses (Pasik and Pasik, 1982) make contact with small dendritic spines, and membrane specializations are asymmetric, i.e. Gray type I junctions. This characteristic has been associated with excitatory synapses, and Pasik and Pasik (1982) suggested that on these sites 5-HT could act as a classical neurotransmitter producing stimulation. This suggestion is consistent with the observation of Adam-Vizi and Vizi (1978), who showed that 5-HT is able to excite cell bodies of cholinergic interneurons in the Auerbach plexus. Therefore it seems very likely that 5-HT can be released from both synaptic and non-synaptic boutons, and can work as a stimulatory transmitter on the dendrites or soma and as an inhibitory modulator on cholinergic axon terminals (see section 4.4).

3.2.4 Enkephalinergic neurons

It is well established that enkephalin released from the intrinsic enkephalinergic interneurons within the dorsal horn (Hökfelt *et al.*, 1977a,c) can play a pre-synaptic modulatory role in pain transmission, by a directly inhibiting substance-P release from primary afferent neurons (Jessel and Iversen, 1977a).

Enkephalin-containing axon terminals most commonly make synaptic contacts with dendritic shafts or spines within layers I and II of the dorsal horn (Fig. 21). Close apposition to other non-degenerating unlabelled vesicle-containing profiles were shown, but synaptic specializations were rare, and in only one case was there direct evidence (an axo-axonic relationship) for a direct interaction between primary afferent fibres and enkephalin-containing axon terminals (Hunt *et al.*, 1980). Neurons within layers I and V have been shown to respond to noxious stimulation and to project through the spinothalamic tract to the higher brain centres. Inhibition of the transmission of noxious stimuli and layer V projection neurons has been demonstrated following iontophoretic application of morphine to layer II, but not when morphine was applied directly to layer V. This suggests that at least part of the physiological effect of enkephalin released from layer II interneurons may be inhibition (cf. Hunt *et al.*, 1980), and no 'morphological substrate for the interaction has yet been found' (Hunt *et al.*, 1980). Using immunocytochemical techniques, Hunt *et al.* (1980) found that following rhizotomy, degenerating primary afferent axon terminals were present throughout layers I and II but a synaptic relationship with a labelled terminal was observed in only one case. Extensive terminal branchings of the primary afferent fibres are characterized by multiple end-terminals and frequent *en passant* enlargements (Fig. 22), in addition to a rarer termination (Réthelyi, 1977; Réthelyi *et al.*, 1982). It is likely that methionin–enkephalin released from terminals acts non-synaptically on opiate receptors located on the varicose axon terminals of fine-diamcter primary afferent fibres (Réthelyi, 1977). There is no direct evidence that substance P-containing axon terminals are identical to those described by Réthelyi (1977) in fact, Priestley *et al.* (1982) have provided evidence that substance-P immunoreactive terminals synapse with dendrites and soma and very rarely receive axo-axonic synapse.

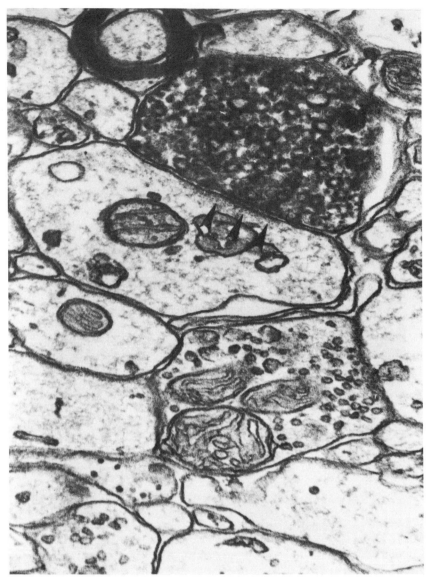

Fig. 21. Enkephalin-positive terminal making asymmetric synaptic contact (arrow-heads) with dendritic shaft in layer of the spinal cord (Hunt *et al.*, 1980, Fig. 4c). (× 66 000). (Reproduced by permission of Pergamon Press Ltd)

In the neostriatum and globus pallidum of the rat all the enkephaline-immunoreactive boutons studied electron-microscopically were found, without exception, to make two distinct types of classical symmetrical or asymmetrical synaptic contacts with cell bodies (Somogyi *et al.*, 1982b). This is at variance with a previous publication (Pickel *et al.*, 1980), in which it was argued that some of the immunoreactive boutons seem to lack synaptic contact, and it was suggested that such axon terminals could influence other

58

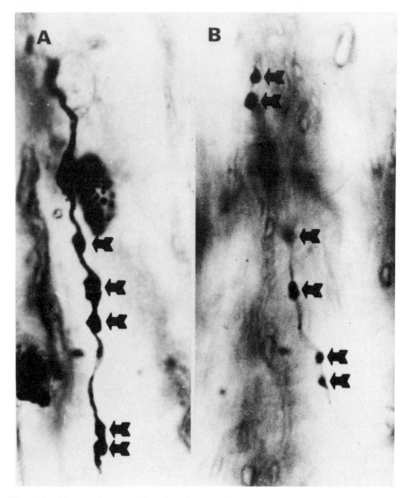

Fig. 22. Photomicrographs showing portions of horseradish peroxidase stained physiologically identified high-threshold mechano-receptor fibres from the spinal marginal zone. *En passant* enlargements of various sizes and shapes (arrows) are interconnected by thin segments of the pre-terminal fibres. Sections were prepared in the sagittal plane from coccygeal segments of the spinal cord. (Magnification: × 2400.) A: cat, B: monkey. (Modified from Réthelyi, Light, and Perl (1982). *J. Comp. Neurol.*, **207**, 382–93)

neurons diffusely while others made axo-axonic contacts: 'axon terminals
. . . form specialized junctions with unlabelled axon and axon terminals'
(Pickel *et al.*, 1980). Each of the almost two hundred varicosities examined
by Somogyi *et al.* (1982b) formed synaptic contacts, but no axo-axonic
contacts were found. On this basis, Somogyi *et al.* (1982b) suggested that
the enkephalin-immunoreactive material may be a synaptic transmitter but
this suggestion is not consonant with those made by Vizi *et al.* (1977) and
Hársing *et al.* (1978), who, using neurochemical methods, showed that enkephalin had two different actions on ACh release, and suggested that enkephalin may be a pre-synaptic modulator on both dopaminergic and cholinergic axon terminals.

3.2.5 Peptidergic neurons

Direct evidence for non-synaptic release of a peptide has been obtained in
Aplysia. Coggershall (1970) demonstrated that neurosecretory bag cells of
the abdominal ganglion terminate in connective tissue, releasing products
directly into the hemolymph for diffusion and thus affect the activity of
various types of target cell in different sites.

3.3 AXO-AXONIC SYNAPSES: A CLASSICAL CORRELATE OF PRE-SYNAPTIC INHIBITION

The synapse is a term now used to describe an interaction between nerve
cells believed to specialize in the transfer of information, and communication
across synapses is now acknowledged to be activated by a chemical agent
released from the pre-synaptic axon terminals.

Pre-synaptic inhibition describes a type of inhibition of chemical transmission which results from a decrease of transmitter release. A conventional
morphological correlate of pre-synaptic inhibition is the axo-axonic synapse,
where an inhibitory input synapses on axon terminals or on the initial segment
of the cell, and the inhibitory transmitter thereby released inhibits the release
of the principal transmitter. It is generally accepted that an axo-axonic
synapse (Figs. 23, 24a,b) is the morphological correlate of the pre-synaptic
type of inhibition (Szentágothai, 1968). Since 1962, axo-axonic synapses have
been described by a number of authors and their nature and possible role as
the structural basis for pre-synaptic inhibition has been discussed by Szentágothai (1968). There have also been speculations on the possibility of a pre-
synaptic inhibition of inhibitory terminals, i.e. a pre-synaptic disinhibition
(Szentágothai, 1968). However, when transmitter release can be pre-synaptically inhibited, although the effect of the inputs of the post-synaptic
membrane is not affected (the number of converging inputs could be 3000!),
the inhibitory effect of the modulator on the axon hillock or on the axon
terminal (varicosity) may completely block the generation of an axon potential, or prevent the axon terminal depolarization.

60

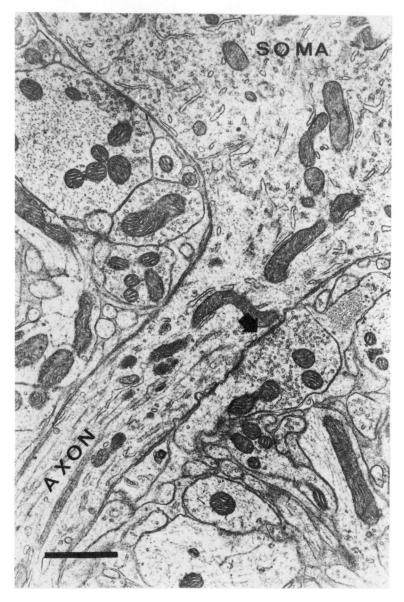

Fig. 23. Synaptic contact located on the axon hillock and the intiial segment of a neuron. Purkinje cell and axo-axonic synapse (arrow). Electron-microscopic study of P. Somogyi and J. Hámori (unpublihsed). Cat cerebellum, 1.5% glutaraldehyde + 2% paraformeldehyde fixation. (Scale 1 μm) (Vizi, 1979) (Reproduced by permission of Pergamon Press Ltd)

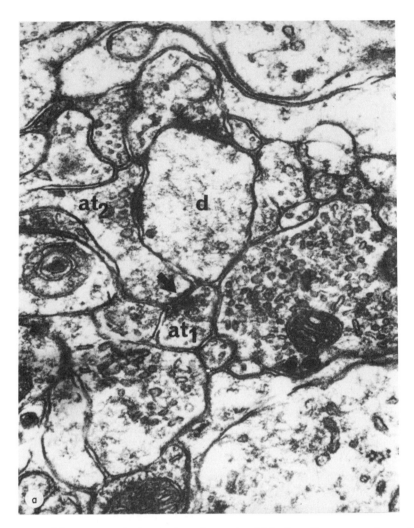

Fig. 24. (a) Axo-axonic (at$_1$–at$_2$) and axo-dendritic (at$_2$–d) synapses from the substantia gelatinosa of the nucleus caudalis in the rat spinal trigeminal nucleus (with the permission of P. Somogyi, unpublished). Note that at$_1$ synapses on at$_2$ at arrow and at$_2$ makes synaptic contact with a dendrite.

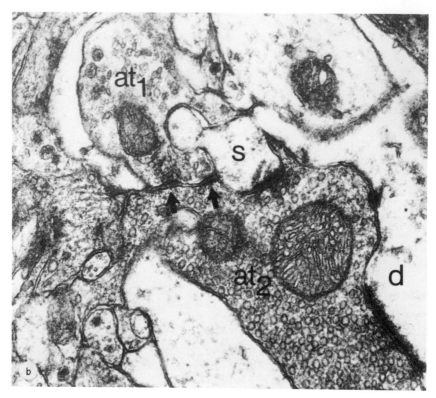

Fig. 24. (b) Axo-dendritic (at₂-d) synapses from the substantia gelatinosa of the nucleus caudalis in the rat spinal trigeminal nucleus (with the permission of P. Somogyi, unpublished). Note that at₂ synapses on dendrite (d) and spine (s)

Somogyi *et al.* (1982a) have provided evidence for the widespread presence of axo-axonic cells in cortical areas, and it has been demonstrated that the axon initial segment of pyramidal cells receives several synapses from a specific interneuron (Somogyi, 1977; Somogyi *et al.*, 1982a). This specific axo-axonal interneuron is identical to the chandelier cell discovered earlier by Szentágothai (Szentágothai and Arbib, 1974). This type of interaction might create a conventional mechanism of pre-synaptic inhibition.

Chapter 4

Interneuronal modulation of transmitter/ modulator release

4.1 MODULATION BY NORADRENALINE

4.1.1. Effect on acetylcholine release

Neurochemical evidence

The gastrointestinal tract

The enteric nervous system (a term introduced by Langley, 1921) can be defined as the intrinsic innervation of the gastrointestinal tract (cf. Gershon, 1981; Wood, 1981). It is comprised of two major plexuses of interconnecting ganglion cells (Schofield, 1968; Gabella, 1976; Furness and Costa, 1980), and these are the submucosal or Meissner's plexus and the myenteric or Auerbach's plexus. The enteric nervous system is a part of the autonomic nervous system, and was originally classified by Langely (1921) as the third division of the autonomic nervous system and essential for the motor control of the gut.

The autonomy of the enteric nervous system was recognized as early as 1899 by Bayliss and Starling. ACh was the first neurotransmitter to be identified in the gut (Dale, 1937), and it was unequivocally shown to be synthesized and released from the myenteric plexus (Paton and Zar, 1968). The structure and functional properties of the myenteric plexus closely resemble those of the central nervous system. Its function is not solely distribution and divergence of parasympathetic outflow to gastrointestinal effector systems: it is perceived as a 'little brain' or 'enteric brain' that also possesses independent integrative circuitry programming and co-ordinating activity (Wood, 1981).

In the nineteenth century it was observed that stimulation of the thoracolumbar, splanchnic, or mesenteric nerves inhibited the movement of the gastrointestinal tract (Pflüger, 1857; Lister, 1858). Later, Finkleman (1930) showed that during sympathetic stimulation of the intestine on adrenaline-like substance was released which inhibited an isolated segment of the gut. Since Finkleman's classic work it has been accepted that inhibition of the rhythmic activity of the gut by sympathetic stimulation is due to release of an adrenaline-like substance, now proved to be noradrenaline. McDougal and West (1952, 1954) found that the peristaltic reflex of the guinea-pig ileum can be inhibited by low concentrations of adrenaline and noradrenaline, which fails to influence the response of the ileum to added ACh.

According to Bayliss and Starling (1899), the intestines are normally under the reciprocal control of vagal excitatory and splanchnic inhibitory fibres. The two types of autonomic nerves are classically considered to be reciprocal in function, and the sympathetic outflow is commonly held to be antagonistic to the parasympathetic outflow. The inhibitory effect of the sympathetic transmitter (noradrenaline) has usually been considered in terms of a direct action on the smooth muscle (Bülbring and Tomita, 1969). However, recent anatomical and physiological findings have caused a revaluation of the classical concepts of the inhibitory innervation of the gut (cf. Vizi, 1976; Gershon, 1981).

The tone and reflex activity of mammalian intestine can be influenced by catecholamines and by a number of other mechanisms: for example, noradrenaline and adrenaline inhibit the reflex of the intestine by stimulating α_2 and β receptors (cf. Vizi, 1979). The location of these receptors was earlier assumed to be in the smooth muscle cells. The old belief that sympathetic nerves terminate on the intestinal smooth muscle layers and that the depressant action of NA released from noradrenergic neurons is mediated by a direct effect on the muscle cells has now been seriously challenged.

In 1968 and 1969 Paton and Vizi (Vizi, 1968; Paton and Vizi, 1969) reported that noradrenaline and adrenaline are capable of reducing the release of ACh from the Auerbach plexus of the guinea-pig ileum. Neurochemical and functional evidence has shown that either added to the tissue or released from axon terminals, reduces the release of acetylcholine (ACh) via a stimulation of pre-synaptic α-adrenoceptors located on cholinergic neurons (Paton and Vizi, 1969).

Resting release. It has been shown (Paton and Vizi, 1969; Paton *et al.*, 1971) that NA in a concentration range of 10^{-6}–10^{-5}M can reduce the resting output of ACh from an isolated longitudinal muscle strip–Auerbach's plexus preparation by an average of 50–80%. Its effect on resting output varies with the base level of the output and is proportionally greater with a high than with a low resting output: isoprenaline has no effect. The order of potency, adrenaline > noradrenaline > isoprenaline > methoxamine, suggests that α_2 adrenoceptors are involved in controlling resting release, and we must first determine how far 'resting release' originates from propagated activity. Hexamethonium, in a concentration of 2.8×10^{-4} M, reduces the resting output by about 53%. The concentration used is five to ten times greater than that required to abolish both the peristaltic reflex and the spontaneous activity of the guinea-pig ileum, and it is sufficient to abolish cholinergic ganglionic transmission in the longitudinal strip. In six experiments tetrodotoxin (5×10^{-7} – 1.6×10^{-6} M) reduced resting output by 55%. From what is now known of the properties of tetrodotoxin it must be assumed that all propagated activity in nerve fibres is eliminated, but that spontaneous quantal release of ACh at nerve terminations is retained. On this basis, the residual output after hexamethonium or tetrodotoxin (about 40% of normal) must represent spontaneous release.

Effect of ACh release stimulated by excess K. The resting output from the longitudinal strip varies with external potassium concentration (K_o) Paton *et al.*, 1971). Tetrodotoxin fails to affect the release of ACh induced by high (K^+_o) at low concentration (23.6 mmol/l), however, it does have an inhibitory action. Noradrenaline and adrenaline, even in high concentrations (10^{-4}M), do not affect ACh release induced by K-excess (49.7 mM), while low K^+ concentrations (23.6 mM 10^{-6} M adrenaline significantly reduced the release. These findings indicate that ACh release induced by high K_o is an enhanced spontaneous release, and is not due to propagated activity of the neurons which are, in fact, not subject to modulation. It is interesting to note that when the release of ACh was enhanced by only 23.6 mM K_o, the release could be reduced by 60% (Vizi, 1979). It seems likely that at lower K_o concentrations a portion of the propagated activity is still maintained and this can be reduced by α_2-adrenoceptor stimulation (Table 7).

Table 7. Effect of noradrenaline/adrenaline on acetylcholine release from varicose cholinergic axons

Type of release	Tetrodotoxin-sensitive	Inhibition by noradrenaline	Note
Resting[a, b]			
(1) Spontaneous	No	No	
(2) Produced by spontaneous electrical activity	Yes	Yes (α_2)	
Induced			
(1) Electrical[b–d]	Yes	Yes (α_2)	Frequency-dependent[a, f]
(2) Ouabain [b, e]	Partly	Partly (α_2)	
(3) K-depolarization[a, b]			
low (<30 mM)	Partly	Partly (α_2)	
high (>30 nM)	No	No	

Conclusion drawn from the data:
[a]Paton *et al.* (1969),[b] Vizi (1979), [c]Drew (1979), [d] Wikberg (1978), [3]Vizi (1974a), [f] Knoll and Vizi (1979, 1971).

Effect on ACh release as a result of neuronal firing. The release of ACh resulting from electrical or chemical stimulation of the neuron or cell bodies and dendrites is tetrodotoxin-sensitive. The first neurochemical evidence of inhibitory interaction between different transmitters was presented when it was shown that NA was able to reduce the field stimulation-induced release of ACh. It has been found that α-adrenoceptor stimulation inhibits the stimulation-evoked release of ACh from the Auerbach's plexus of an ileal longitudinal muscle strip.

In 1969 Paton and Vizi reported that noradrenaline and adrenaline reduced the release of ACh evoked by electrical stimulation from the Auerbach's plexus. Similar observations were made on this preparation (Kilbinger and Wessler, 1979; Alberts and Stjärne, 1982), using guinea-pig colon (Beani et al., 1969), and on human taenia coli (Del Tacca et al., 1970). Since α-receptor blocking drugs prevented the effect of noradrenaline (Paton and Vizi, 1979; Vizi, 1968; Kosterlitz et al., 1970; Kilbinger and Wessler, 1979; Alberts and Stjärne, 1982) and that of sympathetic stimulation (Vizi and Knoll, 1971; Wikberg, 1977; Gillespie and Khoyi, 1977), it was suggested that this effect is mediated via α-receptors. Later it was shown that α_2-receptors are involved in the inhibitory effect (Drew, 1979; Wikberg, 1978a,b).

The inhibitory effect of NA is inversely related to the frequency of stimulation applied: the fewer the shocks delivered and the lower the frequency applied, the higher was the output per volley in the control period and the greater the reduction by noradrenaline (Vizi, 1968; Paton and Vizi, 1969; Del Tacca et al., 1970; Knoll and Vizi, 1970, 1971; Kosterlitz et al., 1970, Alberts and Stjärne, 1982). While adrenaline and noradrenaline were very active, (-)-phenylephrine was comparatively inactive and methoxamine was completely inactive in reducing ACh release following electrical stimulation of Auerbach's plexus (Paton and Vizi, 1969. Phenylephrine and methoxamine are known as α_1 receptor stimulants.

The increased ACh output, at rest as well as in response to stimulation, which occurred after catecholamine depletion by reserpine and guanethidine (Vizi, 1968; Paton and Vizi, 1969; Beani et al., 1969; Kazic, 1971; Vizi and Knoll; 1971) and after 6-hydroxy-dopamine (Knoll and Vizi, 1970) suggests a local permanent control by adrenergic nerves of the cholinergic function. In complete agreement with earlier observations (Vizi and Knoll, 1971; Knoll and Vizi, 1970), prior chemical sympathectomy with 6-hydroxydopamine prevented the reduction of stimulated ACh release by perivascular nerve stimulation (Manber and Gershon, 1979). In additon, it was found that, in the intestine NA, released from the sympathetic nerves by indirectly acting sympathomimetics (Knoll and Vizi, 1970, 1971) or by sympathetic nerve stimulation (Beani et al., 1969; Vizi and Knoll, 1971; Kazic, 1971) can reduce the ACh output. Figure 25 shows the reduction by sympathetic nerve stimulation of acetylcholine release from the nerve terminals of rabbit jejunum. The pendular movement was inhibited by sympathetic stimulation and this effect was prevented by phentolamine, an α-adrenoceptor blocking agent, but not by a β-adrenoceptor block (Vizi and Knoll, 1971, Wikberg, 1978 a,b). These findings indicated that NA can control the ACh output pre-synaptically. When the release of ACh from the intestine was enhanced by gastrin, gastrin-like hormones (Vizi et al., 1972; Vizi, 1973a) or ganglion stimulants (Vizi, 1973a), NA or sympathetic nerve stimulation reduced the output (Vizi et al., 1972; Vizi, 1973a).

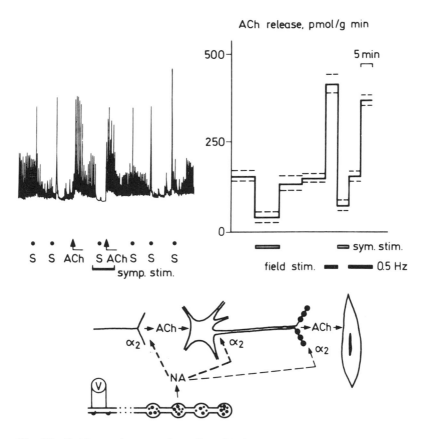

Fig. 25. Evidence that noradrenaline (NA) released from sympathetic axon terminals in response to stimulation can inhibit the release of acetylcholine (ACh) from Auerbach's plexus. Inhibitory effect of sympathetic stimulation on ACh release from rabbit isolated jejunum. Experimental data taken from Vizi and Knoll (1971). Note that in response to sympathetic stimulation, the release of ACh is inhibited. The effect of NA is located on the varicose axon terminals and on the axon hillock both considered to be pre-synaptic (see Section 1.2) since the effect of ACh on the post-synaptic membrane (cell body, dendrites, smooth muscle) is not affected

The cerebral cortex

It has been shown that NA reduces the release of ACh from isolated cortical slices induced either by ouabain (Vizi, 1972) or by electrical stimulation (Beani *et al.*, 1978); however, the resting release was not affected Vizi, 1972; Vizi and Knoll, 1976; Vizi *et al.*, 1977c). Xylazin, a strong α_2-adrenoceptor stimulant, also reduces ACh release enhanced by 2×10^{-5} mol/l ouabain, and phentolamine prevents the inhibitory effect of α-adrenoceptor stimulation (Vizi, 1975a, 1979). Phentolamine by itself enhances the release of ACh induced by ouabain (Vizi, 1975a), and this indicates that there is a permanent control of ACh release by endogenous NA via α-adrenoceptors.

Clonidine has proved to be the most effective and phenylephrine about fifty times less effective than NA in inhibiting the release of ACh induced by ouabain. (The EC_{50} value of NA is 2.6×10^{-6} M) It is interesting to note that the stimulation of α-adrenoceptors does not reduce the release of ACh induced by the membrane ATPase inhibitor, ouabain, when tetrodotoxin is also present in the solution (Vizi, 1975a). The ACh release induced by ouabain is only partially reduced by tetrodotoxin. This indicates that part of the output induced by ouabain is released as a result of action potentials and the consequent calcium influx. Presumably this part of the release is sensitive to α-adrenoceptor stimulation. In addition, amantadine (Beani and Bianchi, 1973), which releases catecholamines from their intraneural storage sites and possibly inhibits their re-uptake, decreases the release of ACh from the neocortex. In the guinea-pig cerebral cortex NA reduces the release of ACh (Beani *et al.*, 1978) evoked by electrical stimulation. The spontaneous and evoked release of acetylcholine was higher in those slices where noradrenergic input was impaired in some way (Vizi, 1980b), either by phentolamine, which blocks the action of noradrenaline, or by 6-OHDA pre-treatment, or by a ipsilateral locus coeruleus lesion. Following a locus coeruleus lesion the spontaneous and evoked release of acetylcholine from slices dissected from the ipsilateral side was higher in comparison with those on the contralateral side. Noradrenaline significantly reduced the resting release of acetylcholine only in those cases where noradrenergic control had been previously removed (Vizi, 1980b). It is suggested that the release of acetylcholine from cholinergic neurons of the cerebral cortex is continuously controlled by noradrenaline released from nerves arising from the locus coeruleus. The removal of this inhibitory system results in an increase of acetylcholine release. In addition, NA has a similar action on isolated human cortical slices (Vizi and Pásztor, 1981), and its effect is mediated via α-adrenoceptors, since phentolamine prevented the effect of NA.

The striatum

In contrast to the findings with slices of cerebral cortex (Vizi, 1972) α-adrenoceptor stimulation either by clonidine or by NA failed to affect the release of ACh from striatal slices (Vizi, 1977a). U'Prichard and Snyder (1979), studying the high-affinity binding of ^{3}H-clonidine, found that there is a fourteen-fold difference between the highest levels in the frontal cortex and the lowest levels in the corpus striatum. This is consistent with the findings of Aghajanian and Bunney (1977), who showed that clonidine, a pure α_2-adrenoceptor stimulant, had no depressent effect on dopaminergic neurons. Furthermore, the nerve terminals of cholinergic neurons in the caudate nucleus probably do not possess inhibitory α-adrenoceptors since A did not affect the release of ACh from caudate nuclei obtained from rats pre-treated with 6-OHDA where DA autoreceptors are absent (Vizi *et al.*, 1977a).

Functional evidence

Guinea-pig ileum, Auerbach's plexus longitudinal muscle preparation, and colon

McDougal and West (1952, 1954) were the first to suggest that NA or A might have an effect on neuronal elements in relaxing the intestine when it is stimulated by indirectly acting drugs. The depressant effect of NA and A on the responses of the guinea-pig ileum to electrical stimulation have been shown by several authors (Ennis *et al.*, 1979; Wikberg, 1977; Andréjak *et al.* (1980).

In the experiments of Kewenter (1965) adrenergic nerves supplying the ileum were tonically active, and only weak excitation occurred when the vagus nerves were stimulated. If adrenergic transmission was blocked by ergotamine or guanethidine the effect of vagal stimulation was markedly enhanced. The adrenergic nerves were not acting on the muscle, and contractions in response to acetylcholine were not dependent on the level of adrenergic nerve activity. Jansson and Martinson (1966) showed that the activation of adrenergic nerves during an intestino-gastric inhibitory reflex did not affect the basic myogenic tone of the stomach, but that it did inhibit contractions in response to stimulation of vagal cholinergic nerves.

Later, this observation was extended to the Auerbach plexus–longitudinal muscle preparation (Kosterlitz *et al.*, 1970; Vizi and Knoll, 1971). α-adrenoceptors antagonists were found to antagonize the inhibitory action of α-agonists (Kosterlitz *et al.*, 1970; Vizi and Knoll, 1971), whereas propranolol (Kosterlitz *et al.*, 1970; Knoll and Vizi, 1971) failed to have any effect.

The isolated longitudinal muscle strip with Auerbach's plexus attached (Paton and Vizi, 1969) is a very good preparation for studying the release of ACh and chemical neurotransmission. The safety margin for neurochemical transmission is one or less than one (Vizi *et al.*, 1977c), indicating that if ACh output per pulse is reduced by a drug, the response of the smooth muscle also becomes reduced. On a longitudinal muscle strip which has not been treated with an anti-cholinesterase, using mechanical contraction as a measure of the ACh released by stimulation, NA (10^{-6} M), A (10^{-6} M), and other α-adrenoceptors stimulants reduced the contraction (Fig. 26) when stimulated at 0.1 and 10 Hz, using a short train. The findings that phentolamine antagonized and pindolol failed to affect the inhibitory action of NA indicates the role of α-adrenoceptors. It was found (Knoll and Vizi, 1971, 1970) that the lower the frequency of stimulation, the greater the inhibitory action of NA, or, at a given frequency, the fewer shocks applied, the higher the inhibitory effect. NA released by methylamphetamine (Knoll and Vizi, 1971) or by guanethidine (Vizi and Knoll, 1971) also reduced the contractions. The contraction produced by adding ACh to the bath or by high-frequency stimulation, however, was not affected, indicating that the reduction observed was not due to a post-synaptic action of methyl-amphetamine

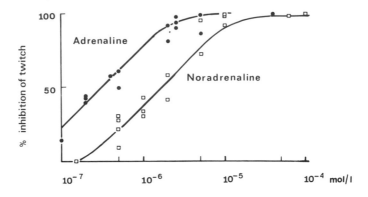

Fig. 26. Inhibitory effect of noradrenaline and adrenaline on twitch responses of the ileal longitudinal muscle strip of guinea-pig (0.1 Hz stimulation)

or guanethidine. The results obtained using an isometric recording system are thus in good agreement with those from direct measurement of ACh release (Table 7). Langer (1974) has suggested that α-adrenoceptors should be subclassified as α_1 (post-synaptic) and α_2 (pre-synaptic). A differential sensitivity to pre- and post-synaptic antagonists has been shown (Drew, 1977; Drew et al., 1979) for the cholinergic nerve terminals in the guinea-pig ileum; α-adrenoceptor stimulants, clonidine (10^{-8}–10^{-7} mol/l), oxymetazoline (5×10^{-7}–5×10^{-6} mol/l), and xylazine (10^{-8}–10^{-7} mol/l) caused concentration-dependent reduction in the twitch response (Fig. 27). Phenylephrine and methoxamine were 1000–10 000 times less potent than clonidine. Phentolamine (10^{-6} mol/l) and piperoxane (10^{-6} mol/l) were potent antagonists of the inhibitory effect of clonidine. In contrast, thymoxamine (1.3 and 10 μg/ml) was only weakly active, and labetalol (0.3 and 1 μg/ml) was inactive against clonidine. These results suggest the pre-synaptic α_2-adrenoceptors located on the parasympathetic nerves in the guinea-pig ileum are of the same type as those located on sympathetic nerve terminals, that is, α_2.

Naloxone did not alter the action of clonidine (Vizi, 1974; Andréjak et al., 1980). Thus the opiate and clonidine receptors are different. Prazosin, which appears to be a relatively specific blocking agent of post-synaptic α_1-adrenoceptors in other preparations, did not antagonize the inhibitory effect of clonidine. Hulten and Jodal (1969) found that low frequencies of stimulation of adrenergic nerves suppressed contractions caused by pelvic or vagus nerve stimulation or by spontaneous activity of enteric neurons, but that neither the myogenic contraction in response to distension of the colon nor the contractions caused by exogenous acetylcholine were antagonized. Watt (1971) has shown that stimulation of mesenteric nerves is more effective in reducing the contraction caused by a single stimulus pulse applied to intrinsic cholinergic nerves than in reducing the contraction to exogenous acetylcholine in the guinea-pig ileum in vitro.

When cholinergic effects are eliminated by muscarinic blocking agents

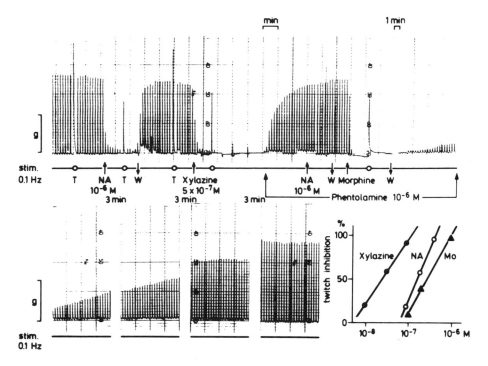

Fig. 27. Pre-synaptic inhibitory effect of xylazine, noradrenaline (NA) and morphine (Mo) on twitch responses of ileal longitudinal muscle strip (guinea-pig). (0.1 Hz stimulation, as indicated). T, tetanic stimulation (10 Hz, 10 shocks). Note that phentolamine, a relatively pure α-antagonist, counteracted the xylazin-induced inhibition of twitch responses, and prevented the effect of NA, but failed to affect that of morphine. The inset shows dose–response relationships

(hyoscine or atropine), stimulation of paravascular nerves with frequencies as low as 1–2 Hz still inhibits intestinal movements *in vitro* (Gillespie, 1960; Burnstock *et al.*, 1966; Campbell, 1966). This implies that noradrenaline release from nerves within the musculature is more effective *in vitro*, and/or noradrenaline diffuses from the terminals in the plexus to the muscle.

The rabbit intestine

In 1930, Finkleman found that in the absence of a cholinesterase-inhibitor sympathetic nerve stimulation inhibited the responses of isolated rabbit ileum to released ACh. This effect was mediated via α-adrenoceptors since phentolamine prevented its action. The contraction caused by high-frequency stimulation (10 Hz, thirty shocks) was not affected. Noradrenaline (10^{-6}–10^{-5} mol/l) behaved in a way similar to that of sympathetic stimulation. Acetylcholine added to the bath increased pendular movement and caused a contraction of rabbit jejunum. This was not affected by sympathetic nerve stimulation, but the contraction to a single 'field' stimulation was still reduced. Gillespie and Khoyi (1977) showed that lumbar sympathetic stimulation inhibited the responses of rabbit isolated colon to pelvic parasympathetic stimulation (Fig. 28). Phentolamine, but not propranolol, prevented the inhibitory effect of sympathetic stimulation on contractions induced by pelvic stimulation. The inhibitory effect of sympathetic stimulation and of NA on the pelvic response is reduced by phentolamine (5×10^{-6} M) but is unaffected by propranolol (5×10^{-6} M), suggesting that the effect is mediated via α-adrenoceptors.

When rabbit jejunum was stimulated electrically, atropine (2×10^{-6} M) almost completely blocked the contractions and clonidine (10^{-7} M) caused an approximately 50% inhibition of the pendular movements. Clonidine is capable of causing relaxation by stimulating α_2-receptors located at the cholinergic neurons and thereby inhibiting ACh release.

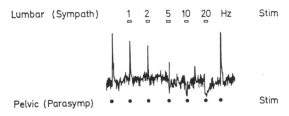

Fig. 28. The effect of lumbar sympathetic nerve stimulation on the response of the rabbit colon to pelvic (parasympathetic) nerve stimulation. (Redrawn from the paper of Gillespie and Khoyi, 1977)

Vagus–stomach preparation

The role of α-adrenergic receptors in gastric function has been recently reviewed (Taylor and Nabi Mir, 1982). Like the intestine, the stomach is under reciprocal control of sympathetic and parasympathetic fibres (Jansson and Martinson, 1966). The existence of two inhibitory andrenoceptors, the prejunctional α_2-adrenoceptor mediating inhibition of acetylcholine release from the vagus, and the post-junctional β-adrenoceptor mediating inhibition of smooth muscle cell excitability, has also been demonstrated in the smooth muscle of the stomach. Jansson and Martinson (1966) have shown the existence of an adrenergic mechanism (in the cat stomach *in situ*) which inhibits smooth muscle responses to excitatory vagal stimulation but not to exogenous acetylcholine. The vagal excitatory responses generally augmented during spinal anaesthesia and during adrenergic blockade by guanethidine, suggests that a sympathetic influence may have been operative throughout. An adrenergic outflow to the stomach can effectively inhibit vagally elicited excitatory motor responses. However, the intestino-gastric inhibitory reflex cannot cause relaxation of the gastric smooth muscle cells when their tone is entirely myogenic in origin. The vagal responses of the stomach are augmented by spinal anaesthesia and by administration of adrenolytic drugs (Jansson and Martinson, 1966). Vizi (1974a), using *in vitro* methods, showed that in a vagus-stomach–smooth muscle preparation from rat, a pure α_2-adrenoceptor agonist inhibited contraction of the stomach induced by pre-ganglionic nerve stimulation but not that induced by exogenous ACh (Fig. 29). Seno *et al.* (1978) have distinguished separation α- and β-receptors which mediate inhibition of gastric smooth muscle in a vagus nerve–smooth muscle preparation isolated from the chick preventriculus, although in their experiments clonidine inhibited only vagally stimulated contraction. Bennett (1969) reported that in the avian stomach the adrenergic nerve fibres terminated in Auerbach plexus. In chicken isolated proventriculus with vagus nerve, adrenaline, in concentrations ranging from 10^{-8} to 10^{-7} M, inhibited the contraction induced by stimulation of the vagus nerve without affecting the response to acetylcholine (Seno *et al.*, 1978). The inhibitory effect increased with increasing concentrations of adrenaline up to 10^{-7} M at which concentration the inhibition reached about 50% of the control response. Noradrenaline (10^{-7}–2.5×10^{-6} M), like adrenaline, produced a concentration-dependent inhibition of the vagally induced response without affecting the response caused by acetylcholine. On the contrary, phenylephrine, in concentrations less than 10^{-6} M, did not inhibit the response induced by either stimulation of the vagus nerve or by acetylcholine. Phentolamine antagonized the inhibitory effect of adrenaline, noradrenaline and clonidine on the vagally induced response. It is interesting to note that it was found that α_2-adrenoceptor stimulation inhibits gastric secretion (Taylor and Nabi Mir, 1982) induced by

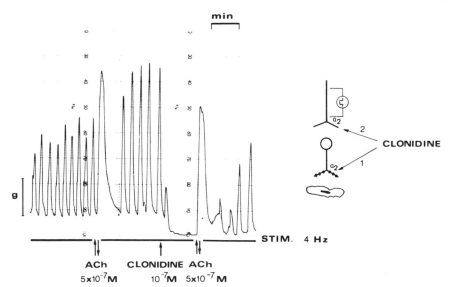

Fig. 29. Effect of clonidine on contractions of vagus–stomach preparation stimulated at 4 Hz (12 shocks) (redrawn from Vizi, 1974). Note that the direct effect of acetylcholine (ACh) on smooth muscle is not affected by clonidine, but the response to vagal stimulation is completely inhibited. The inset shows the possible site of clonidine: (1) on the pre-effectorial axon terminal, (2) on the axon terminal of vagal input, or on both. Since clonidine was active during field stimulation, when the interneuronal axons are also stimulated it seems likely that clonidine can also act on the pre-junctional axon terminal

vagus nerve stimulation in vagi-sectional rats (Jennewein, 1977; Cheng *et al.*, 1981) and the development of gastric lesions in rats in the stress- and reserpine-induced ulcer models (cf. Taylor and Nobi Mir, 1982).

Vagus–heart preparation

The cardiac response to vagal stimulation is inhibited by oxymethazoline and naphazoline (Starke, 1972a). Oxymethazoline, but not naphazoline, decreases the response to exogenous acetylcholine. This effect is quickly reversible, whereas the inhibition of parasympathetic transmission persists. Therefore these α-agonists would seem to inhibit parasympathetic transmission in the rabbit heart not only by a blockade of the action of liberated acetylcholine on ganglion of effector cells but also by a decreasing of the liberation of acetylcholine.

Noradrenaline reduces the negative chronotropic response to vagal stimulation (supramaximal voltage, 1 ms impulse duration, 2, 4, and 8 Hz, ten shocks in each case) in isolated, perfused guinea-pig hearts (Hadházy and Vizi, 1974). The β-blocker, Pindolol, when administered in an amount sufficient to completely prevent the tachycardia produced by NA, exerted an anti-vagal effect. Thus the vagus inhibitory effect of NA is thought to be mediated via β-adrenoceptors. Nevertheless, NA reduces the nicotine-induced ACh release from guinea-pig heart. These results indicate that the anti-vagal effect of NA is mainly due to its action on the effector cells. However, a pre-synaptic inhibitory effect of NA on ACh release may also play some part in this cholinergic–adrenergic interaction.

Guinea-pig oesophagus with parasympathetic nerve

On a preparation of guinea-pig oesophagus with para-sympathetic innervation, NA (5×10^{-6} M) depressed the contractions of smooth muscle (slow component) induced by either pre-ganglionic or post-ganglionic (field) stimulation (Szabolcsi *et al*, 1974) (Fig. 30); however, responses to added ACh were not affected. The contractions of striated muscle (fast component) were not affected by NA (Szabolcsi *et al.*, 1974). Consonant with this observation, Kamikawa *et al.* (1982) observed that in the submucosus plexus longitudinal muscularis mucosae preparation of the guinea-pig oesophagus there are three types of adrenoceptors; inhibitory pre-junctional α-adrenoceptors (α_2), excitatory post-junctional α-adrenoceptors (α_1), and inhibitory post-junctional β-adrenoceptors. Cholinergic neurotransmission is inhibited by catecholamines acting at both pre-junctional α- and post-junctional β-adrenoceptors (Kamikawa *et al.*, 1982).

Electrophysiological evidence

The Auerbach's plexus

Nishi and North (1973a) have obtained electrophysiological evidence

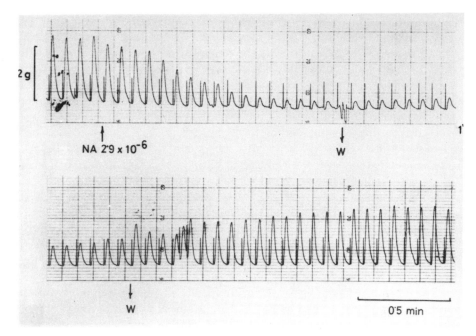

Fig. 30. Inhibitory effect of noradrenaline (NA) on twitch responses of an isolated vagus–oesophagus preparation. Note that there is a slow (atropine-sensitive) and a fast (curare-sensitive) contraction in response to single stimulation (for methods see Szabolcsi *et al.* 1974)

confirming the previous results: NA acts pre-synaptically when it reduces the release of ACh from Auerbach's plexus. Direct intracellular recordings from myenteric neurons in the guinea pig have shown that both NA (Nishi and North, 1973a) and sympathetic nerve stimulation (Hirst and McKirdy, 1974) reduce or suppress excitatory post-synaptic potentials without affecting the electrical properties of the post-ganglionic neurons.

The central nervous system

Noradrenaline-containing axons, arising in the pontine nucleus, coeruleus and subcoeruleus, extend to the cerebral cerebellar and limbic cortics (Unger-stedt, 1971; Olson and Fuxe, 1971; Pickel *et al.*, 1974). Malcolm *et al.* (1967) showed that NA applied to the primary sensory area of the rat cortex inhibits the electrical response evoked by stimulation of the contralateral forepaw. Furthermore, it was also shown that this inhibition is prevented by phenoxy-benzamine, an α-receptor blocking agent. Repeated stimulation of the locus coeruleus induces a strong inhibition (about 50%) of the firing rate of cells of the cingulate rat cortex (Diller *et al.*, 1978), and pre-treatment of the rats with reserpine and α-methyl-p-tyrosine drastically reduces the percentage of cells inhibited (Diller *et al.*, 1978).

Bloom and co-workers (Hoffer *et al.*, 1973; Pickel *et al.*, 1974) in their pioneering work showed that the noradrenergic fibres originating in the locus coeruleus have an inhibitory action on Purkinje cells of the cerebellum and on the pyramidal cells of the hippocampus. The failure of antidromically elicited action potentials to invade the soma was seen more frequently after amphetamine, and is consistent with the view that local extracellular accumulation of noradrenaline hyperpolarizes locus coeruleus neurons to produce an inhibition of firing rate. Changes in the excitability of noradrenergic terminals can be attributed to stimulation of pre-synaptic α-receptors by catecholamines released from the synaptic ending (Nakamura *et al.*, 1981).

Amphetamine infused directly into the frontal cortical terminal fields of locus coeruleus neurons, typically resulted in a decrease in terminal excitability, an effect that could be blocked by infusion of the α-adrenoceptor antagonist, phentolamine, or by pre-treatment with α-methyl-p-tyrosine. These results are consistent with the hypothesis that amphetamine has an indirect effect on terminal excitability by promoting the release and/or blocking the re-uptake of noradrenaline. The NA, in turn, acts back on the presynaptic α-adrenoceptors and leads to a decline in noradrenergic terminal excitability (Nakamura *et al.*, 1981). Iontophoretically applied NA can depress the spontaneous firing of Purkinje cells and hippocampal pyramidal neurons. The inhibition can be blocked by β-adrenergic antagonists (Hoffer *et al.*, 1971; Segal and Bloom, 1974a). β-adrenoceptors have now been visualized on Purkinje cell bodies and apical dendrites in rat cerebellum by means of a new fluorescent analogue of propranolol (Melamed *et al.*, 1976). Stimulation of the nucleus locus coeruleus has an inhibitory action on Purkinje cell and hippocampal neuron-firing which can also be blocked by Sotalol, a β-adrenoceptor antagonist (Hoffer *et al.*, 1973; Segal and Bloom, 1974b). Intracellular recording has revealed that both NA and locus coeruleus stimulation hyperpolarizes the target cells in the cerebellar cortex and the hippocampus, often with an associated increase in membrane resistance (Hoffer *et al.*, 1973). In animals which had been pre-treated with 6-hydroxydopamine, stimulation of the nucleus locus coeruleus failed to produce significant inhibition of Purkinje cell or hippocampal pyramidal cell discharges. Similar results were obtained in animals which were pre-treated with reserpine, i.e. after depletion of catecholamines in brain α-methyltyrosine, a NA synthesis blocking agent had a similar action (Hoffer *et al.*, 1973; Segal and Bloom, 1974b).

Noradrenaline has an inhibitory action on cerebral cortical neurons in several species (Krnjevic and Phillis, 1963; Szabadi *et al.*, 1977) including pyramidal tract neurons (Phillis, 1970; Stone, 1973). Both α- and β-adrenergic blocking agents antagonize its inhibitory action. NA-induced hyperpolarization of feline cortical neurons generally occurs without alteration of membrane resistance.

Studies on identified corticospinal and unidentified deep spontaneously firing neurons have revealed an inhibitory action of nucleus locus coeruleus stimulation (Phillis and Kostopoulos, 1977). The inhibitory effects of NA and locus coeruleus stimulation on these neurons were antagonized by the β-adrenergic blocking agent, Sotalol, and pre-treatment of the cerebral cortex with 6-hydroxydopamine, which destroys the NA-containing nerve terminals, eliminated these inhibitory effects.

The cervical ganglion

Since the initial observation by Marazzi (1939) it has been frequently reported (Marazzi, 1939; Lundberg, 1952; Paton and Thompson, 1953; Christ and Nishi, 1971; Dawes and Vizi, 1973; Kayaalp and McIsaac, 1970), that A and NA are capable of depressing ganglionic transmission, especially at low frequencies of stimulation. However, the mechanism of this inhibition has not been clearly elucidated. It has been suggested that catecholamines hyperpolarize the post-synaptic membrane, thereby decreasing its excitability. However, Christ and Nishi (1971) showed that adrenaline (10^{-5} M) resulted in a rapid blockade of the orthodromic cell body action potential, but that there were only small changes in the resting membrane potential of the cell body.

Adrenaline was also very effective in decreasing the amplitude of EPSPs. These effects were inhibited by phenoxybenzamine (10^{-5} M) but not by propranolol (3×10^{-5} M). On the basis of these results it seems likely that A produces its effect on the isolated ganglion by decreasing the release of transmitter from the pre-synaptic terminals (Paton and Thompson, 1953). Evidence has been obtained from rabbit cervical ganglia that NA reduces the release of ACh when low-frequency stimulation (< 3 Hz) is applied (Dawes and Vizi, 1973), and phentolamine prevents the effect of NA.

Since sympathetic ganglia have been shown to contain and to release adrenaline or an adrenaline-like substance on pre-ganglionic stimulation (Kayaalp and McIsaac, 1970, Bülbring, 1944), there is the possibility that catecholamines have a pre-synaptic modulatory role on ACh release.

It has recently been shown that catecholamines can depress ganglionic transmission. This appears to be an α_2-effect which, in several species, including the rat, has been shown to be of pre-synaptic origin (cf. Brown and Caulfield, 1981). When the extracellular dc potential changes, (unstimulated preparations) and action potentials (supramaximal pre-ganglionic stimulation at 0.2 Hz) were recorded from desheathed rat superior cervical ganglia using a three-chambered bath, there was a clear difference in potency between the hyperpolarizing effect of clonidine in unstimulated preparations (post-synaptic α_2-adrenoceptor effect, mean ED_{50} 2×10^{-9} M) and inhibition of the action potential (pre-synaptic α_2-adrenoceptor effect, mean EC_{50} 1.8×10^{-8} M) (Brown and Medgett, 1982). In addition, phentolamine (10^{-6} M) itself significantly increased the height of the compound action potential by 11% (Brown and Medgett, 1982). This finding suggests that, with continuous stimulation of pre-ganglionic nerves at 0.2 Hz, the acetylcholine release is controlled by an α-adrenoceptor mediated trans-synaptic inhibitory feedback mechanism which is activated by noradrenaline released from the dendrites of the post-ganglionic neurons (Martinez and Adler-Graschinsky, 1980) or from other sites (Noon *et al.*, 1975). Nevertheless, it should be mentioned that propranolol (10^{-5} M) failed to affect the height of the compound action potential, whereas the inhibitory effects of high concentrations of adrenaline (10^{-5}–10^{-4} M) were significantly increased. During an infusion of clonidine (10^{-6} M), adrenaline (10^{-6}–10^{-4} M) and, to a lesser extent, noradrenaline (10^{-5}–10^{-4}M) the height of the compound action potential increased, and these effects were blocked by propranolol (10^{-5} M). These data suggests that facilitatory β-adrenoceptors (possibly β_2) may be present on pre-ganglionic terminals and that activation of these receptors may facilitate transmission when the α_2-adrenoceptor-mediated inhibitory mechanism is fully operative (Brown and Medgett, 1982).

4.1.2 Effect on noradrenaline release

Although strong evidence supports the hypothesis of negative feedback modulation of NA release (see section 5.1), the possibility of interneuronal modulation of NA release by NA secreted from remote axon terminals (Vizi, 1981, a, b) cannot be excluded (Fig. 31).

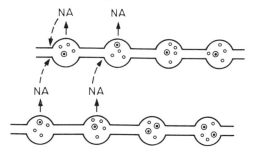

Fig. 31. Interneuronal modulation by noradrenaline (NA) of noradrenaline (NA) release

4.1.3 Effect on adrenaline release

Scatton *et al.* (1979) showed that clonidine decreased the turnover rate of adrenaline and noradrenaline in the region of the nucleus tractus solitarii in the rat and inhibited K^+-induced release of ^{3}H-adrenaline. They suggested that these effects were mediated via α_2-adrenoceptors located on noradrenergic and/or adrenergic neurons. Wemer *et al.* (1982) also found α-adrenoceptors in synaptosomes obtained from the nucleus tractus solitarius and it therefore seems likely that the release of adrenaline from noradrenergic and/or adrenergic neurons is pre-synaptically controlled via α_2-adrenoceptors.

4.1.4 Effect on serotonin release

Terminal serotoninergic fibres may also have pre-synaptic α-adrenoceptors: in coronal brain slices of the rat, the electrically evoked release of ^{3}H-5-

hydroxtryptamine is inhibited by 10^{-5} clonidine (Starke and Montel, 1973a) and slightly increased by 10^{-5} M phentolamine (Farnebo and Hamberger, 1974a). The electrically evoked ^{3}H-5-HT overflow is decreased by noradrenaline and increased by phentolamine, in a concentration-dependent manner. In the presence of tetrodotoxin, throughout superfusion, the Ca^{2+}-induced ^{3}H-overflow from slices superfused with Ca^{2+}-free solution was inhibited by noradrenaline and increased by phentolamine. This indicates that the site of action is on the axon terminal and that secretion-coupling is affected.

In addition, it has been shown that the serotoninergic nerve fibres of the rat brain cortex (Göthert and Huth, 1980) and hippocampus (Frankhuyzen and Mulder, 1980) have α-adrenoceptors. The receptors involved in this modulation have been characterized by Göthert et al. (1981), who found the electrically induced ^{3}H-overflow from rat brain cortex slices pre-incubated with ^{3}H-5-HT was decreased by clonidine and noradrenaline in a concentration-dependent way (the negative logarithms of the IC_{30} values are 6.66 and 6.55, respectively). Yohimbine produced a shift to the right of the concentration–response curves of noradrenaline (apparent pA_2 6.93) and of clonidine (apparent pA_2, 7.06) for their inhibitory effects of induced ^{3}H-overflow. Rauwolscine also shifted the concentration–response curve of noradrenaline to the right, whereas prazosin was ineffective in this respect.

4.1.5 Effect on dopamine release

Scatton et al. (1983), studied the release of DA from the rat striatum, and observed that both yohimbine and rauwolscine enhanced striatal DA turnover, suggesting that the release of DA itself is enhanced. However, when another selective α_2-antagonist RX 781094 was applied (Dettmar et al., 1981) it failed to affect the turnover. These findings contradict the hypothesis that α-adrenoceptor stimulation is involved in brain DA metabolism (cf. Scatton et al., 1983) thereby indirectly influencing ACh release.

4.2 MODULATION BY ACETYLCHOLINE

Brücke (1935) was the first to observe that subcutaneous administration of ACh reduced the pilomotor response of hair to stimulation of sympathetic nerves while the response to exogenous A remained unchanged. He realized that what he had observed was muscarinic inhibition of NA release:

No further attention was paid to this observation and its interpretation until many years later, when an interest in the cholinergic effects on noradrenergic nerve-endings began with the 'cholinergic link' hypothesis of Burn and Rand (1959). They proposed that ACh, which is expected to be present in all post-ganglionic sympathetic fibres, plays a crucial role in noradrenergic transmission: the action potential first releases ACh, which then releases NA from the nerve terminal. Although this proposal did not prove to be valid (cf. Ferry, 1966), it stimulated much thought and experimentation.

Nevertheless, it has been found that ACh in high concentrations releases NA from the noradrenergic axon terminal through the stimulation of nicotinic receptors (cf. Ferry, 1966), but it was found not to be of physiological importance (Table 8).

Table 8. Concentration of ACh affecting the noradrenergic nerve terminal modulating release

Concentration of ACh	10^{-13}–10^{-11} M	10^{-7}–10^{-5}M	10^{-4}M
Change in NA release	Stimulation	Depression	Stimulation
Receptor involved	Neither muscarinic nor nicotinic	Muscarinic	Nicotinic
Importance	?	Physiological importance	Pharmacological importance

4.2.1 Effect on noradrenaline release

In 1967 two examples of pre-synaptic modulation of transmitter release were presented in Cambridge at the Joint Meeting of the German and British Pharmacological Societies. Löffelholz, Lindmar, and Muscholl presented a paper on 'The interference of muscarinic receptors with noradrenaline release from symapthetic nerve endings caused by nicotinic agents' and Vizi presented a talk on 'The inhibitory action of noradrenaline and adrenaline on release of acetylcholine from guinea-pig ileum longitudinal strips' (*Br. J. Pharmc. Chemother*, **31**, 205 (1967).

From the work reported by Lindmar *et al.* (1968) the concept of muscarinic modulation of noradrenergic transmission originated. The theory is comparable to the noradrenergic modulation of cholinergic neurochemical transmission (Vizi, 1968; Paton and Vizi, 1969; cf. Vizi, 1979). It has been proposed (Muscholl, 1980) that there is reciprocal control of transmitter release (Fig. 32).

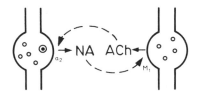

Fig. 32. Reciprocal control of transmitter release, interneuronal modulation. M_1, muscarinic receptor

The occurrence of pre-synaptic inhibitory effects of muscarine receptor stimulation on peripheral adrenergic nerves is now well established (for reviews see Muscholl, 1970, 1980). This was recognized when, in a study of the nicotinic effect of acetylcholine (ACh) on the perfused rabbit heart, it was understood that the muscarinic moiety of the drug greatly decreased the nicotinic noradrenaline relase (Lindmar *et al.* 1968).

After an accidental omission of atropine it was noted that the NA release was dramatically reduced if ACh, but not dimethylphenylpiperazinium (DMPP) was administered (Muscholl, 1980, Fig. 33).

In the absence of a muscarinic antagonist the noradrenaline release by the nicotinic action of ACh was one-tenth of that in the presence of atropine. Furthermore, the effect of ACh was mimicked by a combination of DMPP, a nicotinic drug, and a muscarinic agonist, such as methacholine (MCh), pilocarpine, or ACh itself. In addition, it was found that ACh also inhibited the noradrenaline overflow in response to electrical stimulation of the sympathetic nerves of the rabbit heart (Löffelholz and Muscholl, 1969). These observations provide evidence for the physiological role of muscarinic inhibition.

The hypothesis 'that the peripheral adrenergic nerve fibre contains inhibitory muscarine receptors in addition to the excitatory nicotine receptors mediating NA release' (Lindmar *et al.*, 1968) was soon confirmed by others (Haeusler *et al.*, 1968).

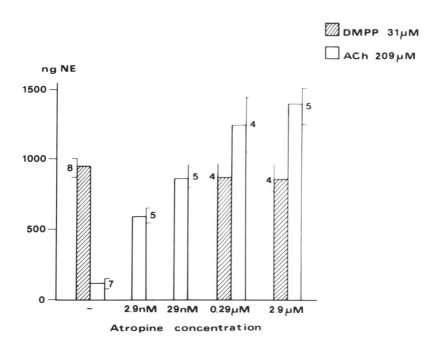

Fig. 33. Effect of atropine on noradrenaline (NE) release induced by dimethylphenylpiperazinium (DMPP) or acetylcholine (ACh) on the isolated rabbit heart [redrawn from the data of Lindmar *et al.*, 1968). Note that atropine does not affect the release of noradrenaline induced by DMPP, but the inhibitory effect of ACh was prevented

The autonomic nervous system

It has been shown that methacholine, ACh, pilocarpine (Lindmar *et al.*, 1968), and oxotremorine, reduce the release of NA from the rabbit heart (Fozard and Muscholl, 1971), the rabbit ear artery (Steinsland *et al.*, 1973) and the guinea-pig ileum longitudinal muscle with Auerbach's plexus (Alberts and Stjärne, 1982). In the absence of eserine, oxotremorine reduced the release of ^{3}H-NA concentration-dependently and reversibly, yielding a EC_{50} of 59 nM. The dissociation constant for atropine was 2 nm. It has been shown (Löffelholz and Muscholl, 1969) that excitation of muscarinic receptors also inhibits NA release elicited by electrical stimulation of the post-ganglionic sympathetic cardioaccelerator nerve running to the rabbit heart. Atropine, but not hexamethonium, completely reversed this inhibition. A similar observation was made when the sympathetic nerve was stimulated, and tritiated NA release was measured from the rabbit central ear artery (Steinsland *et al.*, 1971). In dog cutaneous veins ACh inhibits the release of NA during nerve stimulation but not the release caused by tyramine (Vanhoutte and Shepherd, 1973; Vanhoutte, 1977). Addition of methacholine decreased the NA output induced by 135 mM potassium in a dose-dependent manner; the effect of methacholine was fully reversed by perfusion with atropine (Muscholl, 1973a,b). Methacholine dose-dependently decreased the NA output of the rabbit heart in response to perfusion with high K^+-low Na^+ solution and this effect was also fully reversed by atropine (Dubey *et al.*, 1975). Engel and Löffelholz (1976) showed that in isolated chicken heart there are no nicotinic excitatory receptors. However, convincing evidence has been presented for the presence of muscarinic inhibitory receptors. This observation fails to support the concept of the 'cholinergic link' (Burn and Rand, 1959).

Table 9 shows the dissociation constant of different muscarinic antagonists on pre- and post-synaptic receptors. The decreasing order of potency was quinuclidinyl benzylate (QNB) < ipratropium < scopolamine < atropine < trihexyphenidyl < amitriptyline < gallamine (Muscholl, 1980).

Table 9. Dissociation constants of various muscarine antagonists

Antagonists	Pre-synaptic pA_2	Post-synaptic pA_2
QNB	> 13	11.65
Ipratropium	10.28	10.12
Scopolamine	9.08	9.08
Atropine	8.58	8.70
Trihexyphenidyl	7.40	7.91
Amitriptyline	7.29	7.27
Gallamine	4.85	5.42

Data from Muscholl (1980): pA_2 values against methacholine determined by regression analysis. Note that none of the antagonists possess preferential pre- or post-synaptic effects. Pre-synaptic, noradrenaline release: post-synaptic, effect on heart muscle.

The central nervous system

Acetycholine does not change the NA overflow from rat cerebral cortex slices evoked by electrical stimulation (Starke *et al.*, 1975d), although it reduces the overflow from hypothalamic and cerebellar slices induced by nicotine or high potassium concentrations (Westfall, 1974a,b). In contrast to the observation of Starke *et al.* (1975d), an inhibitory effect of ACh on NA release from the cerebral cortex has been observed (Reader *et al.*, 1976).

4.2.2 Effect of dopamine release

Dopamine-containing nerve-endings seem to possess both nicotinic and muscarinic receptors. Activation of nicotinic receptors elicits a release of DA (Westfall, 1974c; Giorguieff *et al.*, 1976) and stimulation of muscarinic receptors reduces the release evoked by electrical stimulation (Westfall, 1974a,b,c; Reader *et al.*, 1976).

4.3 MODULATION BY DOPAMINE

Although dopamine was discovered as early as 1911 (Funk, 1911), in 1957 Blaschko proposed that, apart from its being a biosynthetic precursor to noradrenaline, dopamine might have a separate, independent peripheral physiological role. Carlsson *et al.* (1957) provided pharmacological evidence that dopamine *per se* is involved in chemical transmission of the brain. In 1959 Bertler and Rosengreen found that the regional distribution of dopamine does not parallel that of noradrenaline, which suggested a separate physiological role for dopamine. Hornykiewicz (1963) presented two significant observations in human neurochemistry: (1) dopamine levels are greatly decreased in the corpus striatum and substantia nigra of sufferers of Parkinson's disease, and (2) l-DOPA improves the symptomatology of these patients. Since Falk's formaldehyde fluorescence technique (Falck, 1962) permitted identification of a specific dopamine neuron system in the brain, biochemical data on the localization of dopamine can be interpreted in terms of morphological observations.

It is now generally accepted that DA is a neurotransmitter/neuromodulator in the central nervous system. However, it has been postulated that DA may also function as an important neuromodulator/neurotransmitter in the peripheral nervous system, affecting both gastrointestinal and cardiovascular activity (Thorner, 1975; Greenacre *et al.*, 1976).

During the 1960s there was an increase in the histochemical, biochemical, and pharmacological evidence supporting the existence of specific DA-containing neurons in the central nervous system (cf. Sedvall, 1975). So far, three main dopaminergic systems have been identified; nigrostriatal, mesolimbic, and tuberoinfundibular fibres. In addition, DA is present in two areas of the rat frontal cortex, the medial prefrontal and the cingulate cortex, where it is believed to be contained in nerve fibres (Lindwall *et al.*, 1974).

4.3.1 Effect on ACh release

The peripheral nervous system

Dopamine in concentrations of 5×10^{-6} M failed to affect the release of ACh from the Auerbach plexus (Paton and Vizi, 1969). In higher concentrations (5×10^{-4} M), however, it reduced the release of ACh in a concentration-dependent manner (Vizi *et al.*, 1974, 1977b,c), whereas apomorphine failed to inhibit ACh release (Vizi *et al.*, 1977b,c).

Figure 34 shows the effect of DA on the contraction of an isolated longitudinal muscle strip induced by field stimulation. The threshold concentration

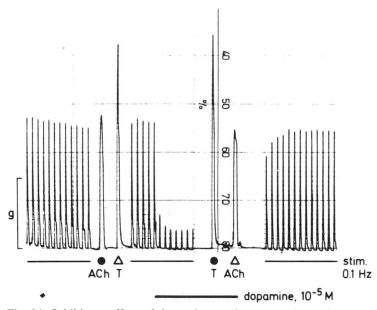

Fig. 34. Inhibitory effect of dopamine on the contractions of isolated longitudinal muscle strip of guinea-pig ileum. Note that DA fails to affect the contractions induced either by exogenous ACh (5×10^{-7}M) or by tetanic stimulation (T) (10 Hz, 20 shocks)

of DA for inducing inhibition was around 5×10^{-7} M, and maximum inhibition was usually achieved by 10^{-5} M of DA. The pre-synaptic nature of DA-induced inhibition of the electrically evoked twitch was indicated when ACh-induced contractions remained unaffected by DA. A similar finding was made by Ebong and Zar (1980). Another DA-agonist, apomorphine, had no effect on neuro-effector transmission (Vizi, 1975b). Bromocriptine (0.1–50 $\times 10^{-6}$ M), a central DA-receptor agaonist, inhibited the electrically induced twitch and ACh-induced contraction. Unlike DA-induced inhibition, the bromocriptine effect was not antagonized by phentolamine, nor was it counteracted by haloperidol, metoclopramide, or pimozide (Ebong and Zar, 1980). Another interesting observation made was that DA ($<10^{-6}$ M), in the presence of phentolamine (5×10^{-6} M), potentiated the electrically evoked twitch without a corresponding potentiation of ACh-induced contraction. Potentiation persisted unchanged after treatment with haloperidol (5×10^{-7} M) or indomethacin (3×10^{-6} g/ml) (Ebong and Zar, 1980).

These findings indicate that the inhibitory action of DA on ACh release is not mediated via DA receptors, but may act via α-adrenoceptors.

The cerebral cortex

Beani *et al*. (1980) showed that DA (5×10^{-5}–2.4×10^{-4} M) failed to affect the release of ACh evoked by electrical stimulation from guinea-pig isolated cerebral cortex slices, whereas apomorphine had an effect. Nevertheless DA (1 mM) reduced the release of ACh from cortical synaptosomes (de Belleroche *et al*., 1982).

The striatum

In *in vivo* experiments it has been shown that DA, added to the superfusion fluid in the cup, had no effect on the release of ACh from the ventricular surface of cat caudate nucleus (Jones *et al*., 1973). Vizi *et al*. (1977a) demonstrated that DA (2.6×10^{-4} M) failed to affect the resting release of ACh in isolated striatal slices of the rat but the opposite was observed by Belleroche *et al*. in 1982. However, when the endogenous DA level was reduced either by single or by repeated 6-OHDA treatments (Vizi *et al*., 1977c) the resting release of ACh more than doubled. In such preparations DA reduced the resting release of ACh (Fig. 35). A similar observation was made by Bianchi *et al*. (1979) on guinea-pig striatal slices. Ouabain (2×10^{-5} M) markedly enhanced the release of ACh from striatal slices, and its effect on the release of ACh was much more marked on striatal slices from 6-OHDA pre-treated rats.

Several lines of evidence indicate that the stimulation of DA-receptors present in the striatum, results in an inhibition of ACh release (de Belleroche *et al*., 1982; Miller and Friedhoff, 1979; Hársing *et al*., 1979a; Bianchi *et al*., 1979; Vizi *et al*., 1977a,b,c, 1979; Vizi, 1979; Stoof *et al*., 1979; Hertting *et al*., 1980) and synthesis (Sethy and Van Woert, 1974). However, there are different presynaptic DA receptors in the striatum and their sensitivity to DA also differs. The stimulation by DA of DA autoreceptors results in a reduction of DA-release and an increase of ACh release from cholinergic interneurons (Vizi *et al*., 1977a). The stimulation of DA receptors present on the cholinergic interneurons may nullify the action of the stimulation of pre-synaptic DA autoreceptors on ACh release. In addition, an increase in DA release induced by a block of pre-synaptic DA autoreceptors and a block of post-synaptic DA receptors by a drug represents opposing functional effects, the net effect being dependent on the relative potency of the drug on the two receptors. Stimulation of pre-synaptic DA receptors situated on the nerve terminals brings about local feedback control of DA release

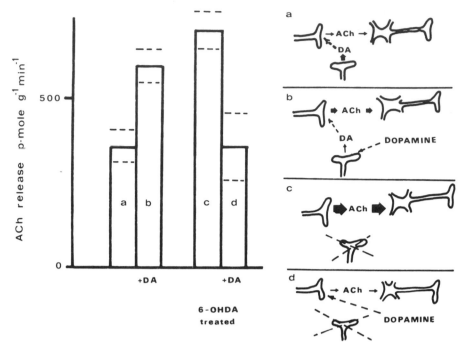

Fig. 35. Effect of dopamine (DA) and 6-OH-dopamine (6-OHDA) pre-treatment on acetylcholine (ACh) release from rat striatal slices (from Vizi *et al.*, 1977c). Insets a, b, c, and d show the possible sites of action of dopamine (DA) and the resulting release of acetylcholine (ACh). Note that when the nigrostriatal dopaminergic pathway is impaired by 6-OHDA pre-treatment the release of ACh is enhanced (c) and the effect of DA is located on the cholinergic interneurons (d)

(Raiteri *et al.*, 1978; Reimann *et al.*, 1979; Westfall *et al.*, 1979b; Brase, 1979; Stoof *et al.*, 1980). Exogenous DA inhibits DA release, and stimulation of post-synaptic DA-sensitive receptors sited on the cholinergic neurons has been shown to inhibit the activity of those neurons.

Apomorphine and piribedil administered in low concentrations to the striatal slices markedly enhances the release of ACh induced by administration of ouabain (Vizi *et al.*, 1977a). The increase of ACh release seems to result from a removal of dopaminergic control by stimulation of pre-synaptic DA receptors, as the effect of the DA agonist to enhance the transmitter output was absent in a preparation where a nigrostriatal pathway had been previously destroyed by 6-OHDA. Evidence has been provided showing that exogenous DA or a DA agonist (N,N-dipropyl-5,6-ADTN) (Stoof, *et al.*, 1980) inhibits the release of DA via stimulation of pre-synaptic DA receptors. It thus appears reasonable to suggest that apomorphine and piribedil in low concentrations activate pre-synaptic DA receptors and that the enhancement of ACh release might be due to a disinhibition of striatal cholinergic interneurons. This latter phenomenon, observed in the release of ACh, antagonized by a very low dose of pimozide, a potent and specific blocking agent of DA receptors (Creese *et al.*, 1976). The blockade of DA receptors by a low dose of pimozide may be localized only at the pre-synaptic level, since no cataleptic activity was observed in response to pimozide administration in rats, an effect mediated via inhibition of DA receptors present on cholinergic interneurons.

The mechanism of the action of DPI (3,4-dihydroxyphenylamino-2-imidazoline), a DA agonist, on striatal release of ACh differs from those of apomorphine and piribedil. Both law and high concentrations of DPI increase release of ACh. The discovery that DPI failed to affect ACh output on striatal slices taken from 6-OHDA pre-treated rats (Vizi *et al.*, 1977a) and that the enhancement by this drug of ACh release could be prevented by a low dose of pimozide suggests that DPI acts only on pre-synaptic receptor areas of nigro-striatal neurons, i.e. its effect is selective. Thus, in the experimental condition, DPI was found to be a 'pure' agonist on presynaptic DA 'autoreceptors' (Vizi *et al.*, 1977a). Nevertheless, these findings (Vizi *et al.*, 1977a,c) provide direct evidence that the sensitivity of pre-synaptic DA receptors (D_2-type) located on nigro-striatal pathways is greater than those located on cholinergic interneurons, and that they can be simulated specifically with DPI. This had several clinical implications: if tardive dyskinesias are related to an enhanced DA release or supersensitivity of DA receptors to long-term treatment with neuroleptic medication (Baldessarini and Tarsy, 1976), it should be possible to treat this disease with DPI. Since DPI reduces DA release it should enhance ACh release, as was the case in our experiments.

If released from the nerve varicosities, DA stimulates the DA receptors located on small cholinergic interneurons.

The differential sensitivity of the two pre-synaptic DA receptors may explain the apparently paradoxical behavioral effects induced by small, compared with large doses of some DA agonists: e.g. in low doses it induces hypomotility and sedation, while in high doses apomorphine causes hypermotility and stereotyped behaviour (Strömbon, 1976). In addition, the fact that apomorphine and piribedil, even in low concentrations, decreased the release of ACh in preparations which had been pre-treated with 6-OHDA might be taken as an indication of post-synaptic DA receptor super-sensitivity.

Bartholini *et al.* (1977) reported that stimulation of cell bodies in the pars compacta of the substantia nigra results in a marked decrease on the rate of release of ACh from the ipsilateral caudate nucleus. These data indicate that the stimulation of nigrostriatal neurons inhibits cholinergic transmission in caudate interneurons.

It has been found that neuroleptics are able to enhance the turnover of ACh in the striatum, but not in the cortex of the rat (Trabucchi *et al.*, 1974), and that apomorphine prevents the effect of neuroleptics (Stalder *et al.*, 1973). (+)-Amphetamine (Hemsworth and Neal, 1968; Beani and Bianchi, 1973) and L-DOPA (Pepeu and Bartholini, 1968) decrease the turnover rate of ACh in the striatum but not in the cortex. Similar results have been obtained with push–pull cannulae, except that ACh release from the cerebral cortex of the guinea pig and the cat was enhanced by (+)-amphetamine and L-DOPA (Beani and Bianchi, 1973; Pepeu and Bartholini, 1968). Stimulation of dopaminergic receptors with apomorphine fails to alter the turnover rate of ACh in rat cerebral cortex but reduces the ACh turnover rate in the striatum (Trabucchi *et al.*, 1974), although it fails to reduce choline uptake in the striatum (Atweh *et al.*, 1975). Using mice and high doses of chlorpromazine, Jenden (1974) demonstrated a decrease in the ACh turnover of the whole brain. In more detailed studies using pharmacological doses of chlorpromazine it was found that this drug and haloperidol increased the ACh turnover rate of N. caudatus and N. accumbens but failed to change that of N. interpeduncularis, amygdala, raphé dorsalis, locus coeruleus, and septum (Racagni *et al.*, 1975). It is well-established that neuroleptics enhance DA release and turnover in the striatum.

It is possible that the enhanced cholinergic outflow via striatonigral fibres stimulates nigrostriatal neurons, and produces an enhanced release of DA. Conversely, the decrease in DA turnover by DA receptor agonists such as apomorphine could be mediated by inhibition of ACh release at the nerveterminals of striatonigral fibres in the substantia nigra.

4.3.2 Effect on NA release

Dopamine has been shown to be a potent inhibitor of NA release in the cat spleen (Langer *et al.*, 1975), nictitating membrane (Langer, 1973; Enero and Langer, 1975b), the guinea-pig hypothalamus (Bryant *et al.*, 1975), the rabbit ear artery (McCulloch *et al.*, 1973), and human blood vessels (Stjärne and Brundin, 1975a), but not very effective in the rat cerebral cortex, the guineapig heart and vas deferens, and the rabbit pulmonary artery (cf. Starke, 1977). The effective concentration of DA was between 10^{-8}–10^{-6} M.

In the isolated was deferens of the guinea pig, contractile responses to adrenergic nerve stimulation at 2 Hz were inhibited by dopamine (5×10^{-6} M), and this effect was prevented by phentolamine (3×10^{-6} M). This finding suggests that the inhibition was due to an agonist action of dopamine on α-adrenoceptors (Bell, 1980).

The depression of dopamine of contractile responses to nerve stimulation was correlated with reduction in amplitude of single excitatory junction potentials (e.j.p.) evoked by nerve stimulation. Phentolamine had no effect on the amplitude of the e.j.p. evoked by a single impulse.

Tayo's proposal (1979), that there is a specific pre-junctional DA receptor has been challenged. In rat vas deferens noradrenaline was approximately forty times more potent than dopamine in producing pre-synaptic inhibition of twitches induced by electrical field stimulation. Yohimbine competitively antagonized the effect of the two agonists to a similar extent (Lcedham and Pennefather, 1982). This finding indicates that both NA and DA act on α_2-adrenoceptors. That DA acts pre-synaptically is confirmed by the finding that DA inhibits the release of ^{3}H-noradrenaline from isolated vas deferens.

Apomorphine and piribedil injected into the femoral artery of the dog induces a dose-dependent increase in femoral blood flow (Laubie *et al.*, 1977). Sections of both femoral and sciatic nerves, ganglionic blockade, section of the spinal cord, guanethidine, and α-adrenergic blockade, eliminated the dilator effect. This suggests that DA agonists act on the noradrenergic nerve-endings. The dilator effect produced by a local injection of apomorphine (Laubie *et al.*, 1977) may have been due to activation of a DA-inhibitory mechanism at the adrenergic nerve-terminals, as was shown for the heart (Long *et al.*, 1975), perhaps acting through inhibition of NA release. Bromocriptine, a DA receptor stimulant, inhibited the pre- and post-ganglionic accelerator nerve-stimulation effect on heart rate but failed to modify the dose-response curve to exogenous A (Scholtysik, 1978), indicating a pre-synaptic action.

4.3.3 Effect on serotonin release

An entirely different action was observed when the effect of DA on 5-HT release was studied.

DA (10^{-5} M) significantly enhanced the release of radioactivity from hippocampal synaptosome preloaded with ^{3}H-5-HT (Balfour and Clasper, 1982). Its effect was mediated via dopamine receptors since haloperidol prevented and apomorphine mimicked the effect of DA. NA had no effect on 5-HT release. DA failed to influence the spontaneous release of ^{3}H-5-HT from synaptosomes prepared from cerebral cortex hypothalamus or pons and medulla.

4.3.4 Effect on neurohormone release

The neurosecretory neurons of the supraoptic and paraventricular nuclei in the hypothalamus and their nerve terminals in the neural lobe of the pituitary have been the subject of many investigations. Using isolated pituitary neural lobe containing only nerve terminals the hypothalamo-inhibitory effect of dopamine on axon terminals was observed by Vizi and Volbekas, (1980a,b,) and it was found that DA (2×10^{-4} M) pre-synaptically controlled the release of oxytocin. A similar effect of DA has been also shown for vasopressin release (Vizi and Volbekas, 1980b). Therefore it seems very likely that DA released from neurons originating from the rostral zone of arcuate nucleus exert a direct modulatory control over the secretion of oxytocin and vasopressin (Fig. 36).

It has been reported that the resting output of vasopressin from the isolated neural lobe is not substantially affected by dopamine (Bridges *et al.*, 1976, Hisada *et al.*, 1977, Wagner *et al.*, 1978). A similar observation was made with oxytocin (Vizi and Volbekas, 1939b). However, when the release was augmented by ouabain, dopamine significantly reduced the release of oxytocin (Vizi and Volbekas, 1980a,b). The finding that the resting release of oxytocin is about two times higher from lobes dissected from α-methyl-p-

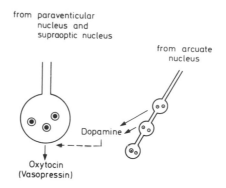

from paraventicular
nucleus and
supraoptic nucleus

from arcuate
nucleus

Dopamine

Oxytocin
(Vasopressin)

Fig. 36. Dopamine released from dopaminergic axon terminals can inhibit the release of neurohypophysial hormones

tyrosine pre-treated (dopamine-deficient) rats than from normal rats indicates that there is a continuous inhibition of oxytocin release by endogenous dopamine (Vizi and Volbekas, 1980a). Dopamine was very very active on DA-deficient neurohypophysis.

Consistent with these observations, Johnson *et al.* (1975) demonstrated a dose-related inhibition of vasopressin and oxytocin release from the isolated neurohypophysis following dopamine administration (0.053–0.53 mM). However, Bridges *et al.* (1976) reported that administration of dopamine (0.0065 mM) to the isolated neurohypophysis of the rat produced an increase in vasopressin and oxytocin release. Nevertheless, Hisada *et al.* (1977) found no change in neurohormone release when DA was administered. The data to the contrary might be attributed to differences in concentration of DA applied and the different sensitivity of DA receptors located on dopaminergic axon terminals (autoreceptors) and on neurosecretory neurons (Fig. 36).

It could be argued that the lower concentration of DA inhibited the release of endogenous DA release from tuberoinfundibular neurons via autoreceptor stimulation and thereby removed intrinsic dopaminergic control (disinhibition), whereas in higher concentrations it acted directly on neuro-secretory neurons (Fig. 37). A similar conclusion has been reached by Passo *et al.* (1981). The effect of dopamine on neurohypophysial hormone release was the result of a change in balance of excitatory and inhibitory inputs to the magnocellular soma of the supraoptic nucleus and paraventricular nucleus of the hypothalamus.

The vasopressin output evoked by high K was unaltered in the presence of dopamine (10 µM) or apomorphine (100–300 µM) (Racké *et al.*, 1982). Bromocriptine had a biphasic effect on vasopressin release stimulation, enhancing it by 20% at 1 nM and inhibiting it by 35% at 3 µM. Sulpiride (10 nM to 1 µM) stimulated the vasopressin release by about 20% and antagonized the inhibitory action of apomorphine and bromocriptine in a concentration-dependent manner. The existence of two dopamine receptors have been proposed (Racké *et al.*, 1982), one mediating inhibition of vasop-

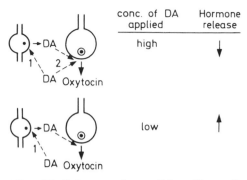

Fig. 37. Scheme of possible effect of exogenous dopamine (DA) or dopamine-receptor agonist in low (1) and high (1 and ¯2) concentrations on neurohypophysial hormone release

ressin release (activated by apomorphine, bromocriptine at high concentrations of endogenous dopamine, and blocked by sulpiride), and the other mediating facilitation of release (activated by endogenous dopamine and blocked by flupenthixol). From these data it can be concluded that dopamine is a pre-synaptic modulator of neurohypophysial hormone release: its effect is on the axon terminal.

It has been observed that dopamine-containing nerve terminals are heavily concentrated in the neural lobe of the pituitary (Saavedra *et al.*, 1975). Therefore it is probable that dopamine released from neurons originating from the rostral zone of the arcuate nucleus exerts a direct modulatory control over the secretion of oxytocin and vasopressin.

The dopaminergic nerve-endings are frequently located in close proximity to the neurosecretory axons and processes of pituicytes (Baumgarten *et al.*, 1972), indicating that there is a synaptic and/or non-synaptic interaction between the nerves. Since noradrenergic fibres are also present in the neurohypophysis, the release of neurohypophysial hormones are possibly mediated via both dopamine and α_2-adrenoceptors. This suggestion is supported by the finding that clonidine, an α_2-adrenoceptor agonist, reduces the release of oxytocin (Vizi and Volbekas, 1980a). Douglas and Taraskevich (1978) have recently reported that action potentials generated in cells cultured from intermediate lobes of the rat can be suppressed by DA. When the rostral arcuate nucleus was electrically stimulated and the subsequent changes in neurohypophysial electrical activity recorded (Passo *et al.*, 1981) neurohypophysial firing rates decreased by approximately 94% during the first 5 s following stimulation. In the presence of a dopamine receptor antagonist (pimozide, 1 μg/ml), stimulation of the rostral arcuate nucleus had no effect on neurohypophysial electrical activity. The results of the experiments of Passo *et al.* (1981) indicate that electrical stimulation of the rostral arcuate nucleus supresses the multi-unit activity recorded from the neurohypophysis.

That this inhibition is mediated by dopamine receptors is suggested by the ability of locally applied pimozide, a dopamine receptor blocking agent, to prevent the response. Moreover, dopamine superfused into the neurohypophysis causes multi-unit electrical activity to decline, whereas α- and β-adrenoceptor agonists have no such effect.

Holzbauer et al. (1978) suggested that the dopaminergic innervation of the pituitary gland may modulate the release of oxytocin and vasopressin and under experimental conditions, when increased amounts of pituitary hormones are released, a decrease in pituitary DA content and an increase in its turnover was observed. In addition, DA has been shown to inhibit the release of ACTH (Fischer and Moriarty, 1977) and melanophore-stimulating hormones (Bower et al., 1974) from intermediate lobe cells in vitro. One of the well-known long-distance modulating effects of dopamine is inhibition of prolactin secretion. Dopamine released from axon terminals of tubero-infundibular neurons is transported by the portal blood to prolactin cells and where it inhibits prolactin secretion by the pituitary through specific receptor sites. This effect of dopamine is a hormonal action, since it reaches it target cells via the bloodstream and it seems very likely that, depending on the circumstances, DA can serve both modulator (synaptic and non-synaptic) and hormone functions (Table 10).

Table 10. Modulatory and hormonal functions of dopamine

Function	Site of action	Mode of action
Transmitter	No evidence	
Modulator	Striatum	Pre-synaptic (synaptic and non-synaptic)
	Neurohypophysis	Pre-synaptic (synaptic and non-synaptic)
Hormone	Adenohypophysis (prolactin release)	Via portal blood

4.3.5 Electrophysiological evidence of the depressant action of DA on neurons

Periphal neurons

Dopamine-containing neurons have been identified in the sympathetic ganglion (Björklund et al., 1970). Libet (1970) postulated that dopamine is the most likely catecholamine that mediates the slow inhibitory post-synaptic potential. (slow i.p.s.p) in the sympathetic ganglion. Dun and Nishi (1974) reported that dopamine depressed synaptic transmission in the cat superior cervical ganglion. Dopamine (10^{-5}–10^{-3} M) also depressed the amplitude of the excitatory post-synaptic potential (ep. sp) and blocked impulse transmission (Dun and Nishi, 1974). The post-synaptic membrane sensitivity to acetylcholine (ACh) applied iontophoretically was not affected by dopamine and reduced the quantal content of the e.p.s.p. in a low CA^{2+} and high Mg^{2+}

solution, but had no effect on the quantal size. The ganglionic blocking effect of dopamine was antagonized by phenoxybenzamine, but not by propranolol, indicating that it is exerted primarily through an α-adrenoceptive site at the pre-synaptic nerve terminal.

The central nervous system

Functionally, evidence has accumulated that DA is able to reduce the unit discharge frequency of most neurons on which it has been tested. It is inhibitory on the cerebral cortex (Krnjevic and Phillis, 1963; Bunney et al., 1973a) and the caudate nucleus (Bloom et al., 1965; McLennan and York, 1966) and similar depressant effects of DA on the neurons of the thalamus (Andersen and Curtis, 1964), the hypothalamus (Bloom et al., 1963), the hippocampus (Biscoe and Straughan, 1966), and on single cortical neurons (Bevan et al., 1975, 1977) has been observed.

The finding that systematically administered apomorphine markedly reduces impulse activity in migral units is compatible with the view that the drug reduces striatal DA turnover by reducing nerve impulse activity in the nigrostriatal pathway (Bunney et al., 1973a,b). The fact that haloperidol and chlorpromazine effectively block the depressant effects of apomorphine provides further evidence that DA receptor stimulation is required for the depressant effect of apomorphine on DA neurons. However, since local application of apomorphine to nigral cells also retards impulse activity, the role of pre-synaptic DA receptors must also be considered in terms of the mechanism of action of apomorphine (Aghajanian and Bunney, 1973).

4.4 MODULATION BY SEROTONIN

Although much circumstantial evidence has accumulated, the transmitter function of serotonin (5-hydroxytryptamine) (5-HT) has not yet been convincingly established. It has been found to be present in the central nervous system (Twarog and Page, 1953; Bogdansky et al., 1956; Brodie and Shore, 1957) in nerve terminals (Zieher and de Robertis, 1963) and is believed to have an important role in the regulation of mood, sleep, excitability of spinal motoneurons, and other physiological processes. Recently, evidence has been presented that it plays a modulatory role both in the autonomic and the central nervous systems.

4.4.1 Acetylcholine release

The gastrointestinal tract

It has been suggested that 5-HT is an excitatory neurotransmitter in Auerbach's plexus (Bülbring and Gershon, 1966), however, 5-HT possess an inhibitory action in the large intestine and in the stomach (Bülbring and

Gershon, 1968). In addition there is histochemical (Gershon *et al.*, 1977) and biochemical (Robinson and Gershon, 1971a,b) evidence for serotoninergic intrinsic neurons in Auerbach's plexus.

5-hydroxytryptamine (5-HT) has both inhibitory and excitatory effects (Gaddum and Picarelli, 1957; Adam-Vizi and Vizi, 1978) on isolated preparations of the intestine. Inhibition of the peristaltic reflex is observed when 5-HT is applied to the serosal surface of the guinea-pig ileum (Kosterlitz and Robinson, 1957), and this is attributed by Bülbring and Crema (1958) to a blockade of ganglionic transmission within the myenteric plexus.

Direct evidence has been presented that 5-HT releases ACh from Auerbach's plexus in amounts large enough to produce contractions (Brownlee and Johnson, 1963; Ádám-Vizi and Vizi, 1978). Morphine, adrenaline, and tetrodotoxin have been shown to completely inhibit the release of ACh induced by 5-HT, indicating that the release of ACh in response to 5-HT is due to the stimulation of enteric ganglion cells. The effect of 5-HT on Auerbach's plexus to release ACh cannot be blocked by hexamethonum, indicating the existence of 5-HT-sensitive receptors in the ganglion cells which are independent of nicotinic receptors (Ádám-Vizi and Vizi, 1978). A direct excitation of myenteric neurons by 5-HT has been reported in electrophysiological experiemnts (Sato *et al.*, 1974; Dingledine and Goldstein, 1975).

There is convincing evidence from the work of Gershon and his colleagues (Robinson and Gershon, 1971b) that the myenteric plexus contains intrinsic neurons which can take up, synthetize, store, and release 5-HT. It is possible that 5-HT released under physiological circumstances from myenteric neurons may act through axo-axonal synapses or simply by diffusion into the vicinity of nerve terminals to inhibit or excite cholinergic neurons within the plexus, and thereby inhibit or induce the peristaltic reflex. Such a functional pre-synaptic inhibitory effect occurs with noradrenaline when the extrinsic sympathetic nerves are stimulated.

5-HT has been detected histochemically in the myenteric plexus following inhibition of MAO (Robinson and Gershon, 1971b; Feher, 1974, 1975).

The central nervous system

A growing bulk of morphological evidence exists on serotonergic inputs into striatal pathways (Dahlström and Fuxe, 1965; Palkovits *et al.*, 1974). Neurochemical studies have indicated an interaction between serotoninergic and cholinergic neurons (Euvrard *et al.*, 1977) by the demonstration that serotonin agonists enhanced the contents of ACh in the striatum. In addition, electrophysiological evidence has shown that the electrical stimulation of raphé nuclei and 5-HT administration results in an inhibition of neuronal firing in the striatum (Miller *et al.*, 1975; Olpe and Koella, 1977; Davies and Tongroach, 1978).

Evidence has been presented (Vizi *et al.*, 1981a) showing that there is a serotoninergic inhibitory influence on striatal cholinergic neurochemical transmission. The reduction of serotonin release by para-chlorophenylalanine (p-CPA) or 5,7 dihydroxytryptamine (5,7 DHT) pre-treatment or administration of serotonin antagonists, mianserine, or methysergide significantly enhances the release of ACh from striatal interneurons when the neurons were activated by ouabain. These data provided direct evidence that serotoninergic input exerts a modulatory effect on striatal cholinergic interneurons inhibiting the release of ACh.

The decrease in ACh release induced by serotoninergic agonists was identical irrespective of how the release was enhanced, either by removal of dopaminergic control, or of serotoninergic control, or of both. It can thus be concluded that in the striatum most of the cholinergic interneurons are under the tonic inhibitory influence of both dopaminergic (see section 4.3) and serotoninergic neurons and the interneurons in the striatum are equipped with both dopamine and serotonin receptors. A more detailed analysis is needed to know which discrete region of the striatum has only one type of modulatory influence.

The substantia nigra contains serotonin and receives a direct inhibitory (serotoninergic) input from the median raphé nucleus (Dray *et al.*, 1976). It is suggested that stimulation of the median raphé nucleus decreases DA release from the nigrostriatal axon terminals, resulting in an increase of ACh release from cholinergic interneurons. These facts indicate that there are two serotonergic pathways which are able to influence the release of ACh from cholinergic interneurons; (1) fibres originating from the dorsal raphé nucleus having a direct inhibitory effect on cholinergic interneurons equipped with serotonin receptors (ACh release is reduced); (2) fibres originating from median raphé nucleus possessing an inhibitory action on nigrostriatal neurons which in turn disinhibits cholinergic interneurons (ACh release increased). Of particular relevance is the significant increase in striatal DA content observed after median raphé lesioning, a fact that indicates a reduction in striatal DA utilization (cf. Roth *et al.*, 1975).

4.4.2 Noradrenaline release

5-hydroxytryptamine has been shown to stimulate NA release from isolated rabbit hearts by activating tryptamine receptors on the terminal sympathetic nerves (Fozard and Mobarok Ali, 1976; Fozard and Mwaluko, 1976). A characteristic feature of this response is the tendency to rapid and reversible tachyphylaxis (Fozard and Mwaluko, 1976). In contrast to this finding, it has been observed that serotonin inhibits the release of 3H-NA induced by electrical stimulation from the dog saphenous vein (McGrath and Shepherd,

1976; Feniuk *et al.*, 1978, 1979), suggesting that activation of a pre-synaptic serotonin receptor is involved. Like the central inhibitory receptor, it cannot be blocked by the classical D-receptor blocking agent, cyproheptadine.

4.4.3 Dopamine release

Ennis *et al.* (1981) observed that serotonin (10^{-7}–10^{-5} M), as well as other indoleamines, reduced the potassium-evoked release of ^{3}H dopamine from superfused striatal slices, an effect blocked by serotonin antagonists such as methysergide, metergoline, methiothepin, cyproheptadine, and mianserin (Ennis *et al.*, 1981). Serotonin produced a concentration-dependent decrease in the field stimulation-induced release of ^{3}H-dopamine from rat striatum, with the threshold concentration being 10^{-6} M or lower (Westfall and Titter-many, 1982). Methysergide prevented the effect of serotonin. By contrast, serotonin stimulated ^{14}C-dopamine release from synaptosomes prepared from striatal tissue (de Belleroche and Bradford, 1980). This effect was additive with the simultaneous application of high potassium (56 mM) and was not blocked by classical serotonin antagonists (de Belleroche and Bradford, 1980).

4.4.4. Electrophysiological findings

Ganglionic transmission

It has been demonstrated that 5-HT inhibits ganglionic transmission (Machova and Boska, 1963; Dun and Karczmar, 1981) and exerts a direct post-synaptic effect on sympathetic neurons (Wallis and Woodward, 1975; Skok and Selyanko, 1979). 5-HT-induced attenuation of the synaptic potential may be due to a reduction in induced ACh release from pre-ganglionic fibres. The findings that the quantal (Dun and Karczmar, 1981) content but not the quantal size is significantly decreased by 5-HT and that the post-synaptic sensitivity to ACh is not diminished by 5-HT are in accord with the suggestion that 5-HT acts pre-synaptically. The membrane depolarization induced by 5-HT was accompanied by a reduction in membrane resistance, suggesting an increase of membrane permeability to certain ions, possibly Na and K (Skok and Selyanko, 1979). The diminution of the EPSP by 5-HT was partially antagonized in high Ca^{2+} solution (Dun and Karczmar, 1981), indicating the possible role of 5-HT on Ca entry.

Auerbach's plexus

Electrical stimulation of the interganglionic fibre tracts of Auerbach's plexus released 5-HT which produced a slow EPSP associated with an increase in input resistance in the pre-synaptic neuron (Wood and Mayer, 1979). Multi-unit extracellular recording from guinea-pig myenteric neurons shows an increase in the rate of spike discharge after addition of 5-HT to the organ bath. In the experiments of Wood and Mayer (1979) concentrations of curare (140–430 μM) which reduced fast EPSPs in other neurons within the myenteric ganglia did not reduce the slow EPSPs and did not affect the response of the cells to exogenously applied 5-HT. However, using intracellular recording it was shown (North *et al.*, 1980) in myenteric neurons of the guinea-pig ileum that 5-HT did not change the amplitude of depolarizations produced by iontophoretic application of acetylcholine. 5-HT also depresses the cholinergic fast EPSP in all myenteric neurons. Since 5-HT does not affect the amplitude of time course of the depolarization produced by an iontophoretic application of ACh, it has been concluded that the action of 5-HT on the fast EPSP is due to pre-synaptic inhibition of ACh release. The mechanism which underlies this pre-synaptic inhibition is now known. It could be a depolarization of pre-synaptic terminals in accordance with classical ideas (Takeuchi and Takeuchi, 1966; Eccles *et al.*, 1963), or it could be a membrane hyperpolarization and conductance increase, as appears to operate for the action of morphine on myenteric neurons (North and Tonini, 1976, 1977).

The locus coeruleus

Segal (1979) produced evidence for a pre-synaptic action of 5-HT in the rat locus coeruleus because it reduced the response of neurons to noxious inputs more than it depressed their spontaneous firing rates. A more convincing case for a pre-synaptic action was provided by Headley *et al.* (1978), who found that 5-HT reduced the response of cat dorsal horn cells to noxious inputs whilst having little effect on the responses of the same cells to non-noxious inputs, an effect which may be functionally related to descending raphé-spinal pathways.

The hippocampus

The hippocampus receives a direct input from the raphé nuclei and its serotonin content is dependent on the integrity of the raphé nuclei. Hippocampal pyramidal cells in the rat are inhibited by serotonin and by electrical stimulation of the dorsal and median raphé nuclei (Segal, 1975). The inhibitory response to raphé stimulation is absent in rats that have been pre-treated with p-CPA, a serotonin synthesis inhibitor, and the effects of p-CPA are alleviated by administration of 5-hydroxtryptophan, the precursor of serotonin. The inhibitory effects of both serotonin and raphé stimulation are blocked by the serotonin antagonists methysergide and cyproheptadine, and potentiated by chlorimipramine, a serotonin re-uptake blocking agent. Serotonin therefore deserves serious consideration as an inhibitory modulator acting on rat hippocampal pyramidal cells.

The cortex

Serotonin depresses neurons of sensory-motor cortex and its effects are selectively antagonized by the drug metergoline (Sastry and Phillis, 1977). Metergoline also antagonizes the inhibitory response generated by raphé stimulation.

Motoneurons

Serotonin, like NA, hyperpolarizes motoneurons (Phillis *et al.*, 1968a) with a concomitant increase in membrane resistance. It inhibits the invasion of ortho- and antidromically evoked spikes into the initial segment and soma-dendritic membrane.

The midbrain raphé

It has been shown that the firing rate of serotonin-containing neurons in the midbrain raphé is modulated by a negative feedback mechanism (Aghajanian, 1972). The micro-iontophoretic application of 5-HT directly into presumed 5-HT-containing neurons consistently produces inhibition, showing that 5-HT-containing neurons *per se* have 5-HT-sensitive receptor sites.

4.5 MODULATION BY γ-AMINOBUTYRIC ACID (GABA)

GABA has been implicated in a pre-synaptic action in the spinal cord of vertebrates (Eccles *et al.*, 1961, 1962; Barker and Nicoll, 1973) and in the neuromuscular junction of crustacea (Dudel and Kuffler, 1960, 1961a,b; Takeuchi and Takeuchi, 1966). There is evidence to suggest a pre-synaptic release of GABA (Jasper *et al.*, 1965; Mitchell and Srinivasan, 1969; Bradford, 1970; Arbilla and Langer, 1978a).

4.5.1 Effect on acetylcholine release

GABA receptor stimulants, diazepam and muscimol, reduce the turnover release of ACh in the cortex and midbrain, does not affect ACh release in the striatum and hippocampus (Zsilla *et al.*, 1976). GABA-containing neurons may impinge on cholinergic neurons, producing their inhibition, since perfusion with GABA (10^{-4}–10^{-3} M) reduces the size of all EPSP's produced by a train of 100 stimuli at 1/s in cholinergic synapse in *Aplysia californica*. However, GABA does not significantly affect the membrane resistance (Tremblay and Plowde, 1975) and fails to affect the iontophoretic potential of ACh. This finding indicates that the effect of GABA is presynaptic.

Minimal stimulation of the right visceropleural connective tissue of the abdomnal ganglion of *Aplysia californica* produces a monosynaptic unitary and cholinergic EPSP in cell R15 (Frazier *et al.*, 1967).

It has been shown that repeated stimulation of RC1–R15 at 0.5–2 stimuli/ s produces changes in the size of the EPSPs. The relative amplitudes are modified by a variety of cholinergic (Woodson *et al.*, 1975), dopaminergic, and serotonergic agents (Tremblay *et al.*, 1976), as well as by opiates (Tremblay *et al.*, 1974).

4.5.2 Effect on noradrenaline release

Arbilla and Langer (1978b) showed that GABA does not modify ^{3}H-noradrenaline overflow nor the post-synaptic responses to sympathetic nerve stimulation. However, Bowery *et al.* (1980) found that GABA facilitated the release of potassium-induced ^{3}H-noradrenaline in rat cerebral cortex slices. This was only observed when low concentrations of potassium were used. Yet in the rabbit pulmonary artery, exposure to GABA reduces the stimulation evoked release of tritiated noradrenaline (Starke and Weitzell, 1980). These inhibitory effects of GABA on noradrenaline release are not blocked by either picrotoxin or bicuculline. In the central nervous system, GABA has a facilitatory effect on the stimulated release and turnover (Biswas and Carlsson, 1977) of noradrenaline from slices of rat occipital cortex. Exogenous GABA (30–1000 μM) increases in a concentration-dependent manner, the stimulation-induced release of ^{3}H-noradrenaline elicited by a 1-min exposure to 20 mMK$^+$. When the release is induced by high potassium concentrations (35 mM) or through a calcium-independent mechanism i.e. exposure to tyramine, GABA had no effect.

4.5.3 Effect on dopamine release

When the striatum was labelled with (^{3}H)-dopamine, and the subsequent basal or K$^+$-induced outflow of tritium was determined, either no effect of GABA (Starr, 1978, 1979; Martin and Mitchell, 1980) or an increase on basal outflow was observed (Kerwin and Pycock, 1979; Roberts and Anderson, 1979; Stoof *et al.*, 1979). An increase in basal (^{3}H) dopamine outflow was also obtained when GABA was added to rat striatal slices superfused with (^{3}H) tyrosine (Giorguieff *et al.*, 1978; Giorguieff-Chesselet *et al.*, 1979). While most investigators report facilitation by GABA of the K$^+$-induced overflow of tritium (Starr, 1978, 1979; Kerwin and Pycock, 1979;

Stoof *et al.*, 1979; Ennis and Cox, 1981), some observed either no effect (Martin and Mitchell, 1980) or inconsistent inhibition (Bowery *et al.*, 1980). Reimann *et al.* (1982) found that γ-amino butyric acid (10^{-4} and 10^{-3} M) increased the stimulation-induced overflow of tritium and of dopamine.

4.5.4 Effect on substance P release

In slices of substantia nigra, administration of GABA at a concentration of 50 μM almost completely prevented the release of substance P normally induced by exposure of 47 mM K+, although GABA had no effect on the resting release of substance P (Jessel and Iversen, 1977b). This effect of GABA seems to be mediated via GABA receptors. An effect, similar to that of GABA could be mimicked by muscimol (10 μM) a GABA agonist, and blocked by picrotoxin (50 μM) and bicuculline (50 μM) (Jessel and Iversen, 1977b). These results suggest that GABA acting on pre-synaptic receptors of a substance P-containing neuron may be able to modulate its release in the substantia nigra. Since GABA and substance P may represent, respectively, inhibitory and excitatory transmitters synapsing on dopaminergic neurons of substantia nigra (Dray and Straughan, 1976), this could represent a modulatory mechanism in the substantia nigra.

4.6 MODULATION BY OPIOID PEPTIDES

Although morphine has been used as a pain-killing drug for many centuries, it is only in the last few years than an endogenous ligand has been discovered in the nervous system which appears to be able to occupy opiate receptors. Since opiates have been known for thousands of years and endorphine-like peptides only for a few years, we call these peptides opioid peptides.

In the last few years evidence that peptides isolated from the brain and pituitary gland (Hughes *et al.*, 1975; Li and Chung, 1976) are capable of mimicking morphine-like effects has been growing. Hughes *et al.* (1975) were the first to describe two pentapeptides: leucine-enkephalin (Tyr–Gly–Gly–Phe–Leu) and methionine–enkephalin (Tyr–Gly–Gly–Phe–Met), extracted from the brain, which produce naloxone-reversible inhibition of electrically induced contraction of mouse vas deferens and of longitudinal muscle strip of the guinea-pig ileum (Hughes *et al.*, 1975).

It was also observed that Met[5] -enkephalin is present in the amino acid sequence of β-lipotropin (LPH 61–65), a polypeptide isolated from pituitary tissue. Longer fractions of lipotropin, such as α-endorphin (LPH 61–76), γ-endorphin (LPH 61–77), δ-endorphin (LPH 61–79), and β-endorphin (LPH 61–91) also produce opiate-like actions. It has been proposed that β-LPH might be a biological source of endogenous substrates with opiate-like activity (Hughes *et al.*, 1975; Gráf *et al.*, 1976).

Since evidence has been obtained that opioid peptides are capable of affecting, pre-synaptically, the release transmitters, it is suggested that they might play both a modulator and a transmitter role. This suggestion is supported by the findings that opioid peptides are synthetized, stored in, and released from both the peripheral and the central nervous system.

The release of enkephalin from the myenteric plexus of the guinea-pig intestine has been demonstrated (Schulz *et al.*, 1977; Hughes *et al.*, 1978). The amount released per pulse was of the order of 0.004 ng/g of tissue for an average tissue content of 250 ng/g. In addition, recent immunohistochemical studies have shown that the enkephalins are localized in neurons in various regions of the central nervous system, with high concentrations in nerve terminals (Elde *et al.*, 1976; Hökfelt *et al.*, 1977a). These studies have indicated that enkephalin-containing nerve terminals are particularly concentrated in the rat globus pallidus, which contains enkephalin in a concentration six to eight times higher than that observed in any other region of the brain (Simantov *et al.*, 1977). It has been found that enkephalins are associated with axons and nerve terminals of striatal neurons (Elde *et al.*, 1976) and that the cell bodies of enkephalin-containing interneurons may reside in striatum (Hong *et al.*, 1977). Opioid peptides are present in relatively high concentrations in the striatum (Hughes, 1975; Yang *et al.*, 1978).

The physiological role of enkephalin and other related peptides is supported by the findings that enkephalin can be released from brain tissues (Henderson *et al.*, 1978; Smith *et al.*, 1976; Iversen *et al.*, 1978; Hársing *et al.*, 1982).

Veratridine (10 μM) was found to enhance the release and tetrodotoxin (1.7 μM) abolish the veratridine-induced output of enkephalin. The ratio of Met-enkephalin to Leu-enkephalin released from the guinea-pig striatum is very similar to the ratio of these peptides present in the tissue. These facts indicate that the source of enkephalin is the nervous tissue.

Leu-enkephalin-containing nerve fibres in the pallidum do not belong to an intrinsic neuronal circuit but to a striopallidal pathway. This was shown by Cuello and Paxinos (1978), who demonstrated that isolation of the globus pallidus from the neighbouring caudoputamen by a knife-cut lesion within the caudoputamen caused an almost complete disappearance of Leu-enkephalin immunofluorescence from the ipsilateral globus pallidus. However, stereotaxic undercutting of the cerebral cortex, stereotaxic injections of 6-OHDA into the substantia nigra (which in parallel experiments resulted in a 60–70% fall in striatal DA content), electrolytic lesions of the medial forebrain bundle (2 mA, 20 s), and stereotoxic knife-cut hemisections at the middle-hypothalamic level did not change the pattern of Leu-enkephalin immunoreactivity in the globus pallidus.

Opiates and opioid peptides can affect chemical neurotransmission pre-synaptically, therefore the different isolated preparations, mainly cholinergic and noradrenergic, have been widely used for studying kinetic parameters of agonists and antagonists, and for identifying receptor selectivity. A summary of this work is given in Table 11.

Table 11. The inhibitory potencies of endogenous opioid peptides in isolated electrically stimulated longitudinal muscle strip of guinea-pig ileum (GPI), mouse vas deferens (MVD), and rabbit ear artery (ART) preparations

Peptide	IC_{50}[a]			References
	GPI	MVD	ART	
Met[5]-enkephalin	183.5	7.0	105.0	Rónai et al. (1978)
	96.2	12.8		Waterfield et al. (1977)
Met[5]-enkephalin-Arg[6],Phe[7]	548	15.4	85.0	Wüster et al. (1981)
	366.5	9.4		Rónai et al. (1982)
Leu[5]-enkephalin	2400	22	630	Knoll (1976)
	463	7, 8		Waterfield et al. (1977)
Dynorphin[1-13]	2, 2	3,6	N.T.[b]	Wüster et al. (1981)
α-neo-endorphin[1-8]	132	107	N.T.[b]	Wüster et al. (1981)
β-endorphin	87	84		Rónai et al. (1977)
			~950	Rónai et al. (1978)
	75	53		Lord et al. (1976)

[a]50% inhibitory concentration expressed in nM.
[b]Not tested.

4.6.1 Acetylcholine release

The autonomic nervous system

Morphine decreases the amplitude of the electrically induced contractions in the myenteric plexus-longitudinal muscle preparation of the guinea-pig ileum (Paton, 1957; Gyang and Kosterlitz, 1966). This preparation has been used to study the basic mechanism of the effects of narcotic analgesics and opioid peptides, since the receptors of the myenteric plexus are similar to those in the central nervous system (Fennessey et al., 1969; Ehrenpreis et al., 1972). It has been shown that the inhibitory action of opiates on neuro-effector transmission of guinea-pig ileum correlates closely with their analgesic potency (Kosterlitz and Waterfield, 1975; Paton, 1958). These results suggest that opiate receptors in both the myenteric-plexus and the central nervous system are similar in nature.

Narcotic analgesics have been shown to reduce the electrically evoked release of ACh (Paton, 1957; Paton and Zar, 1968; Cowie et al., 1968) from Auerbach's plexus. Their effect is frequency-dependent. At high frequencies of stimulation (> 3 Hz) opiates have no effect, but when short trains (2–5 impulses) were applied, morphine was effective (Vizi, 1974c) even at high (10 Hz) frequencies of stimulation. Since Waterfield and Kosterlitz (1975) and Koritsánszky et al. (1977) showed that naloxone, a pure opioid antagonist, increased the stimulated release of ACh from isolated longitudinal muscle strip of guinea-pig ileum it has been suggested that endogenous opioid peptides are involved in the modulation of transmitter release from the intestine.

The central nervous system

Morphine causes a decrease in ACh output from the cerebral cortex in unanaesthetized rabbits (Beani *et al.*, 1968) and in rats (Garau *et al.*, 1975). However, raphé lesions prevented the effect of morphine on ACh release (Garau *et al.*, 1975). Narcotic analgesics inhibit the release of ACh from the cerebral cortex of rats (Pepeu *et al.*, 1975; Beleslin and Polak, 1965) and mice (Jhamandas and Dickinson, 1973) and from the striatum of cats (Yaksh and Yamamura, 1975). Morphine also inhibits the hemicholinium-induced decline of rat brain ACh content and the uptake of choline in rat occipital cortex, but fails to change the uptake of choline in rat striatum (Atweh *et al.*, 1975). Naloxone is capable of restoring the reduction in ACh release from the cerebral cortex of the cat produced by the narcotic analgesics (Jhamandas *et al.*, 1970, 1971; Matthews *et al.*, 1973; Yaksh and Yamamura, 1975) but it fails to affect the spontaneous release of ACh from the surface of cat cerebral cortex (Jhamandas and Sutak, 1976). β-endorphine has no significant effect on the resting release of ACh (Vizi *et al.*, 1977a) from striatal slices of the rat. In agreement with this finding, β-endorphin does not affect the turnover rate of ACh in the caudate nucleus of the rat (Moroni *et al.*, 1977a,b). A similar finding was made by Hársing *et al.* (1978), who showed that Met^5-enkephalin and related pentapeptides did not affect the resting release of ACh from striatal slices of the rat. However, when the release of ACh is induced by administration of ouabain, two opioid peptides (β-endorphin and D-Ala^2-Pro^5-enkephalinamide) enhance this (Vizi *et al.*, 1977c). This increased release of ACh might be due to a pre-synaptic inhibitory effect of added opioid peptides on dopaminergic nerve terminals, leading to a decreased release of endogenous DA (Subramanian *et al.*, 1977) and thus disinhibiting the cholinergic neurons (Hársing *et al.*, 1978). This suggestion, however, has no direct experimental confirmation. Nevertheless, it has been found that high-affinity (morphine-displaceable) binding sites for Leu-enkephalin in the striatum decreased after chemical or mechanical interruption of the nigrostriatal bundle (Pollard *et al.*, 1977). Degeneration of DA terminals can be elicited either chemically by 6-OHDA or by electrocoagulation of the pars compacta of the substantia nigra and neurochemically assessed by the reduction in 'dopa' decarboxylase activity. The effect of such lesions were seen after 6–22 days and resulted in significant decreases in ^{3}H-Leu-enkephalin binding displaceable by morphine. The similar decrease in ^{3}H-naloxone binding following the chemical lesions confirms that the high-affinity binding sites of ^{3}H-Leu-enkephalin represent the 'opiate' receptors. Unilateral microinjections of 6-OHDA into the substantia nigra decreased ^{3}H-Leu-enkephalin binding displaceable by morphine, whereas the high-affinity uptake of ^{3}H-5-hydroxytryptamine in striatal synaptosomes from the same animals was not significantly modified (Pollard *et al.*, 1977).

A disinhibition, of the above-mentioned type might play an important role in the modulation of neurochemical transmission. In contrast to its effects on striatal slices from normal rats, β-endorphin caused an inhibition of ouabain-induced ACh release from striatal slices of 6-OHDA-treated rats, in which the dopaminergic fibres would have been largely destroyed. This can be explained by a direct action of opioid peptides on cholinergic neurons.

Enkephalin, however, had no direct effect on cholinergic neurons (Hársing *et al.*, 1978). These findings suggest that the sensitivity of opiate peptide receptors are different.

Beani *et al.* (1982) showed that morphine at a low dose (3 μM) had no effect on slices of cerebral cortex, but it enhanced the evoked release of ACh in thalamic and caudate slices. At higher doses of morphine (10–30 μM) the ACh release evoked by electrical pulses was significantly inhibited in every area. Met-enkephalin behaved like morphine in thalamic slices. Naloxone antagonized the inhibitory effect of morphine in the cerebral cortex and caudate nucleus slices. Naloxone blocked the releasing effect of morphine in caudate slices. In contrast, naloxone did not affect the increase of ACh release caused by morphine and Met-enkephalin in thalamic slices. A two-fold increase of calcium concentration in the Krebs solution prevented the inhibitory effects of morphine (10 μM).

4.6.2 Effect on noradrenaline (dopamine) release

The autonomic nervous system

Morphine-like drugs inhibit adrenergic neuro-effector transmission in the mouse vas deferens by depressing the induced release of NA from post-ganglionic sympathetic nerve fibres which innervate the smooth muscle (Henderson *et al.*, 1972; Hughes *et al.*, 1975; Henderson and Hughes, 1976). The depression of NA release is stereo-specific and reversed by naloxone. Normorphine and Met-enkephalin markedly depresses the EJP, the twitch response, and NA output to nerve stimulation in mouse vas deferens preparation (Henderson, and Hughes, 1976). It has been shown that morphine prevents the inhibitory effect of sympathetic stimulation on jejunum (Szerb, 1961). This was, in fact, the very first observation of an inhibitory effect of opiates on noradrenergic neruochemical transmission. Henderson *et al.* (1972) later showed that contractions of the mouse vas deferens are inhibited by morphine.

The opioid peptide Met[5]-enkephalin methylester completely inhibits vaso-constrictor responses to nerve stimulation in the rabbit ear artery (Knoll, 1976). Met[5]-enkephalin is more potent than Leu[5]-enkephalin. It has been suggested that enkephalins act pre-synaptically on a neuronal receptor: they inhibit the outflow of ^{3}H-NA from the artery and the vasoconstrictor responses to exogenous NA remain unchanged in the presence of those concentrations of enkephalins, which completely inhibit the effect of nerve stimulation. Interestingly, opiate agonists do not inhibit the vasoconstrictor response to direct electrical stimulation, thus showing the absence of opiate receptors in the neuronal elements of the central ear artery of the rabbit. This, in fact, is the first evidence of differences existing between receptors sensitive to morphine-like substances and endogenous enkephalin-like peptides (Knoll, 1976). Thus, two kinds of opiate receptors have been suggested (Knoll, 1977). Rónai *et al.* (1977) have shown that there is a significant difference in the equilibrium constant (K_e) of the antagonist naltrexone. When studying normorphine and opioid peptides as agonists in the mouse vas deferens; the K_e values for normorphine, Met-enkephalin, and β-endorphin were 0.634, 6.945, and 8.843, respectively. This observation suggests that there is a difference in receptor sensitivity to opiates and to endorphins. Moreover, Lord *et al.* (1977) demonstrated two binding sites for ^{3}H-Leu-enkephalin. Terenius (1978) showed that there are two different binding sites for opiates and enkephalins.

The medial smooth muscle preparation of the cat nictitating membrane has also proved to be a useful preparation for the analysis of opiate receptors (Knoll and Illés, 1978) since opioids inhibit the contraction induced by single-field stimulation.

The central nervous system

Morphine depresses the potassium-induced release of ^{3}H-NA in the rat cerebral cortex (Montel *et al.*, 1974a,b, 1975; Celsen and Kuschinsky, 1974) and of ^{3}H-DA in the rat striatum (Loh *et al.*, 1976). While Met-enkephalin mimicked the effects of morphine in rat cortex slices (Taube *et al.*, 1977), it failed to reduce the release of DA from the rat striatum (Loh *et al.*, 1976). Like narcotic analgesics, enkephalin reduces the electrical stimulation evoked overflow of tritium from rat occipital cortex slices pre-incubated with ^{3}H-NA (Taube *et al.*, 1976a) and this depression can be antagonized by naloxone. Methionine-enkephalin (10^{-6}–10^{-5} M) significantly reduced the potassium-induced release of DA from isolated striatal slices and of ACh from hippocampal slices of the rat (Subramanian *et al.*, 1977), while naloxone (10^{-5} M) prevented this effect. In contrast to these findings, opiate receptor agonists were unable to inhibit the release of ^{3}H-DA from rat striatal slices, even when tested in concentrations higher than those found to be effective in the cerebral cortex (Arbilla and Langer, 1978b).

The failure of morphine and of β-endorphin to reduce the release of ^{3}H-DA from the striatum conflicts with the evidence presented in a previous report (Loh *et al.*, 1976). It should be noted, however, that in the experiments of Arbilla and Langer (1978b) a 1-min exposure time and 20 mM K^+ were used, while in their previous publication, 53 mM K^+ for 20 min was used (Loh *et al.*, 1976). Arbilla and Langer (1978b) claimed that their experimental conditions (1-min short pulse of a moderate potassium concentration) mimic nerve stimulation more closely than a prolonged exposure to a rather high concentration of K^+.

Naloxone by itself failed to affect the release of noradrenaline induced by electrical stimulation (Montel *et al.*, 1974a,b, 1975). On the other, naloxone increased the potassium-induced release (Taube *et al.*, 1976a). Perhaps potassium depolarization provides a better stimulus for the release of endogenous morphine receptor ligands than do electrical pulses. Thus it cannot be excluded that one physiological role of enkephalin or related compounds in pre-synaptic inhibition of central noradrenergic synaptic transmission.

4.6.3 Effect on the release of substance P

It has been suggested (Lembeck, 1953; Otsuka and Konishi, 1976) that the undecapeptide substance P acts as an excitatory transmitter released from the terminals of primary sensory nerves in the spinal cord. It depolarizes the motorneuron and is concentrated mainly in the primary afferent terminals.

On a molar basis, substance P is about 200 times more potent than L-glutamate. It seems to selectively excite those cells in the dorsal horn that respond to noxious stimuli. To understand the role of substance P in analgesia it is worth mentioning that, in the spinal cord and brain stem, the region described by Atweh and Kuhar (1977a,b,c) as containing high densities of opiate receptors, corresponds to those containing substance P terminals.

It was shown that substance P-like immunoreactivity is released from the mammalian spinal cord *in vivo* (Yaks *et al.*, 1980) following chemical stimulation of sensory neurons with reduced responsiveness of CNS tissue to nociceptive stimuli. The release of substance P induced by high-intensity stimuli was completely inhibited by intrathecally administered morphine (Yaks *et al.*, 1980).

In vitro experiments have shown that morphine (10^{-6} M) and normorphine (5×10^{-6} M) inhibit the release of substance P (Fig. 38). Induced by K-excess (47 mM) from superfused spinal trigeminal nervi nuclei (Jessel and Iversen, 1977a,b). Naloxone (1 μM), given 2 min before the addition of 10 μM morphine, caused an almost complete reversal of the effect of morphine on substance-P release. The opiate effect was stereospecific in that levorphanol produced an almost complete inhibition of the release, whereas the inactive enantiomer dextrorphan did not significantly inhibit the release. Opiate binding sites (Atweh and Kuhar, 1977a,b,c; Larsson *et al.*, 1976; Pert and Snyder, 1974) of the dorsal horn appear to be on the terminals of primary afferent fibres, since dorsal root rhizotomy decreases the number of binding sites by about 50% (Larsson *et al.*, 1976).

The finding that the enkephalin content of the spinal cord is not changed by dorsal root rhizotomy indicates that enkephalin is contained mainly in short-axoned interneurons.

The Met-enkephalin analogue (D-Ala^2-met-enkephalin-amide), reported to be metabolically stable and to possess high analgesic potency in low concentrations (3 μM), almost completely suppressed the K$^+$ evoked release of substance P, producing a 94.33 $\pm$8.1% inhibition. However, no significant inhibition of the spontaneous or K$^+$-evoked release of substance P from nigral tissue was observed with morphine or β-endorphin (Jessel and Iversen, 1977a).

Substance P probably represents one of the major transmitters involved in the transmission of nociceptive information in primary sensory fibres (cf.

Primary afferent fibre

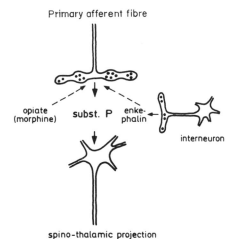

opiate (morphine)   subst. P   enke-phalin   interneuron

spino-thalamic projection

Fig. 38. Effect of opioid peptides on substance P release from the primary afferent fibre (Jessel and Iversen, 1977)

Henry 1976, 1980), the interaction of opiate receptors with substance P in sensory pathways seems likely to represent an important target as a primary site for opiate-induced analgesia. There have been previous reports that administration of substance P can antagonize the analgesic actions of morphine in experimental animals (Zetler, 1956), which would be consistent with the present hypothesis of Iversen and his group. On the other hand Stewart *et al.* (1976) found that intracerebral, and even intraperitoenal injections, of substance P caused a morphine-like analgesia in mice. The analgesia effects seen after substance P administration were very slow in onset by comparison with those elicited by morphine. One possible explanation for these results might be that administration of relatively large amounts of substance P could cause a progressive and persistent desensitization of substance P receptors.

4.6.4 Electrophysiological evidence

It has been shown that the effect of opiates and opioid peptides on neurons is to decrease excitation (cf. Zieglgänsberger and Frey, 1978). There are at least two proposed mechanisms by which the depressant effects are produced. One is a pre-synaptic effect, where the release of an excitatory transmitter is decreased by opiates, the second is a post-synaptic effect, by which the efficacy of an excitatory neurotransmitter is decreased by opiates.

Using electrophysiological techniques, it has been observed that morphine blocks transmission within ganglionic sites of Auerbach's plexus of guinea-pig ileum (Sato *et al.*, 1973; Dingledine *et al.*, 1974; Ehrenpreis *et al.*, 1976). The authors suggested that the site of inhibition of ganglionic transmission appears to be at the pre-synaptic site. This is indicated by the lack of effect on exogenous ACh under conditions in which the opiate blocks transmission. Dingledine and Goldstein (1976) concluded that the primary site of morphine blockade of synaptic transmission is at the post-ganglionic nerve terminal, as postulated by Ehrenpreis *et al.* (1973). At variance with this suggestion, North and Henderson (1975) concluded that morphine has no action on ganglionic transmission. In their study, intracellular electrodes were used to explore morphine's action on single ganglion cells within Auerbach's plexus. It was reported that morphine had no effect on the evoked EPSP or on the electrochemical properties of membranes of various types of ganglion cells. However, morphine can inhibit the increased firing rate produced by serotonin, caerulein, sodium picrate (Sato *et al.*, 1974) or ACh (Ehrenpreis *et al.*, 1976). Narcotic analgesics inhibit neuronal firing in the myenteric plexus (Sato *et al.*, 1973, 1974; Ehrenpreis *et al.*, 1976; North and Williams, 1977). The spontaneous and pharmacologically enhanced (by serotonin and nicotine) firing rates of neurons in the guinea-pig myenteric plexus are found to be inhibited by morphine (Dingledine *et al.*, 1974). Since the inhibition on spontaneous electrical activity was also seen under conditions of ganglionic blockade (Dingledine and Goldstein, 1976), these results suggest that morphine acts on the final cholinergic (pre-effectorial) neuron. This action could have been exerted on the cell soma and/or on the nerve terminal.

Convincing electrophysiological evidence has been obtained by Illés and North (1982) to show that the site of action of normorphine in mouse vas deferens is pre-synaptic and it reduces the induced EJP, in a naloxone-reversible way (Fig. 39). Naloxone in itself has no effect on EJP amplitude (Illés, personal communication). What is interesting is that the administration of naloxone to tissue which had been exposed to normorphine (Fig. 39) results in 'overshoot', i.e. a sudden increase of amplitude. This could be a sign of acute 'withdrawal', frequently seen even in isolated preparations.

In addition, Jurna (1966) showed that morphine selectively blocks the repetitive discharge activity of neurons, an effect possibly exerted somewhere on the axon hillock. Intracellular recording from some of the cells in the myenteric plexus has shown that the stimulation of opiate receptor produces hyperpolarization (North and Tonini, 1976).

A recent pharmacological study carried out by Konishi *et al.* (1979) has shown that enkephalin causes a pre-synaptic inhibition of cholinergic transmission in sympathetic ganglia. The occurrence of enkephalin in pre-ganglionic neurons (Dalsgaard *et al.*, 1982) suggests the possibility of a co-existence of enkephalin and acetylcholine in these neurons.

It has also been shown that pre-ganglionic stimulation of lumbar splanchnic nerves produces pre-synaptic inhibition of cholinergic excitatory post-synaptic potentials in the guinea-pig inferior mesenteric ganglion, which can be antagonized by naloxone (Konishi *et al.*, 1980).

Zieglgänsberger and his associates (1976, 1978, 1979) provided evidence that opiates (morphines and enkephalins) depress the excitability of different neurons. They were the first to realize that opiates may interefere with

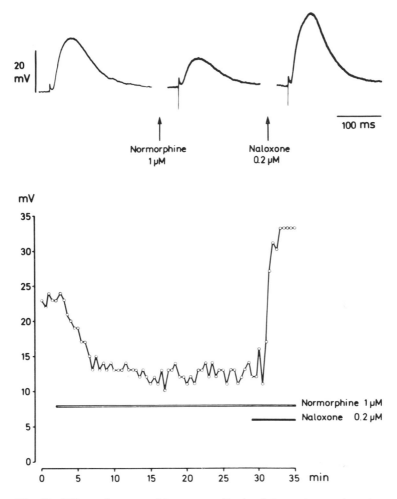

Fig. 39. Effect of normorphine on amplitude of the excitatory junction potentials induced by a single pulse (for method, see Illés and North, 1982) in the mouse vas deferens (intracellular recordings, amplitude in mV) (with the permission of Dr P. Illés)

inhibitory circuits in the hippocampus. They suggested, on the basis of single-unit recordings, that excitation of pyramidal cells may activate inhibitory interneurons via recurrent collaterals of their axons and that this inhibitory circuit may be interrupted by the application of opiates; the resulting disinhibition becomes evident as strong excitation. The inhibitory interneuron may be the basket cell, which releases γ-aminobutyric acid (GABA) in the neighbourhood of the pyramidal cell. The experimental proof for this hypothesis was that both opiates and GABA antagonists (e.g. bicuculline) excited pyramidal cells, although only the effect of the opiates could be antagonized by the specific opiate-antagonist naloxone. The inhibition of on-going synaptic transmission by Mg^{2+} antagonized the excitation elicited by opiates. Therefore, a direct effect of opiates on the pyramidal cell membranes is unlikely. This hypothesis was confirmed by experiments carried out on the hypocampal slice preparation. The pyramidal cell population spike, registered by extracellular glass microelectrodes, increased during opiate administration and naloxone reversed this effect (Dunwiddie *et al.*, 1980; Corrigall and Linseman, 1980). Opiate affects are mimicked by GABA antagonists and are partially or fully reversed by GABA itself. Intracellular studies on pyramidal cells in tissue culture or in the hippocampal slice preparation led to similar conclusions (Gahwiler, 1980; Nicoll *et al.*, 1980; Siggins and Zieglgänsberger, 1981).

4.7 MODULATION BY HISTAMINE

In the last few years evidence has been obtained that histamine and L-histidine decarboxylase are present in the central nervous system. Most of the amine and its synthesizing enzyme are associated with the synaptosomal fraction of brain tissue. Transection of the rat medial forebrain bundle results in a reduction in histamine levels and L-histidine decarboxylase activity in the ipsilateral cerebral cortex, suggesting that a histaminergic pathway may ascend through this tract.

The finding of Phillis' group is very significant in that it showed that histamine inhibits the firing of cerebral cortical neurons (Phillis *et al.*, 1968b) and studies with H_1 and H_2 agonists and antagonists have provided evidence for the existence of both H_1 and H_2 receptors on cerebral cortical neurons (Phillis *et al.*, 1968b; Sastry and Phillis, 1976). These results suggest that an inhibitory histaminergic pathway of unknown origin projects to the cerebral cortex. However, it has been found that histamine releases ^{3}H-catecholamines from isolated cat nictitating membrane (Enero and Langer, 1975a) and from hypothalamic, striatal, hippocampal, and cortical slices, but failed to release ^{14}C-GABA and ^{14}C-ACh (Subramanian and Mulder, 1977). It is probable, therefore, that histamine depresses neuronal firing via NA and/or DA release, and H_1 and H_2 receptors are involved in the release mechanism. It is also shown in the periphery using isolated dog saphenous vein strips that histamine depresses contractions in response to electrical stimulation as well as the release of labelled NA (McGrath and Shepherd, 1976b). Any effect of histamine on ACh release can be excluded, since histamine (10^{-6} M) fails to affect the release of ACh from Auerbach's plexus in guinea-pig ileum at rest and at 0.2 Hz stimulation (Vizi, unpublished). Ouabain (2×10^{-5} M), an inhibitor of Na^+-K^+-activated ATPase, which otherwise is able to release transmitters (cf. Vizi, 1978), failed to release histamine from isolated rat skin, where 48/80 was very effective (Vizi, 1975b).

It has been shown that, in the superior cervical ganglion, where ACh is the transmitter, histamine has a facilitatory effect on neurochemical transmission which is mediated via H_1 receptors, and an inhibitory effect mediated via H_2 receptors (Brimble and Wallis, 1973). A similar observation was made on bullfrog sympathetic pre-ganglionic nerve terminals; low concentrations of histamine (1–3 μM) increased the amplitude of fast, excitatory post-synaptic potentials and increased ACh quantal content, while high concentrations (100–300 μM) decreased the amplitude and content. Since the amplitudes of miniature EPSPs and ACh potentials were not affected by histamine (0.3–300 μM), these results suggest that histamine has facilitatory and depressant actions on ACh release. The facilitatory action is probably mediated by the H_1-receptor and the depressant action by the H_2-receptor (Yamada et al., 1982).

4.8 MODULATION BY EXTRACELLULAR POTASSIUM

Neurons and glia are surrounded by narrow intercellular clefts with an ionic composition. Transmembrane change of Na^+ and K^+ during nervous activity can lead to a transient increases in K^+_o to more than five times the resting level.

Frankenhauser and Hodgkin (1956) found that a train of impulses results in a depolarization of the membrane, and conluded that the potassium ions liberated by nerve impulses do not diffuse freely but accumulate in an extra-cellular space in the order of 300 Å wide (Vyklicky et al., 1972, 1975; Kriz et al., 1974; Erulkar and Weight, 1974, 1977), modifying the activity of discharging neurons (Frankenhauser and Hodgkin, 1956) and impulse trans-mission (Baylor and Nicholls, 1969). It is far from clear what the physiological role of extracellular potassium accumulated in the close vicinity of the fibre and nerve terminal is. It was established that, during high-frequency ortho-dromic stimulation, extracellular potassium increases from 3 to 9 mM. When potassium accumulates in the synaptic gap, transmitter release from the nerve terminals is reduced (Weight and Erulkar, 1976; Stjärne, 1976, 1977). However, the post-synaptic membrane sensitivity to the transmitter (Stjärne, 1977) is enhanced by potassium. Depending on the predominance of pre-junctional depression or of post-junctional facilitation, an increase in concen-tration of potassium may produce either reduction or potentiation of neuro-effector transmission. Stjärne (1977) suggested that potassium released *in loco* in the close synaptic gaps may modulate transmission. In the guinea-pig vas deferens a rise of about 0.5 mM per nerve impulse of potassium was calculated (Stjärne, 1977), which seems to be enough to affect transmission.

Chapter 5

Auto-modulation of transmitter/modulator release (negative and positive feedback modulation)

Neurochemical evidence of pre-synaptic autoreceptors was first proposed in 1971, simultaneously with evidence for noradrenergic, dopaminergic (Farnebo and Hamberger, 1971a,b; Starke, 1971; Langer *et al.*, 1971; Kirpekar and Puig, 1971), central cholinergic (Polak, 1971), serotonergic (Farnebo and Hamberger, 1971b), and GABA neurons (Johnson and Mitchell, 1971).

5.1 NORADRENALINE

Knowledge of pre-synaptic autoreceptors began with the original observation that the overflow of noradrenaline in the venous blood from the cat spleen or colon, as a result of adrenergic nerve stimulation, was greatly increased by dibenamine or phenoxybenzamine (Brown and Gillespie, 1956, 1957). Gillespie (1980) has described this discovery; 'Since the known effect of dibenamine and phenoxybenzamine was to block α-receptors and since inhibition of monoamine oxidase [the chief enzyme known at that time to destroy noradrenaline] did not alter the overflow of transmitter in these experiments, the protective effect was attributed to α-receptor blockade. It was suggested that either combination with the α-receptor or some event subsequent to combination was responsible for inactivation.' Brown and Gillespie (1957) originally supposed that a combination of NA with post-synaptic α-receptor were the 'sites of loss' of NA released, since attachment to these sites is a prelude to uptake into the effector cells and subsequent metabolic destruction. This idea has been challenged by Paton (1960), when, in discussion of Sir Lindor Brown's paper (1960) at the CIBA Symposium on Adrenergic Mechanism in 1960, he argued that, in adrenals, receptor binding could hardly be inactivating the transmitter. NA should be 'sucked back . . . returned to store, when the events of excitation is over'. This suggestion, mainly based on experiments carried out in the adrenal medulla, was, in fact, the first indication of a functional role of pre-synaptic nerve terminal receptors in neurochemical transmission. Paton's idea was soon proved to be correct. Radioactive noradrenaline was taken up and stored in the adrenals and in tissues with a dense adrenergic innervation, and this was abolished by sympathetic denervation (Hertting *et al.*, 1961a) and blocked by phenoxybenzamine (Hertting *et al.*, 1961b). The ability to increase transmitter overflow still seemed associated with α-receptor blocking drugs rather than with drugs blocking neuronal uptake (Hertting *et al.*, 1961b). A similar conclusion from observations on the cat spleen was reached by Kirpekar and Cervoni (1963). These results were interpreted as being due to the blockade of the post-synaptic α-andrenoceptors by phenoxybenzamine, preventing released noradrenaline from combining with the receptors and thus increasing the transmitter collected in the venous effluent. A similar hypothesis was postulated several years later by Häggendal (1970), namely that the post-synaptic α-andrenoceptors are involved in a trans-synaptic regulatory mechanism for transmitter release and that blockade of the post-synaptic α-adrenoceptors would decrease the response of the effector organ and lead to an increase in noradrenaline release from the pre-synaptic nerve-endings (Häggendal, 1970; Farnebo and Malmfors, 1971). Häggendal (1969, 1970) showed that drugs such as desmethylimipramine effectively blocked uptake but increased overflow only slightly, whereas subsequent addition of phenoxybenzamine increased overflow greatly with no increase in uptake blockade. The extent of the increase in overflow with phenoxybenzamine was related to the reduction in the post-junctional response, indicating a trans-synaptic modulation. The trans-synaptic regulation of noradrenaline release was challenged by the finding that in tissues (e.g. the heart) where effector cells possess a predominantly β-receptor population, α-antagonists increased transmitter overflow without altering noradrenaline uptake or metabolism and without reducing the response (Starke *et al.*, 1971a; Starke 1972a,b; Farah and Langer, 1974; Langer *et al.*, 1971; McCulloch *et al.*, 1972). In addition, Enero *et al.* (1972) have shown that phenoxybenzamine in a concentration of as low as 10^{-9} g/ml increased NA overflow. This concentration did not affect uptake inhibition.

Iversen and Langer (1969) suggest that the effect of phenoxybenzamine on the overflow of NA following nerve stimulation cannot be explained only by the ability of phenoxybenzamine to prevent uptake of NA nerve terminals. They have concluded that prevention of metabolism of the released transmitter by uptake might be an important contributory factor.

In 1968 and 1969 (Vizi, 1968, and Paton and Vizi, 1969) published the first neurochemical evidence of the inhibitory role of α-adrenoceptors located on a neuron. They demonstrated that stimulation of α-adrenoceptor, present in the neurons of the Auerbach plexus, by NA or A leads to a reduction of transmitter release of ACh.

In 1971, in different laboratories (Starke, 1971; Kirpekar and Puig, 1971; Langer *et al.*, 1971; De Potter *et al.*, 1971; Farnebo and Hamberger, 1971) it was suggested that '. . . neuronal α-receptors mediate . . . inhibition of NA liberation in the presence of α-blockers; this restriction is attenuated' (Starke, 1971); or in other words 'the NA released . . . acts on . . . presynaptic sites to inhibit its own release by a negative feedback mechanism. . . . Adrenoceptor blocking agents enhance . . . NA overflow . . . because they remove this autoinhibition by blocking the presynaptic sites' (Kirpekar and Puig, 1971). During the same period, xylazine (BAY-1470), a hypotensive drug, was found to have α-agonistic activity and reduce the overflow of noradrenaline from nerve stimulation in the cat spleen (Heise and Kroneberg, 1970; Heise et al., 1971). Starke and his colleagues provided direct evidence that α-agonists such as clonidine, oxymetazoline, and naphazoline are able to produce a dose-dependent depression of NA release (Werner et al., 1972; Starke, 1972b; Starke et al., 1975a,b). A new hypothesis was developed (for reviews, see Langer, 1974; Stjärne, 1975a; Starke, 1977; Westfall, 19711; Vizi, 1979; Gillespie, 1980; Rand et al., 1980; Starke, 1981): noradrenaline released from sympathetic axon terminals acts not only to mediate transmission to the effector cells but also acts on pre-synaptic/pre-junctional α2-adrenoceptors to inhibit its own subsequent release from the varicosity where it originated (Starke, 1977; Langer, 1974) (Fig. 40). This concept has recently been challenged (cf. Kalsner, 1982a,b; Alberts et al., 1981).

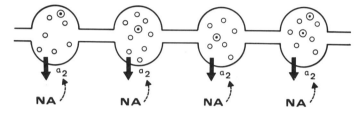

Fig. 40. Negative feedback modulation of noradrenaline (NA) release. Noradrenaline released inhibits its own subsequent release from the same varicosity (cf. Starke, 1977; Langer, 1974). Note the auto-inhibition of secretion coupling of NA release

5.1.1 Neurochemical evidence

The notion that the release of NA can be modulated pre-synaptically/pre-junctionally via stimulation of α2-adrenoceptors located on noradrenergic axon terminals is now widely accepted. Neurochemical evidence has been obtained showing that both in the peripheral and central nervous systems pre-synaptic/pre-junctional α2-adrenoceptors are involved in the modulation of NA release.

The peripheral autonomic nervous system

α2-adrenoceptor agonists

It has been recognized that in several tissues pre-synaptic α2-adrenoceptors differed from those located on post-synaptic sites (Starke et al., 1974, 1975a,b; Langer, 1974). Experiments carried out on the main pulmonary artery of the rabbit showed that clonidine, tramazoline, and α-methylnoradrenaline preferentially activated the pre-junctional α2-receptors, that methaxamine and phenylephrine preferentially activated post-junctional α1-adrenoceptors, and that NA and A activated either group at similar concentrations. It has been shown that the α_2-adrenoceptor agonists xylazine and clonidine reduce the stimulation-induced overflow of NA from the cat spleen (Heise and Kroneberg, 1970) and the rabbit heart (Werner *et al.*, 1972), respectively. The inhibitory effect of NA added to the tissue on NA release has been shown on sympathetic nerves in guinea-pig atria (McCulloch *et al.*, 1972), in the vas deferens of the mouse (Farnebo and Malmfors, 1971), the rat (Vizi *et al.*, 1973), and the guinea-pig (Stjäne, 1974), in the human omental artery (Stjärne and Brundin, 1975a), in the rabbit ear artery (Rand *et al.*, 1973), and in the cat spleen (Kirpekar *et al.*, 1973). A wide range of α-agonists have been shown to reduce the release of the NA induced by electrical stimulation (see review of Starke, 1977). Table 12 shows the effect of different α-agonists on pre-and post-junctional α_2-adrenoceptors. It was shown that, in contrast to $(-)$-NA, the $(+)$ enantiomer fails to depress release (McCulloch *et al.*, 1973; Rand *et al.*, 1973; Stjärne, 1974).

Table 12. Pre- and post-synaptic potencies of different α-adrenoceptor agonists in the rabbit pulmonary artery: post-synaptic α-adrenoceptors mediate smooth muscle contraction, pre-synaptic α-adrenoceptors mediate inhibition of NA release

	EC_{20} post-(M)	Relative potency[a]	EC_{20} pre-(M)	Relative potency[a]	Pre-post-ratio
Tramazoline	5.6×10^{-8}	0.13	3.7×10^{-9}	3.24	0.07
α-methylnoradrenaline	6.1×10^{-8}	0.12	8.1×10^{-9}	1.48	0.13
Clonidine	6.5×10^{-8}	0.11	1.0×10^{-8}	1.2	0.15
Oxymetazoline	1.8×10^{-8}	0.4	3.1×10^{-9}	3.8	0.17
Naphazoline	3.8×10^{-8}	0.19	1.6×10^{-8}	0.75	0.41
Adrenaline	3.2×10^{-9}	2.28	1.9×10^{-9}	6.3	0.58
Noradrenaline	7.3×10^{-9}	1	1.2×10^{-8}	1	1.6
Phenylephrine	5.4×10^{-8}	0.13	1.7×10^{-6}	0.007	30.9
Methoxamine	7.4×10^{-7}	0.0098	2.4×10^{-5}	0.0005	32.5

Data of Starke *et al.*, 1975. (Reproduced by permission of Pergamon Press Ltd).
EC_{20} post-, concentration of α-adrenoceptor agonist needed to cause a contraction of 20% of the maximal contraction. EC_{20} pre-, concentration of α-adrenoceptor agonist needed to reduce transmitter release.
[a]Taking the potency of NA as unity.

α_2-adrenoceptor antagonists

Evidence has been obtained showing that there are antagonists which preferentially inhibit either pre- or post-junctional α-adrenoceptors. In analogy to the observation made with agonists, some antagonists, such as prazosin and E-643 (2-4-/n-butyryl/-homopiperazine-l-yl-4-amino-6,7-dimethoxy-quinazoline), preferentially inhibit post-junctional α_1-adrenoceptors. At the other extreme, yohimbine or rauwolfscine preferentially block pre-junctional α_2-adrenoceptors.

Brown and Gillespie (1956, 1957) were the first to observe that α-adrenoceptor blocking agents, such as phenoxybenzamine and dibenzamine, enhanced the release of NA in response to electrical stimulation. Later, evidence was obtained showing that α_2-adrenoceptor antagonists enhance the release of NA (Starke, 1972b; Starke et al., 1971b; Starke et al., 1975a; Vizi et al., 1973). Starke (1972a, b) showed that the inhibitory effect of NA (10 ng/ml) was prevented by 1 μg/ml phenoxybenzamine in the rabbit heart. Recently, Kalsner (1982a,b) questioned the discussion and hypothesis drawn from these data. He claimed that in his experiments there was no interaction between oxymetazoline and yohimbine when he measured the release of NA from guinea-pig ureter. Yohimbine (3×10^{-7} M) did not prevent the inhibitory effect of oxymetazoline (10^{-8} M), a relatively pure α_2-agonist, on NA release when he expressed the effects as a percentage. Since the effect of oxymetazoline was in no way antagonized by yohimbine, he concluded that the increased release of ^{3}H-NA induced by antagonists and the reduced efflux produced by exogenous agonists may represent actions at different loci. However, even in these experiments it seems that yohimbine is able to reduce the effectiveness of both oxymetazoline and noradrenaline on ^{3}H-NA release when the effect of agonists is calculated in absolute values rather than as a percentage.

Effect on potassium-induced release

Kirpekar and Wakade (1968) were unable to show unequivocally an enhancing effect of phenoxybenzamine on potassium-induced release of NA from the perfused cat spleen. Starke and Montel (1974) showed that, at a concentration of 10^{-5} M, oxymetazoline reduced the overflow of the transmitter from the perfused rabbit heart by only 40%. On the other hand, when a higher potassium concentration (80 mM) was used, the release of NA was not changed by oxymetazoline. The authors concluded that released NA and exogenous α-adrenoceptor stimulating drugs induce a common inhibitory mechanism which is saturated when the NA concentration in the biphase is high (i.e. when a high frequency of nerve stimulation or very high potassium concentrations are used). The blockade of α-adrenoceptors by phenoxybenzamine (10^{-5} M) enhanced the potassium-induced release of NA by 50%. Lower concentrations of phenoxybenzamine did not modify the NA release, a finding different from those observed with electrical nerve stimulation (Enero et al., 1972). It has also been reported that ^{3}H-NA release from rat vas deferens by high potassium concentrations (80 nM) was not appreciably affected by exogenous NA or phentolamine (Wakade and Wakade, 1977). In addition, phenylephrine and clonidine (10^{-7}–10^{-3} M) were shown not to modify significantly the potassium-induced NA release (Garcia et al., 1978). In recently formed nerve endings from cultured rat superior cervical ganglia, phenoxybenzamine enhances the release of ^{3}H-noradrenaline induced by potassium-depolarization (Vogel et al., 1972). Under these experimental conditions α-receptor blockade enhanced the stimulation-induced release of ^{3}H-NA in the absence of an effector post-synaptic cell. It was suggested that the failure of sympathomimetic amines to depress, and of α-adrenoceptor blocking agents to enhance the release of NA by high potassium concentrations may be related to the length of time the nerve membrane remains depolarized. Stjärne (cf. Alberts *et al.*, 1981) concluded that the ineffectiveness of an α_2-agonist on potassium-excess-induced release of NA was evidence that via α_2-adrenoceptor stimulation not the secretion-coupling is affected.

The central nervous system

Effect of α-adrenoceptor agonists and antagonists

It has been shown that noradrenaline and other α-agonists reduce the release of ^{3}H-monoamine from field-stimulated rat brain slices (Farnebo and Hamberger, 1971b; Starke *et al.*, 1965d; Taube *et al.*, 1977; Wemer and Mulder, 1981; Hedler *et al.*, 1981). Oxymetazoline reduced and phentolamine increased depolarization, and the potassium- or electrical field stimulation-evoked release of ^{3}H-NA from superfused neocortical slices of the rat in a dose-dependent manner (Dismukes and Mulder, 1976). Exogenous NA strongly inhibited (^{3}H)-NA release from cortex slices of newborn rats, indicating that pre-synaptic α-receptors already exist at birth (Wemer and Mulder, 1981). During the first post-natal month, the noradrenergic system develops rapidly and approaches adult levels with respect to high-affinity NA uptake, vesicular storage capacity, biosynthetic and metabolizing enzyme activities, and NA content. Up to 17 days of age the inhibitory effect of $0.1–1.0 \times 10^{-6}$ M NA on (^{3}H)-NA release was significantly greater with neonatal than with adult rats. No obvious differences were observed (Wemer and Mulder, 1981) in the NA concentrations causing half-maximal effect ($IC_{50} = 10^{-7}$ M). The inhibitory effect of NA at a concentration of 10^{-7} M was considered as (near) maximal. Reinbacher *et al.* (1982) provided clear evidence that in the brain cortex of the rabbit, NA, α-methylnoradrenaline, clonidine, oxymetazoline, xylazine, and quanabenz all inhibited the stimulation induced release of ^{3}H-NA. Conversely, yohimbine, corynanthine, phentolamine, tolazoline, and azapetine all enhanced the release.

Inhibition of depolarization-induced NA release by α_2-receptor agonists has been demonstrated using brain slices labelled with radioactive catecholamine (Farnebo and Hamberger, 1971; Dismukes and Mulder, 1976; Dismukes *et al.*, 1977; Mulder *et al.*, 1978; Wemer *et al.*, 1979). K^+-induced release of ^{3}H-NA from cortex synaptosomes was inhibited by different α-adrenoceptor agonists. These findings substantiate the hypothesis that release-modulating α_2-receptors are localized on noradrenergic nerve terminals in the brain. Exogenous NA concentration-dependently inhibited ^{3}H-NA release from synaptosomes obtained from the cortex or the hypothalamus but not from striatal synaptosomes (Langen *et al.*, 1979). In these experiments adrenaline, clonidine, and oxymetazoline dose-dependently ($10^{-8}–10^{-6}$ M) inhibited ^{3}H-NA release evoked by 15 mM K_o from cortical synaptosomes. At a concentration of 10^{-8} M these α_2-adrenoceptors agonists significantly reduced ^{3}H-NA release to 45% (adrenaline), 58% (clonidine), and 71% (oxymetazoline) of control. Phentolamine by itself did not significantly influence synaptosomal ^{3}H-NA release from cortical or from hypothalamic synaptosomes (Langen et al., 1979). It was, however, observed that α-receptor antagonists enhanced ^{3}H-NA release from brain slices (Taube et al., 1977; Starke and Montel, 1973a,b; Dismukes et al., 1977).

Frequency-dependence

There is an inverse correlation between the inhibitory effect of α-agonist and the frequency of stimulation applied to the tissue (Vizi et al., 1973; McCulloch et al., 1973; Stjärne, 1974; Karlsner, 1982a). Neither NA nor phentolamine has any effect on ^{3}H-NA release at high frequencies of stimulation.

This frequency-dependent characteristic of the regulation mediated by pre-synaptic α-adrenoceptors is in contrast to the effects of neuron blocking agents like guanethidine or bretylium, which decrease the release of the neurotransmitter at both low and high frequencies of nerve stimulation (Armstrong and Boura, 1973).

In superfused strips of rabbit pulmonary artery the inhibition caused by clonidine was greater at low (2 Hz) than at high (8 Hz) stimulation frequencies (Starke et al., 1975c). Phenoxybenzamine has been shown to cause a large increase in NA output (five to ten times control) from the cat spleen in response to nerve stimulation at low frequencies (Brown and Gillespie, 1956; Kirpekar and Cervoni, 1963). It has been reported (Starke and Montel, 1974) that phenoxybenzamine enhanced output in a guinea-pig artery at 5 Hz but not at 25 Hz, and a similar relationship was reported by others with 2 Hz and 50 Hz in rat vas deferens (Vizi et al., 1973) and with 5 Hz and 30 Hz (Langer et al., 1975) and 10 Hz and 30 Hz (Brown and Gillespie, 1956) in the cat spleen. The frequency-dependence of the inhibition of transmitter release by α-receptor agonists like clonidine has also been observed under in vivo conditions in the rat and in the dog (Armstrong and Boura, 1973; Scriabine and Stavorski, 1973; Robson and Antonaccio, 1974; Yamaguchi et al., 1977). This ineffectiveness at very high frequencies was attributed by some to a disengagement of the feedback system (Langer, 1977) and to stimulation of Na,K-ATPase (Vizi, 1979) and the consequent hyperpolarization (Vizi, 1979).

5.1.2 Electrophysiological evidence

It has been shown that the nucleus locus coeruleus consists of NA-containing neurons (Dahlström and Fuxe, 1964) and it has been suggested (Fuxe et al, 1974) that these cells receive a tonic, inhibitory input from other brainstem catecholaminergic neurons. Direct microiontophoretic application of NA or clonidine into the vicinity of locus coeruleus neurons inhibits the spontaneous firing (Svensson *et al.*, 1975; Cedarbaum and Aghajanian, 1977), α-adrenergic blocking drugs antagonize these inhibitions, whereas β-adrenergic blocking agents do not (Cederbaum and Aghajanian, 1977), indicating the role of α-adrenoceptors. The α-receptor of the locus coeruleus is similar to 'pre-synaptic' or 'α_2' receptors of peripheral sympathetic nerves, since phenylephrine does not affects its activity.

5.1.3 Neuroeffector transmission

The first reports on the inhibitory effect of NA on twitch responses to electrical stimulation in the isolated vas deferens of the rat (Somogyi *et al.*, 1970) and of the guinea-pig (Ambache and Zar, 1971) appeared in 1970 and 1971. It was claimed that α-blocking agents such as phentolamine, tolazoline, yohimbine, and phenoxybenzamine markedly potentiated the contractions of the vas deferens to a single stimulation (Somogyi *et al.*, 1970) and prevented the inhibitory effect of NA and tyramine. These observations were confirmed by others and were shown to be true for the mouse vas deferens by Jenkins *et al.* (1977). In tissues where the effector organ responses are mediated by α-adrenoceptors, blockade of the pre-synaptic α-adrenoceptors enhances NA release by nerve stimulation, but decreases the response by blockade of the post-synaptic receptors. Since α-blocking agents with pre-synaptic receptor effects are also post-synaptic blockers, the post-synaptic actions would presumably counteract the effect of increased transmitter synthesis and release caused by removal of the pre-synaptic dampening mechanism, just as they counteract the effects of the neuronally mediated increase in impulse flow.

In contrast, for tissues in which the responses are mediated by post-synaptic α-adrenoceptors, blockade of pre-synaptic β-adrenoceptors should enhance NA release and response to sympathetic nerve stimulation. Nevertheless, in the perfused cat heart and in isolated guinea-pig astria there is no significant increase in the positive chronotropic responses to nerve stimulation in conditions in which the release of NA is increased by phentolamine (Vizi *et al.*, 1973). However, it was also observed in guinea-pig atria with their accelerans nerve that phentolamine (3×10^{-5} M) produce an almost four-fold increase in ^{3}H-NA release and a significant increase in the positive chronotropic effect in response to nerve stimulation (Langer *et al.*, 1977). The first observation regarding the differences between pre- and post-synaptic α-adrenoceptors was made for the α-adrenoceptor agonists in the cat spleen (Langer, 1973). It was shown in the perfused cat spleen that phenoxybenzamine is at least thirty times more potent in blocking the post-synaptic α-receptors than it is in blocking the pre-synaptic α-receptors. This finding was confirmed in perfused cat spleen: phenoxybenzamine was thirty to a hundred times more potent in blocking the post-synaptic α-receptors than the pre-synaptic receptors (Cubeddu *et al.*, 1974). On the other hand, these authors found only a very small difference for the potency of phentolamine in blocking and pre- and post-synaptic α-adrenoceptors in the prefused cat spleen. It was also shown that the α-receptor agonist clonidine is more potent in reducing noradrenaline release during nerve stimulation than in stimulating the postsynaptic α-adrenoceptors (Starke *et al.*, 1974b). These results led to the concept (Langer, 1974b) that there were actually two types of α-adrenoceptors; α_1- and α_2-pre-synaptic receptors. Starke *et al.* (1975b,c) showed that phenylephrine preferentially stimulated post-synaptic adrenoceptors, whereas the pre-synaptic receptors were preferentially stimulated by clonidine (Starke *et al.*, 1974b). Yohimbine has been shown to block pre-synaptic, α_2-adrenoceptors in concentrations lower than are required to block postsynaptic α_1-adrenoceptors (Starke *et al.*, 1975c).

The rabbit pulmonary artery contains post-synaptic α_1-adrenoceptors which mediate smooth muscle contraction: its noradrenergic nerves contain presynaptic α_2-adrenoceptors which mediate inhibition of the release of the transmitter induced by nerve impulses. Starke *et al.* (1975c) have demonstrated that there are large differences in the relative pre- and post-synaptic potencies of various agonists, there being a nearly five hundred- fold variation in the pre- and post-synaptic effects. The post-synaptic rank order of potency was: A < NA < oxymethazoline < naphazoline < phenylephrine < tramazoline < methylnoradrenaline < methoxamine (Starke *et al.*, 1975c). According to the concentrations, which reduced the stimulation-induced overflow by 20% (EC_{20} pre-), the rank order of potency was: A < oxymethazoline < tramazoline < α-methylnoradrenaline < NA < naphazoline < phenylephrine < methoxamine. Some of the references for pre-synaptic effects of α-adrenoceptor agonists and antagonists on isolated preparations in which noradrenaline is the presumed transmitter are given in Table 13.

The order of potency of the agonists for twitch inhibition of the rat vas deferens was: clonidine < oxymethazoline < dopamine < apomorphine < noradrenaline (Tayo, 1979). Yohimbine readily blocked the inhibitory effects of clonidine, oxymetazoline, and NA.

5.1.4 Is auto-inhibition a true auto-inhibition?

Auto-inhibition by single pulse

The hypothesis that NA released from axon terminals inhibits its own release has been challenged (Kalsner, 1979a,b; cf. Kalsner, 1982a; Stjärne *et al.*, 1981).

The physiological importance of apparent negative feedback control, where a transmitter released from a varicosity and auto-inhibits further release from the same varicosity, has been questioned (Stjärne, 1978). It has been suggested (Vizi, 1979, 1981b) that the role of negative feedback modulation is just to keep constant the output per volley, i.e. it is not a real modulation. On theoretical grounds it is very difficult to understand the physiological meaning of 'negative feedback' inhibiton.

Table 13. Effect of different α_2-adrenoceptor agonists and antagonists on neurochemical transmission presumed to be mediated by noradrenaline

Tissue	Species	References
Vas deferens	Rat	Somogyi et al. (1970), Ambache and Zar (1971), Vizi et al. (1973b), Drew (1977), Dozey (1979)
	Mouse	Farnebo and Malmfors (1971), Jenkins et al. (1977)
	Guinea-pig	Stjärne (1978), Kalsner (1979 a,b, 1980b), Ambache and Zar (1971), Alberts et al. (1981).
Pulmonary artery	Rabbit	Starke et al. (1974, 1975a,b), McCulloch et al. (1973), Weitzell et al. (1979), Borowski et al. (1977), Hope et al. (1976)
Ear artery (perfused)	Rabbit	Drew (1979), Hope et al. (1976) Hieble and Pendleton (1979), Holman and Surprenant (1980)
Mesenteric artery (perfused)	Rat	Doxey and Roach (1980)
Saphenous vein	Dog	Drew and Sullivan (1980)
Facial artery	Bovine	Kalsner and Chan (1979)
Renal artery	Bovine	Kalsner and Chan (1979)
Anococcygeus muscle	Rat	Doxey (1979), Drews and Sullivan, (1980), Idowu and Zar (1978), Gillespie and McGrath, (1975)
Cardiac tissue	Rat	Idowu and Zar (1977)
	Guinea-pig	Vizi et al. (1973b), McCulloch et al. (1972), Commarato et al. (1978), Medgett et al. (1978), Langer et al. (1977), Angus and Korner (1980)
	Rabbit	Starke and Altmann (1973)

The α_2-adrenoceptor-mediated inhibition of transmitter release would not be expected in response to a single pulse. Blockade of a feedback regulatory system by antagonists should not increase transmitter efflux with one pulse as the released NA cannot retroactively inhibit its own release (Kalsner, 1979a).

The negative feedback hypothesis has been challenged by the observation that the release of NA and the mechanical response to a single pulse are both enhanced by phenoxybenzamine in the guinea-pig vas deferens (Kalsner, 1979b). Since a transmitter released by a single pulse cannot retroactively affects its own release, it was suggested that the antagonist exerts increasing actions on transmitter release other than by inhibition of a 'negative feedback' system activated by in loco-released noradrenaline (Kalsner, 1979b). It has been shown (Angus and Korner, 1980) that phenoxybenzamine markedly potentiates the chronotropic response to a single pulse and also responses up to maximum stimuli. In contrast, phentolamine and yohimbine were totally without effect, even at high concentrations. The action of phenoxybenzamine was largely accounted for by its effect on uptake blockade. The above facts are held as evidence against a physiological role of pre-synaptic α_2-adrenoceptor in the heart.

It was concluded by Angus and Korner (1980) that 'presynaptic α-adrenoceptor modulation by synaptically released noradrenaline plays no part in cardiac sympathetic transmission'. This conclusion conflicts with the work of Langer *et al.* (1977). The study of Angus and Korner (1980) was carried out with guinea-pig right atria in Krebs solution which contained 1.2 mM Mg^{2+} but Langer *et al.* (1977) obtained their results with Locke's solution, which does not contain Mg^{2+}.

The choice of phenoxybenzamine was not the best for disproving the validity of negative feedback because first, phenoxybenzamine is not a selective, pre-synaptic α-adrenoceptor antagonist; second, it is a strong uptake inhibitor and third, its effect on transmitter release might be affected by its inhibitory effect on Na,K-ATPase (Hexum, 1978; Schaefer *et al.*, 1979. Svoboda and Mosinger, 1981; Serfózó and Vizi, 1983) a mechanism generally accepted to lead to an increase of transmitter release (cf. Vizi, 1978). The date with phentolamine seems to be much more convincing. Holman and Suprenant (1980), in an electro-physiological analysis of the effects of noradrenaline and α-adrenoceptor antagonists on neuromuscular transmission in mammalian muscular arteries, also provided evidence against negative feedback modulation of noradrenaline release. In a large proportion of their preparations (9 out of 14 rabbit ears and 8 out of 9 tail arteries) an excitatory junction potential induced by single stimulus was increased above control values when either phentolamine or (2.5 μM) phenoxybenzamine was added (Holman and Suprenant, 1980). No increase in the amplitude of the EJP induced by a single stimulus was observed in any artery when prazosin or labetolol was added to the bathing fluid.

No support for Kalsner's results were obtained in the mouse vas deferens using yohimbine (Baker and Marshall, 1982) and RX781094 (Baker and Marshall, 1982). The twitch response and the fractional release of ^{3}H-noradrenaline induced by a single pulse (2 ms impulse duration) were not increased by RX781094 (30, 100 and 300×10^{-9}M, 20 min equilibrium), but the release by 10 pulses (1 Hz) was increased after 20 min equilibrium. In addition, the work of Rand *et al.* (1975) and Markiewicz *et al.* (1979) showed the opposite result to that of Kalsner (1979a,b; 1980a,b). In their experiments α-blockade enhanced the secretion induced by two or more impulses, but not that caused by a single pulse. Nevertheless, when the secretion of ^{3}H-NA induced by a single pulse was calculated by extrapolation from data Stjärne (1980), the results yielded support for Kalsner's data, indicating that α-adrenoceptor blocking agents enhance secretion by single nerve impulses. Stjäne (1980) suggested that auto-inhibition does not necessarily imply that NA released from a varicosity inhibits further release from the same varicosity, but rather that it lowers the probability for secretion from other varicosities in the same ('some auto-inhibiton') (Fig. 41a) or in a neighbouring ('lateral inhibition') terminal branch (Fig. 41b) by blocking centrifugal propagation or nerve impulses. A similar conclusion was drawn by Vizi (1981a,b, 1982a), who suggested that, like the noradrenergic cholinergic interneuronal non-synaptic interaction (see section 4.1), noradrenaline release from a varicosity of a neuron inhibits the release of NA released from another axon terminal (Fig. 41c).

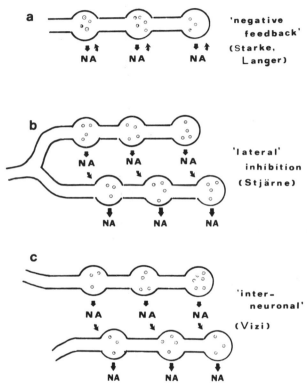

Fig 41. Different kinds of 'auto'-inhibition of noradrenaline (NA) release: (a) 'negative feedback' inhibition (Starke, Langer) when NA inhibits the subsequent release of NA from the same varicosity: (b) 'lateral' inhibition (Stjarne, 1980); (c) 'interneuronal' modulation (Vizi, 1980, 1981a, b) of NA release

5.1.5 Positive feedback modulation of noradrenaline release

It has been observed that isoprenaline enhances the stimulation-induced release of ^{3}H-NA from isolated guinea-pig atria, cat aorta, nictitating membrane, spleen (Langer *et al.*, 1975; Adler-Graschinsky and Langer, 1975; Celuch *et al.*, 1978), and rat portal vein (Dahlöf *et al.*, 1978.)

It has been suggested that there are presynaptic β-adrenoceptors in noradrenergic axon terminals (Adler-Graschinsky and Langer, 1975) and that released NA, acting on presynaptic β-adrenoceptors, causes its own release by a positive feedback mechanism. Pre-synaptic β-receptors may be selectively activated by low concentrations of NA, for instance during low rates of impulse flow, so that release per pulse increases. As the NA concentration rises, the α-adrenergic negative feedback may be triggered, so that release per pulse falls (Adler-Graschinsky and Langer, 1975). In agreement with this hypothesis, propranolol reduces release of NA in some tissues (Adler-Graschinsky and Langer, 1975). On the other hand, in cat spleen, propranolol fails to reduce release, even though isoprenaline causes significant facilitation, and the same may apply to human blood vessels (Stjärne and Brundin, 1975). Baumann and Koella (1980) failed to observe a positive feedback mechanism in isolated cortical slices of the rat, as propranolol did not affect stimulation-induced ^{3}H-NA overflow even when a low-stimulation rate (0.25 Hz) was applied.

The effects obtained *in vivo* with $β_1$-and $β_2$-selective antagonists suggest that facilitation of NA release elicited by axonally conducted nerve action potentials is mediated by pre-synaptic $β_1$-adrenoceptors (Dahlöf *et al.*, 1978; Dahlöf, 1982). According to Dahlöf *et al.* (1975, 1978), the pre-synaptic receptors are of the $β_1$-type because they are blocked by *meteprolol*, a selective $β_1$-receptor blocking agent. In the isolated rat portal vein *in vitro* experiments, facilitation of NA release elicited by transmural nerve stimulation was mediated by pre-synaptic $β_2$-adrenoceptors (Dahlöf, 1981).

In an isolated portal vein preparation where NA uptake mechanisms and α-adrenoceptor-mediated effects had been inhibited, the transmitter release elicited by transmural nerve stimulation was increased dose-dependently by isoprenaline, adrenaline (A), and NA, the relationship for equipotent concentrations being 1:2:900,respectively (Dahlöf, 1981). Practolol decreased the inhibitory effect of sympathetic nerve stimulation on the Finkleman preparation (Szenohradszky *et al.*, 1980). Since practolol did not influence the effect of exogenously administered NA, the site of action must be pre-synaptic. The action of practolol depended on the number of shocks and on the frequency of stimulation (Szenohradszky *et al.*, 1980). Nevertheless $β_2$-adrenoceptors were proposed (Westfall *et al.*, 1979a) in the positive-feedback control of NA release. The secretion of NA from human vasoconstrictor nerves appears to be under pre-synaptic feedback regulation: NA release is increased by stimulation of β-adrenoceptors (Stjärne and Brundin, 1975, 1976).

In the superfused human oviduct, 0.5 nM isoprenaline increased the release of ^{3}H-NA by transmural stimulation (Hedqvist and Moawas, 1975). Enhancement of evoked transmitter release by isoprenaline was inhibited by propranolol. Similarly, the increase in nerve stimulation. induced secretion of ^{3}H-NA from isolated human omental vessels caused by 0.04 μM isoprenaline was prevented by 1 μM propranolol (Stjärne and Brundin, 1976).

It was suggested that the secretion of NA by human vasoconstrictor nerves may be modulated by $β_2$-adrenoceptors, sensitive to circulating A. Physiologically, this could provide a mechanism for a co-operative increase in NA release in various conditions associated with increased adrenomedullary secretion (Stjärne, 1975b).

Even the positive feedback regulation of noradrenaline release from sympathetic nerve has been questioned (Kalsner, 1980a, 1982c). In the left atria, isoprenaline (1.2 × 10^{-8} M) enhances the efflux of ^{3}H-NA with 50 pulses at all four test frequencies, exogenous noradrenaline (1.8 × 10^{-8} M) inhibits efflux, and isoprenaline is ineffective. Propranolol (1 × 10^{-7} M) did not decrease the stimulation-induced efflux of NA.

5.2 DOPAMINE

The first evidence of the negative feedback control of central dopaminergic activity was presented in 1971 by Farnebo and Hamberger, who reported that the release of ^{3}H-dopamine *in vitro*, induced by electrical field stimulation of neostriatal slices, was increased by pimozide and chlorpromazine, and was decreased by apomorphine when added to the incubation medium.

Although it is now widely accepted that pre-synaptic autoreceptors control dopamine (DA) release, (Farnebo and Hamberger, 1971b; Starke, 1977; Westfall *et al.*, 1976; Plotsky *et al.*, 1977) recent observations argue against this interpretation (Raiteri *et al.*, 1978). Superfusion of ^{3}H-tyrosine-treated synaptosomes with apomorphine (10^{-6} M) have been found to lead to a decrease in the spontaneous as well as the veratridine-induced release of ^{3}H-DA. Superfusion with α-methyl-tyrosine, an inhibitor of catecholamine synthesis, leads to a similar reduction. If DA agonists did have an effect on release they should be detectable due to labelling with ^{3}H-tyrosine when DA synthesis is stopped. ^{3}H-DA release caused either by veratridine or by high K$^+$, is not inhibited by apomorphine in the presence of α-methyl-tyrosine (Raiteri *et al.*, 1978).

However, there is another possibility, which is that the DA receptors of dopaminergic neuron terminals could be activated by DA molecules released from other dopaminergic neurons. Perhaps this type of mechanism can modulate neurochemical transmission at the dopaminergic synapse (interneuronal modulation), although neither morphological nor electrophysiological evidence has been presented to prove this.

Important evidence for the existence of DA autoreceptors is the so-called paradoxical response to apomorphine (Carlsson, 1975). At least one order of magnitude below that required for producing stereotyped behaviour in the rat apomorphine can inhibit the DA synthesis in the striatum and in the limbic system in a dose-dependent manner. These low doses also cause an inhibition of locomotor activity, compatible with a reduction in the release of DA. In man, low doses of apomorphine have been shown to have antipsychotic and antidyskinetic effects. (Tamminga *et al.*, 1978; Smith *et al.*, 1977) whereas higher doses stimulated, for example, locomotor activity via post-synaptic dopamine receptors. A DA analogue, 3-(-hydroxyphenyl)-N-n-propldiperidine (3-PPP), produced all the effects observed after low doses of apomorphine (Nilsson and Carlsson, 1982). 3-PPP appears to be the first apparently selective dopaminergic autoreceptor agonists to have been reported (Nilsson and Carlsson, 1982). Hopefully, this will open up possibilities of achieving an antipsychotic agent without extrapyramidal side-effects.

The most striking behavioural action of 3-PPP in rat and mice is a marked inhibition of exploratory activity. No stimulatory effects of stereotyped behaviour occur at any dosage, and the action of reserpine is not antagonized. Gooddale *et al.* (1980) reported 6,7-dihydroxy-2-(dimethylamino) tetralin (TL-99) to be a selective agonist for DA autoreceptors.

DA is not uniformaly present throughout the brain. The striatum and the limbic system are rich in DA nerve terminals and are therefore considered to be under the control of DA neurons. Those antipsychotic drugs (neuroleptics) with a central DA receptor blocking activity are able to increase the synthesis and release of striatal DA and the firing of dopaminergic neurons in the substantia nigra. These effects have been postulated to be secondary to the drug-induced blockade of striatal post-synaptic DA receptors and mediated by a negative feedback loop impinging on nigral DA neurons (Carlsson and Lindqvist, 1963). Alternatively, it has been suggested that neuroleptics stimulate dopaminergic firing by blocking nigral DA receptors (autoreceptors), thus relieving DA neurons from the inhibitory action of DA released from dopaminergic neurons.

5.3 ACETYLCHOLINE

McIntosh and Oborin (1953) observed that atropine, a muscarinic blocking agent, increased the release of acetylcholine (ACh) from the cerebral cortex of the cat. McIntosh (1963) postulated that atropine stimulates the release of ACh from the cerebral cortex in *in vivo* experiments by blocking cholinergic synapses which form a part of inhibitory neuronal circuits controlling the activity of the intracortical neurons from which the ACh is released. It was suggested by Koelle (1962) that cholinoceptors are located not only post- but also pre-synaptically at cholinergic transmission sites. Koelle also suggested that there is a positive feedback mechanism, a hypothesis that has since been challenged as evidence presented by Polak and his colleagues (cf. Vizi, 1979; Molenaar and Polak, 1980) demonstrates that the stimulation of muscarinic receptors on the cholinergic neuron leads to a reduction of ACh release. The results of Szerb and co-workers called the attention of scientists to the feedback inhibition of ACh release as a modulating process in neurochemical transmission and provided direct neurochemical evidence.

5.3.1 Negative feedback modulation via muscarinic receptors

The central nervous system

McIntosh and Oborin (1953) observed that atropine, in the presence of a cholinesterase inhibitor, enhanced the release of ACh from the cerebral cortex of the cat. Similar observation were made by others on the cortex, the hippocampus, and the caudate nucleus (Mitchell, 1963; Szerb, 1964; Polak, 1965; Celesia and Jasper, 1966; Bartholini and Pepeu, 1967; Phillis, 1968; Dudar and Szerb, 1969; Dudar, 1977).

It has been observed that some degree of activity is required of the cholinergic nerve fibres for atropine to exert its stimulating action: for instance, atropine has no effect on spontaneous ACh release from the cerebral cortex of the cat (Dudar and Szerb, 1969). Polak and Meeuws (1966) found that the release and synthesis of ACh was greatly stimulated when the KCl concentration of the medium was raised from 4.7 to 24 mM. Atropine and other antimuscarinic substances further enhanced ACh release and synthesis provided that the medium contained a high KCl concentration. The atropine effect is linked to the antimuscarinic property, since the l-stereo isomer of atropine, which has a much stronger antimuscarinic activity than its optical antipode, was much more effective in stimulating ACh release (Polak and Meeuws, 1966). *In vitro* experiments with cortical slices have shown that atropine enhances the release of ACh by high concentrations of potassium in the presence of tetrodotoxin, a compound known to block nervous conduction. This excludes the involvement of neuronal loops in the atropine effect (Molenaar and Polak, 1970). These results led Polak (1971) to propose that there is a local regulation of ACh release from central cholinergic neurons through a negative feedback mechanism.

The muscarinic agents, oxotremorine and acetyl-β-methylcholine, competitive antagonize the stimulating action of atropine (Polak, 1971) on ACh release. In addition, it has been shown that the muscarinic agonists oxotremorine, carbamylcholine, and arecoline reduce the rate of induced release of ^{3}H-ACh (Szerb *et al.*, 1977) from septohippocampal cholinergic fibres with an ED_{50} similar to the ED_{50} required to displace specific ^{3}H-quinuclidinyl benzylate (QNB) binding, as found by Yamamura and Snyder (1974a,b).

The inhibitory effect of a cholinesterase blocking agent on induced ACh release from cerebral cortical slices has been reported (Szerb and Somogyi, 1973). A quaternary cholinesterase inhibitor (neostigmine) and an organophosphate inhibitor (echothiophate) depressed the induced ACh release by about 70%, as did physostigmine. In these experiments atropine by itself failed to potentiate the induced release of ACh. In the absence of a cholinesterase inhibitor, oxotremorine (10^{-6}–10^{-5} M) reduced the release of ACh from rat cerebral cortex slices following electrical stimulation (Szerb and Somogyi, 1973). The presence of a cholinesterase inhibitor caused a similar inhibition (Bourdois *et al.*, 1974). In the isolated perfused rat brain, only a small and insignificant elevation of the level of endogenous ACh occurred when the brain was perfused with 10^{-7} M oxotremorine (Kilbinger and Krieglstein, 1974).

Hadházy and Szerb (1977) showed that in hippocampal and caudate slices atropine not only prevented the depressant effect of the ACh accumulated under the influence of cholinesterase inhibition but that it also potentiated

ACh release in the absence of a cholinesterase inhibitor. This finding supports the hypothesis that this phenomenon is of physiological importance. Yamamura and Snyder (1974b) questioned the existence of pre-synaptic muscarinic receptors in the hippocampus because destruction of the septum was not followed by a reduction in the binding of (^{3}H) quinuclidinyl benzylate (QNB), a potent selective muscarinic antagonist, in homogenate of hippocampus. This argument was, however, refuted by Szerb *et al.* (1977), who compared the concentrations in which muscarinic agonists and antagonist altered the rate constants of induced ACh release from hippocampal slices with the concentration in which these drugs displaced specific (^{3}H)-QNB binding in the hippocampus, as reported by Yamamura and Snyder (1974a). Much higher concentrations were needed from the antagonist to displace ^{3}H-QNB-binding than to affect ACh release.

In the isolated frog spinal cord preparation, the application of hyoscine (2.2×10^{-5} M) was followed by a large rise in the unstimulated ACh output (Nistri, 1976). However, the release of ACh induced by orthodromic stimulation was not enhanced.

Szerb and Somogyi (1973) suggested that the reduction of induced ACh release in the presence of a cholinesterase inhibitor was correlated with the extracellular accumulation of ACh which inhibits the release of ACh via muscarinic receptors. The importance of extraneuronal ACh in ACh release seems to be confirmed by the observation (Vizi, 1974) that the rate of ACh release depends on the concentration of ACh in the surrounding medium. It was shown that exogenous ACh added to the bathing fluid reduces the release of ACh. A similar observation was made by Fosbraey and Johnson (1978), who showed that the increase in efflux of ^{3}H-ACh on electrical field stimulation of ileal strips pre-treated with ^{3}H-choline for 2 h was inhibited by exogenous ACh (1.7×10^{-6} M) in Krebs solution containing physostigmine (0.18×10^{-6} M) and hemicholinium (1.7×10^{-5} M). This inhibition of ^{3}H-ACh efflux was reversed by atropine (3.5×10^{-5} M) (Fosbraey and Johnson, 1978).

Neuromuscular junction

Using an intracellular recording system, quantal release was determined by the ratio of end-plate potentials and miniature end-plate potentials in a rat hemidiaphragm phrenic nerve preparation. Eserine sulfate (3×10^{-6} M) and neostigmine (6×10^{-6} M) reduced and curare (1.4×10^{-5} M) enhanced the quantal release (Wilson, 1982). Although it has been suggested that anticholinesterase drug treatment could have a direct inhibitory effect on the nerve terminal (cf. Wilson, 1982), Wilson (1982) suggested that residual ACh inhibits transmitter release and that blockade of ACh receptors enhances transmitter release.

According to Wilson, a direct inhibitory effect of a cholinesterase inhibitor can be excluded since it was found to be effective only with a higher frequency of stimulation (> 10 Hz). This could explain the observation that the ACh output per stimulus (volley output) is high at low frequencies of stimulation but is greatly reduced as the frequency of stimulation is increased (Potter, 1970; Földes and Vizi, 1980), i.e. at high frequencies the negative feedback inhibition reduces the volley output. But this result is entirely different from those obtained in other tissues (cortex, hippocampus, Auerbach's plexus), where it was shown (Szerb and Somogyi, 1973; Kilbinger, 1977 cf. 1982) that negative feedback inhibition operates only at a lower frequency of stimulation (< 5 Hz).

Torpedo marmorata

The hypothesis of pre-synaptic autoreceptors has also been tested (Dunant and Walker, 1981) at the nerve electroplaque junction of a Torpedo marmorata preparation in which the post-synaptic receptors are known to be purely nicotinic. Using single stimuli, it was observed that oxotremorine decreased the EPP by $56\pm3\%$ at a concentration of 10^{-6} M and by $77\pm2\%$ at 10^{-5} M. At 20 Hz for 1 s, oxotremorine reduced the release of ACh by $44\pm10\%$ at 10^{-6} M and by $60\pm9\%$ at 10^{-5} M concentration. Atropine (10^{-9} to 10^{-6} M) antagonized all actions of oxotremorine completely. In contrast, synaptic transmission was not detectably affected by other muscarinic agonists such as muscarine (10^{-6}–10^{-5} M), pilocarpine (10^{-7}–10^{-5} M) or bethanecol (10^{-6}–10^{-3} M). Nevertheless, the release of ACh was markedly reduced in the presence of physostigmine and neostigmine (10^{-4} M). However, atropine alone (10^{-9} – 10^{-6} M) had no action on ACh release or on the EPP (Dunrant and Walker, 1981).

Auerbach's plexus

It has been observed that oxotremorine, a strong muscarinic stimulant, reduces the release of ACh from the isolated longitudinal muscle strip of guinea-pig ileum (Cox and Hecker, 1971; Vizi and Knoll, 1972; Kilbiner and Wagner, 1975; Kilbinger, 1977; Kilbinger and Wesler, 1980). Kilbinger (1977), studying in detail the negative feedback modulation of ACh release from Auerbach's plexus, showed that it operates at a lower frequency of stimulation (< 5 Hz). Oxotremorine inhibits only that fraction of the ACh resting release that is calcium-dependent, whereas the spontaneous release from an isolated longitudinal muscle strip in a calcium-deficient medium is not affected (Kilbinger and Wagner, 1975). *In vitro* experiments demonstrated that the release of ACh from Auerbach's plexus induced by 25 mM potassium is potentiated by small concentrations of atropine and of other antimuscarinic agents when measured following the inhibition of cholinesterase (Kilbiner and Wagner, 1975; Dzieniszewski and Kilbinger, 1978; Sawynok and Jhamandas, 1977).

The DMPP-induced release of ACh from Auerbach's plexus can be dose-dependently inhibited by oxotremorine and increased by atropine. In addition, the ACh output induced by either 45 mM or 108 mM potassium was enhanced by atropine. Oxotremorine does not affect the ACh release by high potassium, but it prevents the facilitatory effect of atropine (Dzieniszewski and Kilbinger, 1978).

Johnson (1963) found that when the ACh from such ileal segments was collected during 5-, 10-, 20-, or 40-min periods the total amount recovered increased with the increasing collection time, but the rate of spontaneous release fell progressively when longer collection periods were used. The accumulated ACh in the synaptic cleft might have been responsible for the inhibition of output by the transmitter. A similar observation was made by Vizi (1974a).

Other tissues

No such pre-synaptic inhibitory receptors were found in the heart of the guinea-pig or the chicken or in the urinary bladder of the guinea-pig (Kilbinger, 1982).

5.3.2 Positive feedback

In contrast to these results, it has been suggested that ACh facilitates its own release by a positive feedback mechanism on pre-synaptic muscarinic and oxotremorine-sensitive receptors (Das et al., 1978; Ganguly and Das, 1979). These authors reported that stimulation of these receptors by 10.5×10^{-6} M oxotremorine caused an almost sixty-fold increase of the stimulated (1 Hz) ACh release from a rat phrenic·nerve–diaphgram preparation (Ganguly and Das, 1979). The effect of oxotremorine on ACh release was removed by 0.5×10^{-6} M atropine. Oxotremorine (10.5×10^{-6} M) also caused a significant increase of the spontaneous ACh release and produced a significant neuro-muscular block (Das et al., 1978).

Investigation of the effect of oxotremorine on the rat phrenic nerve–hemidiaphragm preparation and its interactions with d-tubocurare and atropine at the newomuscular junction (Földes, Chaudry, Nagashima, Sherman, Vizi, unpublished data) gave results that are difficult to reconcile with the report by Das et al. (1978). It was observed that oxotremorine, d-Tc and atropine mutually increased the neuromuscular blocking effect of one another (Földes and Vizi, 1980). If oxotremorine produced a manifold increase of stimulated ACh release, then, instead of augmenting it, it should antagonize the neuromuscular blocking effect of d-tubocurare.

It has been suggested by Koelle (1962) that cholinoceptors are present not only post- but also pre-synaptically at cholinergic transmission sites. According to his hypothesis, the relatively small amount of ACh initially released by the nerve impulse stimulates, by a positive feedback mechanism on pre-synaptic nicotine cholinoceptors, the release of much larger quantities of ACh, which then depolarizes the post-junctional membrane. The ACh-induced release of ACh is especially important for the maintenance of transmission during relatively high rates of stimulation (Bowman and Webb, 1976). This positive feedback mechanism of ACh release can be blocked by compounds capable of blocking ACh-induced depolarization at various cholinergic transmitter sites. The decreasing order of the relative potency of pre-junctional nicotinic receptors is hexamethonium > d-tubocurarine (d-Tc) > pancuronium. In agreement with this, the intravenous injection of a dose of hexamethonium that had little effect on the twitch (0.1 Hz) or peak tetanic (100 Hz) tension produced pronounced tetanic fade. Pancuronium affected both twitch and peak tetanic tension and tetanic fade about equally and the effects of d-Tc were mid-way between hexamethonium and pancuronium (Bowman and Webb, 1976). Additional evidence in favour of the existence and physiological role of pre-junctional nicotinic cholinergic receptors has been provided by the finding that after intravenous injection of ACh antagonists there was little fade of tetanus (Bowman and Webb, 1976), indicating that pre-junctional receptors, which are less accessible to or combine more slowly with ACh antagonists, are involved in this phenomenon.

Evidence has been presented showing that nicotinic receptors are also involved in a feedback inhibition of evoked ACh release (Miledi *et al.*, 1978). α-Bungarotoxin, a neurotoxin obtained from the venom of the snake *Bungarus multicinctus* which blocks neuromuscular transmission by combining irreversibly with the ACh receptors, increased the release of ACh from the rat diaphragm, whether induced by KCl or by electrical stimulation of the nerve.

5.4 SEROTONIN

Farnebo and Hamberger (1971b, 1974a) first suggested that serotonin is involved in the regulation of 5-hydroxytryptamine (5-HT) release from central serotoninergic neurons. In addition, the existence of 5-HT autoreceptors has been proposed on the basis of results obtained with brain slices utilizing LSD or ergocornine as 5-HT receptor agonists (Hammon *et al.*, 1974; Göthert and Winheimer, 1979; Göthert, 1980).

The depolarization-induced release of serotonin (5-HT) from hypothalamic synaptosomes was inhibited by extracellular 5-HT. The inhibition was counteracted by the central 5-HT-receptor blocking agent, methiothepin (Cerrito and Raiteri, 1979a,b). 5-HT (0.5×10^{-6} M), strongly, inhibited the release of ^{3}H-5-HT induced by high K^+ (15 mM). The inhibition was dose-related. The serotonin receptor blocking agent, methiothepin (0.5×10^{-6} M), inhibited the effect of 5-HT. Other 5-HT antagonists (cyproheptadine, methysergide, and mianserin) tested at 10^{-6} M were ineffective. Also, the spontaneous release of ^{3}H-5-HT was inhibited by 5-HT and methiothepin antagonized this

132

inhibition. These results provide evidence for a modulation of 5-HT release through pre-synaptic autoreceptors, which appear to be located pre-synaptically on the terminal serotoninergic nerve fibres themselves (Cerrito and Raiteri, 1979b).

The inhibitory effect of unlabelled 5-HT on electrically induced ^{3}H overflow increases (Fig. 42a,b) with decreasing Ca^{2+} concentrations (1.3–0.65 mM, 3 Hz) or with a decreasing frequency of stimulation (10–0.3 Hz, 1.3 mM Ca^{2+}) (Göthert, 1980). Methiothepin causes a substantial shift to the right of the concerntration response curve of unlabelled 5-HT for its inhibitory effect. In contrast, metergoline produces only a very slight shift of this curve (Göthert, 1980).

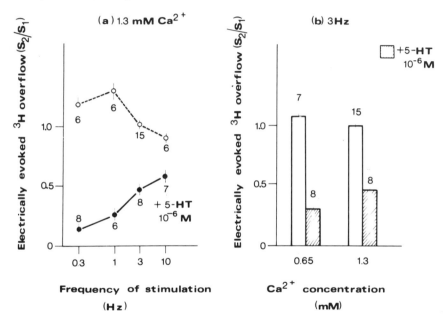

Fig. 42. Inhibitory effect of 10^{-6} M cold serotonin (5-HT) on the electrically induced tritium release from rat brain cortex slices at various frequencies (a) and at two (0.65 and 1.3 mM) Ca^{2+} concentrations (b) (redrawn from Göthbert's (1980) paper: for method and details see Göthert, 1980). Field stimulation 20 mA, 2 ms, 0.3–10 Hz. Note the frequency-dependence of the inhibitory effect of 5-HT. (Reproduced by permission of Springer-Verlag)

A superfused synaptosomal preparation seems to be particularly suitable for studying the role of pre-synaptic receptors in negative feedback modulation. Nevertheless, so far, no-one has succeeded in preparing a synaptosome which contains homogenous axon terminals, i.e. only one type of transmitter/modulator present in the synaptosomes. Therefore the data obtained with this technique seems to be very convincing, but caution is needed in the interpretation. In the light of this discussion it is highly probable that the site of action of serotonin is on the axon terminal.

5.5 γ-AMINOBUTYRIC ACID (GABA)

Johnston and Mitchell (1971), observed that GABA antagonists enhanced the release of ^{3}H-GABA from rat brain slices. Evidence has shown that γ-aminobutyric acid (GABA) inhibits the potassium-induced release of ^{3}H-GABA from the substantia nigra (Kamal et al., 1978) and from the frontal cortex (Mitchell and Martin, 1978). There was no similar mechanism in the occipital cortex (Kamal et al., 1978). Muscimol (19^{-7} M) caused similar GABA-like effects, and bicuculline or pictrotoxin (10^{-6} M) prevented its action (Mitchell and Martin, 1978). Bicuculline alone, however, failed to facilitate the release of GABA. This is not consistent with the finding of Johnston and Mitchell (1971), who succeeded in showing an increase of release. In the rat substantia nigra it was demonstrated that pre-synaptic GABA autoreceptors regulate the calcium-dependent release of ^{3}H-GABA from GABA-ergic nerve terminals (Arbilla et al., 1979). These authors showed that GABA receptor agonists such as muscimol and GABA reduced the potassium-induced release of ^{3}H-GABA (Arbilla et al., 1979). These findings indicate that the release of GABA is subject to a negative feedback control.

5.6 HISTAMINE

Although histaminergic neurons have not yet been histochemically visualized, there is convincing evidence that histamine is released from neurone. Arrang et al. (1983) have shown that histamine inhibits its own release from depolarized slices of rat cerebral cortex. 10^{-6} M histamine inhibited by about fifty percent of ^{3}H-histamine release evoked by 30 mM K, while the basal release was not affected. The inhibitory effect of histamine was mimicked by its two N and N,N-methyl derivatives which showed the same maximal effect but with about three- and two-fold higher relative potency, respectively, while they are slightly less potent than histamine at typical H_1 and H_2 receptors. Since impromidine (half-maximal effect at 3×10^{-8} M) enhanced the release and antagonized the effect of agonists, therefore it has been suggested (Arrang et al., 1983) that the negative feedback modulation of histamine release is mediated via a class of receptor (H_3) pharmacologically distinct from those previously characterized, that is, the H_1 and H_2 receptors.

5.7 ENKEPHALIN

No evidence has been found for pre-synaptic autoreceptors at enkephalin neurons (Richter et al., 1979; Osborne and Herz, 1980; Sawynok et al., 1980).

Chapter 6

Trans-synaptic (trans-junctional) modulation

6.1 ADENINE NUCLEOTIDES

Recent experimental evidence concerning the effects of adenine nucleotides (ATP, ADP, AMP, and adenosine) (Fig. 43) on different nerve structures emphasizes the possibility that a mechanism involving the release of ATP and/or related substances in different nervous systems, either centrally or peripherally (autonomic or motor), may modulate the output of transmitters to some extent. Nevertheless, one of the still-unsolved problems is where adenine nucleotide is released from.

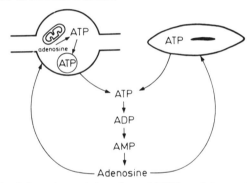

Fig. 43. Adenosine–triphosphate (ATP) and its metabolites

The purinergic nerve concept (Burnstock, 1972), i.e. that the nerves release ATP or a related substance, stimulated much research. The release of ATP together with noradrenaline from adrenergic nerves (Su *et al.*, 1971, Su, 1975; Langer and Pinto, 1976; Westfall *et al.*, 1978) and from cholinergic nerves (Silinsky, 1975) had been previously claimed but, it has always been difficult to ascertain whether the source of nucleotide release is from nerves or from muscle as a result of changes in tension or of receptor stimulation.

Further, it had previously been shown that the release of nucleotides from guinea-pig taenia coli and bladder during non-adrenergic, non-cholinergic nerve stimulation was from nerves and not muscle (Burnstock, *et al.*, 1978, Rutherford and Burnstock, 1978). However, convincing evidence was available (Fredholm *et al.*, 1982) showing that adenine nucleotides were released in response to depolarization of the target cells and that they acted trans-synaptically, reducing the release of the transmitter that had released it from the effector cell. Since adenine and its nucleotides can be also formed by and released from the post-synaptic elements it was suggested that they can act as 'trans-synaptic modulators' (Vizi, 1979; Fredholm, 1981; Hedqvist, 1981) and 'retrograde transmitters' (Israel et al., 1976). Nevertheless, the amount of adenosine and/or ATP released from pre-synaptic nerve-endings cannot be overlooked. They are also released from neurons and can act either on other neurons ('interneuronal modulation') or on the neuron from which it is released ('negative feedback').

6.1.1 Effect on ACh release

Burnstock (1972, 1978b, 1979b) has postulated the presence of non-adrenergic and non-cholinergic inhibitory neurons in addition to the adrenergic fibres in the mammalian intestinal tract. He proposed that the substance in the non-adrenergic neuron may be ATP or a closely related nucleotide. It was found that the nature of smooth muscle response to ATP is very similar to non-adrenergic nerve stimulation. Both are characterized by a rapid onset of hyperpolarization, and this effect is transient, being maintained for no more than 20–30 s (cf. Burnstock, 1972).

In addition to the post-synaptic effect of ATP and closely related nucleotides it was shown that adenosine and other adenine nucleotides are able to act on neurons present in the gastrointestinal tract, in the vas deferens, in the neuromuscular junction, and in the central nervous system.

Gastrointestinal tract

The classical concept of the innervation of the gut is that the cholinergic excitatory neurons are opposed, mainly pre-synaptically, by post-ganglionic sympathetic noradrenergic fibres. However, it has also been shown that there are non-adrenergic, non-cholinergic inhibitory neurons (cf. Burnstock, 1979b) supplying the smooth muscle of the gastrointestinal tract (Burnstock, 1972). Neurochemical evidence has been obtained showing that adenosine and related nucleotides are able to inhibit ACh release from the axon terminals of Auerbach's plexus (Vizi, 1975b, 1976; Vizi and Knoll, 1976; Cook et al., 1978; Gustafsson et al., 1978; Fredholm and Hedqvist, 1979; Hayashi et al., 1978; Vizi et al. 1981b, 1983b).

Resting release

Adenosine (10^{-6}–10^{-4} M) and ATP dose-dependently reduce the resting release of ACh (Vizi, 1976; Vizi and Knoll, 1976; Cook et al., 1978; Gustafsson et al., 1978) from Auerbach's plexus of guinea-pig-ileal longitudinal muscle. When the concentration of ATP is increased to 10^{-3} M, no further decrease in ACh release is seen. No tachyphylaxis was observed to the action of ATP when it was administered repeatedly. Much of the resting release of ACh was due to the spontaneous activity of neurons, since it was reduced by 60% in the presence of tetrodotoxin (10^{-6} M). Under the latter conditions the residual release of ACh was not affected by adenosine (10^{-4} M). Therefore it seems very likely that adenosine only inhibits the release of ACh induced by spontaneous firing of cholinergic interneurons.

Stimulated release

Evidence has been obtained (Vizi, 1975b; Vizi and Knoll, 1976; Fredholm and Hcdqvist, 1979; Hayashi *et al.*, 1978; Vizi *et al.*, 1983b; Gustafsson *et al.*, 1978; Gustafsson, 1980) that adenine nucleotides are able to inhibit the release of ACh from the ileal Auerbach plexus. The inhibitory effect of adenosine and ATP on the release of ACh induced by electrical stimulation depends on the frequency of stimulation applied (Vizi and Knoll, 1976; Gustafsson *et al.*, 1978; Fredholm and Hedqvist, 1979; Vizi *et al.*, 1983b).

The fact that adenosine, ATP, ADP, and AMP were active while adenine, guanosine, and inosine were inactive indicates that both the 6-amino and a sugar moiety are necessary for the effect. The order of potency (adenosine < ATP=AMP=ADP) (Vizi and Knoll, 1976) indicates that P_1 purinoceptors are involved in the pre-synaptic effect of adenine nucleotides. The release of ACh enhanced by cholecystokinin was also inhibited by adenosine (Vizi and Knoll, 1976).

It has been shown (Cook *et al.*, 1978) that Co-enzyme A (2×10^{-6} M is able to reduce the release of ACh from Auerbach's plexus and theophylline (10^{-4} M) antagonizes these inhibitory effects. Any discussion of the effect of CoA should take into account that it may be metabolized to simpler purine derivatives such as adenosine, which are the active factors.

Theophylline is a blocking agent of adenosine receptors (McIlwain, 1972). By itself, it enhances the release of ACh induced by stimulation (Vizi and Knoll, 1976; Cook *et al.*, 1978) at either low (0.2 Hz) or high (10 Hz) frequency. At low frequencies of stimulation (< 3 Hz) the potentiation was seen to be more pronounced than at high frequencies. Adenosine, which otherwise reduced the release of ACh, failed to affect the release in the presence of theophylline. The inhibitory effect of ATP on the ACh released by stimulation at 0.2 Hz could be also partly prevented by theophylline. In line with these observations, it was observed (Moritoki *et al.*, 1976) that methylxanthines increased the release of ACh from the whole ileum preparation during stimulation and in the resting state.

Effect of endogenous adenosine

Theophylline, in a concentration which did not influence phosphodiesterase activity by itself, enhanced the release of ACh from Auerbach's plexus (Vizi and Knoll, 1976; Hayashi *et al.*, 1978; Gustafsson *et al.*, 1978). This finding indicates that the release of ACh may be continuously controlled by an endogenous substance via P_1-receptor stimulation.

The neuromuscular junction

Adenine nucleotides exogenously administred (in concentrations similar to those estimated as being released in rat phrenic nerve–diaphragm preparation) decreased the spontaneous release of ACh as measured by the ratio of the mean amplitude of evoked end-plate potentials (MEPP) and the induced release of ACh as measured by the ratio of the mean amplitude of evoked end-plate potentials to the mean amplitude of MEPPs (Ginsborg and Hirst, 1972; Ribeiro and Walker, 1975; cf. Ribeiro, 1981). The action of adenosine and/or ATP seems to be confined to the nerve terminals, since purinergic substances do not affect the action potential of the axon (Ribeiro and Dominquez, 1978).

The central nervous system

Both adenosine (3×10^{-5} M) and ATP (10^{-5} M) failed to significantly reduce the resting release of ACh (Vizi and Knoll, 1976) from isolated cortical slices of the rat. However, theophylline by itself significantly enhanced the release by 36%. This finding indicates that the release is controlled via P_1 purinoceptors.

6.1.2 Effect of noradrenaline release

Peripheral nerves

Intra-arterial infusion of ATP into the isolated canine subcutaneous adipose tissue inhibited NA release in response to sympathetic nerve stimulation (Fredholm, 1974). As with acetylcholine release, evidence has been presented (Hedqvist and Fredholm, 1976) showing that adenosine also blocks the release of NA from axon terminals: adenosine (10^{-6}–10^{-5} M) was able to cause a concentration-dependent and reversible inhibition of the efflux of ^{3}H-NA resulting from transmural nerve stimulation of a superfused guinea-pig vas deferens. Treatment with phenoxybenzamine (3×10^{-6} M) or indomethacin (6×10^{-6} M) in order to block the α-adrenoceptors and the prostaglandin synthetase respectively, did not significantly alter the inhibitory effect of adenosine on ^{3}H-NA release. A very similar observation was made on rat vas deferens (Wakade and Wakade, 1977a, 1978): adenosine inhibited the release of ^{3}H-NA and aminophylline prevented its action even when the tissue was pre-treated with phenoxybenzamine (Wakade and Wakade, 1978).

Several vascular vessels are known to possess a dilator innervation of a non-adrenergic nature. Acetylcholine or unknown substances have been shown to mediate sympathetic vasodilation in the skeletal muscle and skin of several species. Hughes and Vane (1967) considered that ATP or a related purine compound might account for the relaxation of the rabbit portal vein. Purinergic innervation, has been shown for the portal vein (Hughes and Vane,

1967; Su, 1974, 1975). ATP inhibited the tritium overflow from the rat portal vein pre-labelled with ^{3}H-NA at the rather high concentration of 3×10^{-4} M. When the pre-junctional α-adrenoceptors were blocked with phenoxybenzamine (3×10^{-6} M), the inhibitory effects of ATP on the tritium overflow was increased (Enero and Saidman, 1977). Since the inhibition by ATP increased when the pre-junctional α-adrenoceptors were blocked, prejunctional purinergic inhibition may be a process which operates in parallel with the negative feedback mechanism through α-adrenoceptors. The ATP metabolites ADP, AMP, and adenosine were approximately equipotent with ATP itself in inhibiting tritium overflow from the rat portal vein induced by stimulation (Enero and Saidman, 1977). Further support for this hypothesis was obtained when uptake (dilazep, 3×10^{-5} M) and metabolism (erythro-9-2-hydroxy-3-nonyl-adenine 3×10^{-5} M) of adenosine were inhibited. Under these conditions, the vascular contractions and the tritium overflow in response to transmural stimulation were significantly reduced. This also indicates a possible role of adenine nucleotides released by nerve stimulation in the modulation of NA release.

These findings are consistent with numerous reports in the literature showing that adenosine produces vasodilation in most vascular beds (cf. Haddy and Scott, 1968). Concentrations of 10^{-6}–3×10^{-4} M of exogenous ATP were needed to inhibit noradrenergic transmission. Su (1975) postulated that ATP overflow, produced by stimulation, was due to the nucleotide release from adrenergic and non-adrenergic nerves. Since, in the adrenergic terminals, the molar ratio of NA and ATP has been estimated to be between 4:1 and 20:1 (Geffen and Livett, 1971; Lagercrantz and Stjärne, 1974), Su (1975) calculated that in the rat portal vein transmural stimulation releases enough ATP from the adrenergic terminals to raise the concentration in the biophase to up to 1 μM ATP. Guanethidine, in a concentration which blocks NA release, reduces the ATP release in the portal vein to only 50% suggesting that part of the ATP released by stimulation originated from non-adrenergic terminals (Su, 1975). However, the actual concentration reached in the vicinity of nerve-endings during nerve stimulation may be different from the concentration of the purine compounds added to the incubation specimen.

In contrast to other species, adenosine does not inhibit noradrenergic neurotransmission in the cat.

The central nervous system

It has been found (Wakade, 1980) that adenosine is able to reduce NA release from the central noradrenergic neurons. Adenosine inhibited the ^{3}H-NA release from the isolated hypothalamus and cortical slices of the rat when electrical stimulation was applied. The concentration of adenosine (10^{-4} M) required to inhibit NA release in the central nervous system was considerably higher than those needed for peripheral neurons.

6.1.3 Inhibitory effect on neuroeffector transmission

Cholinergic transmission

Longitudinal muscle strip of the guinea-pig ileum

It has been reported that adenosine inhibits the intestinal movements of the anaesthetized cat (Drury and Szentgyörgyi, 1929) and produces relaxation of an isolated rabbit intestine and reduction of the spontaneous contractions. Evidence has been found of a neuron type in the intestine which is neither cholinergic nor adrenergic (cf. Burnstock, 1972) yet which causes inhibition of the gut. It has been shown (Burnstock *et al.*, 1970b) that a purine nucleotide, probably ATP, is the transmitter released from inhibitory nerves present in Auerbach's plexus. The inhibitory effect of ATP on intestinal motility was attributed to its direct action on the smooth muscle cells (cf. Burnstock, 1972).

However, it has also been demonstrated that adenosine (10^{-6}–10^{-5} M), AMP (10^{-6}–10^{-4} M), ADP (10^{-6}–10^{-4} M), and ATP (10^{-6}–10^{-4} M) reduced the size of contractions of the electrically stimulated guinea-pig myenteric plexus longitudinal muscle preparation (Vizi and Knoll, 1976; Sawynok and Jhamandas, 1976; Gustafsson *et al.*, 1978) without influencing the sensitivity of the smooth muscle cells to added ACh (Vizi and Knoll, 1976). This finding indicates that the inhibitory effect of nucleotides is pre-synaptic. The effects of nucleotides were rapid in onset and after washing out, a fast recovery was observed. The reduction depended on the concentrations of nucleotides used and the frequency of stimulation applied: the contractions induced by high-frequency stimulation (>5 Hz) were not affected by the nucleotides. Figure 44 shows the effect of adenosine on an electrically stimulated longitudinal muscle strip of the guinea-pig ileum. The site of action is pre-synaptic since the direct effect of ACh on smooth muscle is not affected. ID_{50} values of different adenine nucleotides are in the range of 10^{-6} M. The isopropyl derivative of adenosine proved to be sixty-six times more effective than adenosine (Vizi *et al.*, 1983b). Figure 45 shows the effect of N^6-isopropyl-adenosine on the twitch responses of ileal longitudinal muscle strip; the interaction between N^6-isopropyl adenosine and theophilline is competitive (cf. Fig. 45).

No tachyphylaxis was seen with P_1 receptor agonists as repeated exposure to adenine nucleotides did not result in a diminution of the inhibitory action (Vizi and Knoll, 1976). However, it was observed that at higher concentrations (10^{-4} M) ATP caused a contraction of the smooth muscle. Theophylline, in concentrations of up to 2×10^{-4} M, potentiated the contractions of muscle strip to stimulation at a frequency of 0.2 Hz, and reversed or prevented the inhibition produced by each of the nucleotides. Theophylline competitively antagonized the inhibitory effect of adenosine on cholinergic nerve-mediated contractions but phentolamine (Vizi and Knoll, 1976) and phenoxybenzamine

140

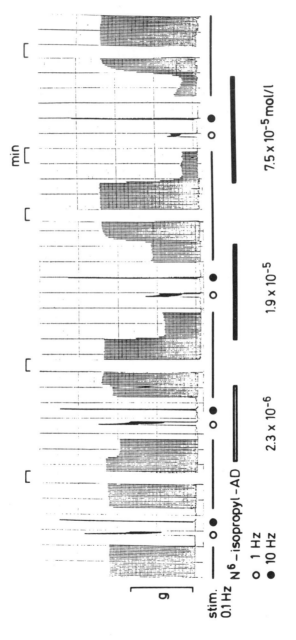

Fig. 44. Inhibitory effect of N⁶-isopropyl-adenosine (N⁶-isopropyl-AD) on contractions of guinea-pig ileal longitudinal muscle strip (redrawn from Vizi *et al.*, 1983)

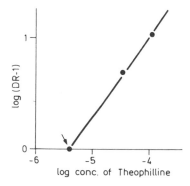

Fig. 45. Competitive nature of the antagonism of theophylline on the agonistic effect of N^6-phenylisopropyladenosine on P_1-receptors (Vizi *et al.*, 1983). Longitudinal muscle strip of guinea-pig ileum, 0.1 Hz stimulation, pA_2 value at the arrow

(Sawynok and Jhamandas, 1976) did not antagonize the effect of adenosine.

Guanosine triphosphate, in concentrations of up to 2×10^{-4} M, failed to affect the contractions of the strip in response to stimulation at 0.1 Hz. Unlike adenosine, the dibutyryl derivative of adenosine, 3,3,-monophosphate, in concentrations ranging from $10^{-6} - 10^{-4}$ M did not reduce the size of the contractions. Dipyridamole and dilazep were found to enhance the inhibitory action of adenosine in contrast to the action of theophylline (Gustafsson *et al.*, 1980). Two of these drugs, dipyridamole and dilazep, are also known inhibitors of adenosine uptake and inactivation, which would fit well with their ability to cause a marked leftward shift in threshold dose for adenosine, especially since it could be shown that the smooth muscle preparation had a substantial capacity to inactivate adenosine. Dipyridamole and dilazep in concentrations higher than 10^{-7} M inhibited transmurally induced responses at 10^{-6} M, producing a 50% inhibition. Dilazep at this concentration did not alter contractions induced by acetylcholine, indicating that its action is presynaptic. The effect of dipyridamole and dilazep could be effectively antagonized by theophylline (Gustafsson *et al.*, 1980), indicating that inhibition of adenosine uptake was their major mechanism of action. This suggestion is supported by the finding that dilazep and dipyridamole enhanced the action of exogenous adenosine, most likely via inhibition of adenosine inactivation, since they had their action in common with erythro-9-9 (2-hydroxy-3-nonyl)-adenine (ENHA) an inhibitor of adenosine deaminase, and since dipyridamole increased the overflow of endogenous adenosine (Gustafsson *et al.*, 1980).

Theophylline itself, at concentrations ineffective on cAMP phosphodiesterase, enhanced and dipyridamole and dilazep inhibited responses to transmural nerve stimulation, indicating an endogenous release of adenosine having an inhibitory effect on neurotransmission. Since endogenous adenosine is spontaneously released in amounts sufficient for modulation of the cholinergic neurotransmission in the isolated guinea-pig ileum and this release can be augmented by nerve stimulation (Gustafsson, 1980), it is suggested that adenosine may also have an inhibitory influence *in vivo* on cholinergic neurotransmission, although responsiveness of circular gastric smooth muscle to stimulation may be enhanced by adenosine (Gustafsson, 1980).

Whole guinea-pig ileum

Adenosine and ATP produced a substantial inhibition of the isometric contractions induced by transmural field stimulation (Gintzler and Musacchio, 1975; Moritoki *et al.*, 1976; Hayashi *et al.*, 1978). This was supported by the experiments of Ally and Nakatsu (1976), who showed that the spontaneous contractions of isolated rabbit ileum could be inhibited by the addition of adenosine to the incubation medium. The ED_{50} proved to be 6×10^{-7} M. Adenine, which differs from adenosine by the absence of a ribose moiety, had no effect at concentrations as high as 10^{-4} M. Similarly, the purine nucleotides, inosine monophosphate and guanosine monophosphate (10^{-4} M), and the pyrimidine nucleotides, cytidine monophosphate and thymidine monophosphate (10^{-3} M), were also without effect. Both theophylline and caffeine effectively antagonized the inhibitory responses elicited by adenosine and adenine nucleotides on the isolated rabbit ileum. The data obtained with the isolated rabbit ileum are consistent with the action of adenosine on NA release. This indicates that this nucleotide does not induce an inhibitory response by releasing endogenous catecholamine from adrenergic nerves. Tomita and Watanabe (1973) showed that adenosine has a direct effect on smooth muscle: tetrodotoxin did not abolish the adenosine-induced hyperpolarization.

Methylxanthines (10^{-5}–10^{-3} M) were found (Moritoki *et al.*, 1976) to increase the amplitude of contractions of guinea-pig ileum induced by transmural stimulation but to inhibit those induced by ACh or histamine. Imidazole (3×10^{-5}–10^{-3} M) had an essentially similar effect to the methylxanthines. The order of the abilities of methylxanthines to augment contractions of the ileum induced by transmural stimulation was theobromine > caffeine > theophylline, whilst their inhibitory effects on phosphodiesterase have been reported to be in the order, theophylline > caffeine > theobromine (Butcher and Sutherland, 1962). Thus, their abilities to potentiate the contractile responses are not correlated with their abilities to inhibit phosphodiesterase.

The neuromuscular juction

Of more direct interest, is the possibility that ATP may be involved in controlling the release of ACh from the motor nerve-ending. Ribeiro and Walker (1973) have demonstrated that ATP in concentrations of 10^{-4} M depresses MEPPs, an effect independent of the calcium chelator action of ATP. Adenosine in doses which increased intercellular cAMP in brain slices decreased MEPPs at the neuromuscular junction (Ginsborg and Hirst, 1972; Branisteanu *et al.*, 1979). Both effects are antagonized by theophylline. Adenosine, in a concentration of 2×10^{-4} M, decreased the amplitude of the EPPs (Ribeiro and Walker, 1975; cf. Ribeiro, 1978) and 400–800 μg/kg, i.a. dose-dependently inhibited neuromuscular transmission of neurally stimulated cat soleus muscle (Standaert *et al.*, 1976). It also prevented the facilitatory effect of dibutyryl cAMP and theophylline (Standaert *et al.*, 1976). Pre-treatment with the phosphodiesterase inhibitor, theophylline, and conditions expected to increase the intraneuronal concentration of cAMP (nerve stimulation, NaF, and dibutyryl cAMP), increased the response to drugs whereas pre-treatment with the phosphodiesterase activator, imidazole, decreased the response to these agents. In addition, phosphodiesterase inhibitors antagonized the blockade of neuromuscular transmission produced by D-tubocurarine (Goldberg and Singer, 1969).

Cyclic AMP and dibutyryl cAMP were found to facilitate neuromuscular transmission by enhancing the quantum content of end-plate potentials from the sciatic nerve (Breckenridge *et al.*, 1967) or phrenic nerve (Goldberg and Singer, 1969).

A post-synaptic action of ATP has also been suggested: evidence has been presented (Ewald, 1976) showing that the cholinergic sensitivity of rat diaphragm muscle, measured as the magnitude of depolarization responses to repetitive, iontrophoretic pulses of ACh onto neuromuscular end-plates, is increased by addition of ATP to the perfusion medium.

The urinary bladder

In studies on parasympathetic ganglia of the urinary bladder of the cat, for which an adrenergic inhibitory mechanism has been described, deGroat and Theobald (1976) observed no significant reduction of ganglionic transmission with cAMP or dibutyryl cAMP injections. However, non-cyclic nucleotides (AMP, ADP, and ATP) in smaller doses, produced considerable depression of ganglionic transmission. These substances also depressed bladder contractions elicited by electrical stimulation of the pre-ganglionic nerves, but did not alter the responses to injected ACh. This indicates that non-cyclic nucleotides have a pre-synaptic effect and depress bladder contractions by preventing the release of motor transmitter ACh.

Noradrenergic transmission

The terminal ileum

In the terminal ileum, the neurochemical transmission at 30-Hz stimulation is believed to be of adrenergic origin. Adenosine, cAMP, and dbcAMP have been found (Kazic and Josavljevic, 1976) to inhibit nerve-mediated contractions of the terminal ileum caused by electrical stimulation. Adenosine, cAMP, and dbcAMP did not affect the contractile responses of the preparation to histamine, potassium and NA: therefore the observed inhibition of nerve-mediated contractions was thought to be due to an action on pre-synaptic sites. The following order of relative potencies have been established: adenosine > cAMP > dbcAMP. The calculated ID_{50} for adenosine was 1.4×10^{-6} M, for cAMP 2.1×10^{-6} M, and for dbcAMP 2×10^{-5} M. Aminophylline and calcium were found to counteract the inhibitory action of adenosine and cAMP.

The vas deferens

It has been shown (Clanachan et al., 1977; Paton et al., 1978) that ATP, ADP, AMP, and adenosine (10^{-6}-10^{-4}M) are able to inhibit the responses of the rat vas deferens to field stimulation in a dose-dependent manner. The inhibition was pre-synaptic in origin as motor responses evoked by exogenous NA were not altered by these nucleotides. Theophylline (10^{-4} M) was able to antagonize their pre-synaptic action. The common structural requirements for activity include a primary or secondary amine function at C_6 of the purine ring with little tolerance for major steric changes or substitutions on the sugar moiety. None of the analogues studied prevented the pre-synaptic inhibitory action of adenosine (Paton et al., 1978).

Blood vessels

In vitro experiments have demonstrated that ATP and related compounds inhibit the contractile amplitude of the rat portal vein (Hughes and Vane, 1967) and the concentration needed to produce a 50% inhibition was as high as 3×10^{-4} M. Vergaeghe et al. (1977) showed that there is an inhibition of sympathetic neurotransmission in canine blood vessels by adenosine and adenine nucleotides. In rabbit kidney, adenosine (10^{-7}–10^{-5} M) produced a concentration-dependent increase in vascular resistance and in vasoconstrictor response to nerve stimulation and exogenous NA (Hedqvist et al., 1978). In the adipose tissue, adenosine also increased the vasoconstrictor responses, but it decreased vascular resistance (Hedqvist and Fredholm, 1976). It was concluded that adenosine affects adrenergic neuro-effector transmission by two discrete mechanisms, pre-synaptic inhibition and junctional stimulation.

In vivo experiments have led to similar conclusions (Hom and Lokhand-wala, 1981). Infusion of adenosine (1 mg/kg/min i.v.) to phenobarbital-anaesthetized dogs decreased the blood pressure and significantly attenuated the femoral vasoconstrictor responses to lumbar sympathetic nerve stimulation. The vasoconstrictor action of exogenous noradrenaline was not altered during adenosine infusion. The administration of theophylline (5 mg/kg i.v.) prevented the inhibitory action of adenosine on responses to sympathetic nerve stimulation. Sympathetic denervation of the femoral vascular bed did not alter the vasodilatory action of adenosine, which was antagonized by theophylline. These results suggest that adenosine causes inhibition of sympathetic neurotransmission to the femoral vasculature via an action on pre-synaptic purinergic receptors. However, this pre-synaptic inhibitory action of adenosine is not involved in the femoral vasodilatation produced by the compound. It has been reasoned (Hom and Lokhandwala, 1981) that if the vasodilatation produced by adenosine did possess a neurogenic component, then sympathetic denervation should attenuate this action? However, the same authors demonstrated that the increase in femoral blood flow produced by adenosine was still of the same magnitude in the denervated hind-limb, indicating that inhibition of sympathetic vasoconstrictor discharge was not completely responsible for the vasodilator action of adenosine. Most of the evidence favoring the presence of an adenosine-mediated inhibitory mechanism on organs innervated by the autonomic nervous system has been obtained from *in vitro* experiments (Hughes and Vane, 1967; Clanachan *et al.*, 1977; Hedqvist *et al.*, 1978), although there are *in vivo* studies reporting the presence of this mechanism in the saphenous vein of the dog (Verhaeghe *et al.*, 1977; Mey *et al.*, 1979).

It can be concluded that adenosine is a potent vasodilator agent and since it also causes inhibition of sympathetic neurotransmission, it is possible that this sympatho-inhibitory action may lead to passive vasodilatation and an increase in blood flow.

Inhibitory effect on neuronal firing

It has been found that adenosine has a potent depressant action on cerebral cortical neurons, including identified corticospinal cells (Phillis *et al.*, 1974, 1975; Kostopoulos *et al.*, 1975; Phillis and Kostopoulos, 1975; Phillis and Kostopoulos, 1977). In experiments recording neuronal activity and ionto-phoresing drug solutions, it has been shown that adenosine 2,3 and 5-phosphates, including adenosine 5-imidodiphosphate, have comparable depressant actions and 2-chloroadenosine is an even more potent depressant. Inhibitors of adenosine uptake, hexobendine and papaverine, potentiate the actions of adenosine and adenosine 5-monophosphate while theophylline and caffeine antagonized the depressant actions of adenosine and adenosine 5-monophosphate. A similar observation was made by Stone and Taylor (1977), who showed that adenine derivatives such as adenosine and its nucleotides have powerful depressant actions on the firing rate of single neurons in the brain when applied by microiontophoresis (Stone and Taylor (1977). 4-Aminopyridine blocked their depressant effects on 24 out of 26 cells without affecting inhibitions produced by aminobutyric acid (GABA). The 4-aminopyridine molecule bears some structural similarities to adenosine, and it is well known that it blocks membrane potassium channels, both calcium-

146

dependent and voltage-dependent, whether applied extracellularly or intra-cellularly (Meves and Pichon, 1977). Two possible explanations for this effect were suggested by Stone (1982): (1) adenosine and ATP depress neurons by increasing their permeability to potassium, which uses channels blocked by 4-aminopyridine: (2) 4-aminopyridine may be an antagonist at purine receptors on central neurons.

The amount of the electroplaque discharge in the electric organ of *Torpedo* was reduced by 5 mM ATP, ADP, AMP, and adenosine (Israel *et al.*, 1977). The less phosphorylated nucleotides appeared to be more efficient in blocking ACh release as shown by the reduction of the discharge. GTP and CTP were inactive.

Source of adenine nucleotides released

The source of adenine nucleotides involved in pre-junctional inhibition of transmitter release may be either purinergic nerves (cf. Burnstock, 1972, 1979) or cholinergic and noradrenergic neurons which have been claimed to release ATP together with ACh (Dowdall *et al.*, 1974) and NA. (see Fig. 46). Convincing evidence has also been obtained showing that the effector cells could be also a source of adenine nucleotides.

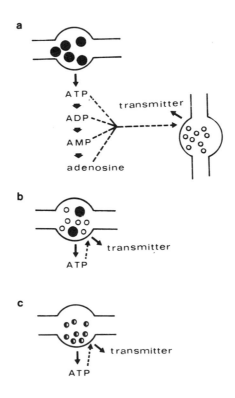

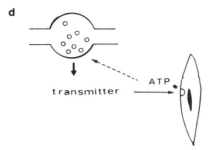

Fig. 46. Possible sources of ATP (taken from Vizi *et al.*, 1983b, Fig. 4)

Pre-synaptic release

The possibilities of the origin of ATP (or related nucleotide) release from nerve terminals are the following: transmitter and ATP released from (1) different nerve terminals (interneuronal modulation); (2) from the same nerve terminal; (3) from different vesicles; or (4) from the same vesicles. (Fig. 46a–d).

The release of ATP during nerve activity was described long ago, when Holton and Holton (1953) demonstrated the release of the nucleotide at sensory nerve-endings. It has been shown that prolonged depolarization of excitable membranes causes the outflow of adenine nucleotides (Abood *et al.*, 1962). More recently, Silinsky and Hubbard (1973) showed that ATP was released from the diaphragm after stimulation of the phrenic nerve.

Purine compounds are released by nerve stimulation in several adrenergically innervated organs (Su, 1975; Fredholm, 1976; Westfall *et al.*, 1978; Fredholm and Hedqvist, 1978; Luchelli-Fortis *et al.*, 1979). KCl (30 mM) produced a Ca-dependent purine release in arterial segments (Katsuragi and Su, 1980).

It has been shown that purine compounds are taken up by (Rowe *et al.*, 1978) and released from (Westfall *et al.*, 1978) the guinea-pig vas deferens and that this release is enhanced by electrical field stimulation. Evidence has been obtained (Fredholm *et al.*, 1982) that after labelling of the vasa deferentia of intact rats with ^{3}H-adenine there was a spontaneous outflow of labelled purines, mainly as adenosine. The basal rate of release of labelled and endogenous purines was markedly increased by dipyridamole, a potent inhibitor of adenosine uptake, from 0.1 to 1.0 pmol min^{-1} mg^{-1}. Westfall *et al.* (1978) reported a tetrodotoxin-sensitive release of purines following electrical stimulation in the guinea-pig vas deferens even when smooth muscle contraction was prevented. These results demonstrate that there may be purine release independent of smooth muscle contraction in the vas deferens, and indicate that purine release is not necessarily linked to smooth muscle contraction.

Adenine nucleotides released from cholinergic neurons

Since it has been shown that ATP is a constituent of cholinergic synaptic vesicles (Dowdall *et al.*, 1974) it is possible that ATP is released together with ACh. Silinsky (1975) succeeded in providing direct evidence for the release of ATP and ADP from motor-nerves using a sensitive modification of the firefly luciferase method (Silinsky, 1974). The presence of tubocurarine and normal (2 mM) calcium, nerve stimulation caused the appearance of ATP and/or ADP from an isolated rat phrenic nerve–hemidiaphragm preparation. It was shown that this nucleotide efflux was not produced by a secondary action of released ACh but was derived from the cholinergic vesicle. Zimmermann and Whittaker (1974) have demonstrated that prolonged stimulation of Torpedo nerve-endings produced a parallel decrease in vesicular ACh and vesicular ATP. Whether ATP release necessarily accompanies the secretion of ACh from motor nerve-endings in all circumstances, or whether the association between ACh and nucleotide is a general property of cholinergic systems is by no means certain.

Nevertheless, White et al. (1980) provided evidence that the depolarization-induced release of ATP from cortical synaptosomes is not associated with ACh release. Moreover, this release is highest from synaptosomes prepared from regions of the brain which contain the relatively greatest amounts of ACh (see Table 14).

Table 14. Depolarization-induced release of ATP from synaptosomes prepared from different parts of rat brain (data taken from Potter and White, 1980)

| | ATP release (10^{-13}M/LDH unit) | |
	KCl (23 mM)	Veratridine (50 μM)
Cerebellum	1.0	12.9
Medulla	6.5	90.0
Hypothalamus	5.2	40.4
Striatum	12.0	52.2
Cortex	9.1	20.1

For details, see Potter and White (1980).

Adrenine nucleotides released from adrenergic neurons

It has been reported that ATP is released together with catecholamines from adrenal medullary vesicles in perfused adrenal glands (Douglas ahd Poisner, 1966) and stored with catecholamines (Smith, 1972, 1973). There are certain similarities between the catecholamine storage granules in nerves and in the adrenals, and the former may also contain ATP, although in small amounts (Lagerantz, 1976). This means that catecholamines and ATP might be released concomitantly. Nevertheless, direct evidence for the release of ATP from adrenergic nerve terminals is still lacking.

It appears that one molecule of ATP compared with over 50 or 100 molecules of NA is released (cf. Fredholm et al., 1982). If ATP is released as a co-transmitter with NA, the concentration of ATP is likely to be small in comparison with NA, which also exerts a pre-synaptic inhibitory effect. Since NA is equi-effective with adenosine in inhibiting pre-synaptic release of transmitter, it seems very unlikely that ATP/adenosine released together with NA would contribute significantly to the overall effect of stimulation.

Adenine nucleotides released from unidentified neurons

In the majority of experiments, field stimulation was used to release transmitters and ATP. In this case it is very difficult to say where ATP comes from; it is often maintained that purines and transmitter are released together, but these findings do not mean that the transmitter and the purine originate from the same source. On the basis of these experiments the existence of purinergic nerves cannot be excluded.

Geffen and Livett (1971) and Su *et al.* (1971) have suggested that ATP might be released with NA. Langer and Pinto (1976) presented evidence that there is a substantial residual contraction of the cat nictitating membrane following depletion of NA. They suggest that the transmitter causing this is stored in the same nerve-endings from which NA is released in the untreated tissue, and it is thought to be ATP.

The evidence that adenosine is released from synaptosomal beds prepared from cerebral cortical tissue (Kuroda and McIlwain, 1974) and from intact cerebral cortex (Sulakhe and Phillis, 1975) strongly suggests that adenosine may be involved in intracellular mediation of synaptic events. A release of labelled adenosine and adenine nucleotides from an intact cerebral cortex which can be enhanced by direct electrical stimulation of the cortex has now been demonstrated in the laboratory by Dr Phillis (Sulakhe and Phillis, 1975), adding strength to the proposal that adenosine serves as a modulator in the central nervous system.

Kato *et al.* (1974), in experiments on cat superior cervical ganglion, have concluded that ATP is not released together with ACh from the pre-ganglionic nerve endings. These studies do not, however, exclude the possibility that ATP might be released from the pre-ganglionic nerve-endings and subsequently degraded before entering the vasculature, since a perfusion technique was used.

The depolarization-induced release of a number of putative neurotransmitters from nervous preparations has been shown to be Ca-dependent (Israel *et al.*, 1976; Baldessarini and Kopin, 1967; Blaustein *et al.*, 1972; DeBelleroche and Bradford, 1972; Osborn *et al.*, 1973). 4.7 mM EGTA reduced the release of ATP by elevated extracellular K^+, indicating a similarity of ATP to putative transmitters. It therefore seems that the release of ATP from synaptosomes is at least in part Ca-dependent (Israel *et al.*, 1976).

Depolarization of rat brain synaptosomes caused a release of ATP (White, 1977) and increasing the extracellular KCl concentration by 34 mM caused a rapid release of ATP from synaptosomes. The response to elevated extracellular K^+ was not prevented by the specific Na^+ channel blocker, tetrodotoxin (Narahashi *et al.*, 1964; Kao, 1966). Veratridine also caused a rapid release of ATP which could be prevented by the prior administration of tetrodotoxin (White, 1977, 1978). As to the origin of the ATP released, it seems most likely that pre-synaptic nerve terminals are involved, although it is also possible that glial contaminants of the synaptosomal preparation (Henn *et al.*, 1976) might respond to depolarizing influences by releasing ATP. On the other hand, the veratridine-induced release of ATP from synaptosomal preparations indicates an involvement of Na^+ channels, which glial elements may not possess.

Recent experiments have described a Ca^{2+}-dependent release of ATP when varicosities isolated from the myenteric plexus of the guinea pig ileum were depolarized with veratridine or elevated extracellular K^+ (White and Leslie, 1982).

Briggs and Cooper (1982) have reported a nicotinic release of ^{3}H-acetylcholine from myenteric varicosities isolated from the guinea-pig ileum. However, the possibility of co-release of ATP with acetylcholine seems unlikely, since Briggs and Cooper (1982) demonstrated a muscarinic inhibitory influence on the release of ^{3}H acetylcholine by nicotinic agonists, and no inhibitory muscarinic influence was observed for the nicotinic receptor-mediated release of ATP from myenteric varicosities (White, 1982).

Exogenously applied ATP caused mechanical responses which resemble those produced by electrical field stimulation of these preparations (Burnstock *et al.*, 1970). Moreover, tetrodotoxin-sensitive release of radio-labelled adenosine derivatives has been demonstrated during field stimulation of these tissues following incubation with ^{3}H-adenosine (Su *et al.*, 1971). It is likely that much of the ATP is very rapidly degraded extracellularly following its release (Satchell and Burnstock, 1971). Potter and White (1980) reported that depolarization induced a significant release of ATP from synaptosomes prepared from different parts of the rat brain. Table 14 shows the release of ATP induced by either KCl (23 mM) or veratridine (50 μM). Veratridine, which specifically depolarizes excitable cells by opening the Na^+ channels, caused a rapid release of ATP from myenteric varicosities. In contrast to the release from rat brain synaptosomes (White, 1978), the veratridine-induced release of ATP from myenteric varicosities was shorter in duration and required extracellular Ca^{2+}. Veratridine-induced release of ATP from guinea-pig brain synaptosomes was increased rather than decreased in the absence of extracellular Ca^{2+}.

Morphological evidence suggests that both adrenergic and cholinergic as well as possible purinergic varicosities are present in myenteric preparations (White and Leslie, 1982). It seems possible that ATP could be released during the release of these other neurotransmitters. If this is so, then the released ATP might function as a co-transmitter or modulator rather than as a primary neurotransmitter.

Adenine nucleotides released from purinergic nerves

Another possibility is that adenosine or related nucleotides are released from specific nucleotide-containing axons (Burnstock *et al.*, 1970b; Su *et al.*, 1971;

cf. Burnstock, 1972), i.e. it is released as a purinergic neuromodulator/trans-
mitter in its own right. This hypothesis has been challenged, although there
is no convincing evidence for excluding this possibility.

Post-synaptic release

ATP could possibly be released from the post-synaptic membrane as a consequence of depolarization produced by a transmitter (Fig. 47). Meunier *et al.*
(1975) have shown that ATP is released from the electric organ of Torpedo
after the motor nerve has been stimulated for 5 s, and that curare reduces this
release, indicating that ATP might come from the post-synaptic structures.

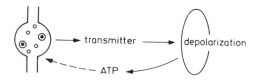

Fig. 47. Trans-synaptic inhibitory effect of ATP on transmitter release.
ATP is released in response to the effect of the transmitter on the effector
cell (cf. Fredholm, 1981; Hedqvist, 1981)

Israel *et al.* (1976, 1977) showed that ATP is released from the electric organ
of *Torpedo marmorata* even after a single nerve stimulus. High K^+ caused
release of ATP even in the presence of curare, and, therefore, the release
of ATP could be a consequence of the depolarization of the post-synaptic
membrane (cf. Abood *et al.*, 1962). It was suggested that the occurrence of
ATP during nerve activity might be a consequence of the downhill ion flux
triggered by the transmitter. The released ATP could potentiate the action
of ACh or modulate its release, or it may be involved in the uptake of ACh
precursors. Fredholm and Hedqvist (1978) provided evidence that tissues
(rabbit kidney, guinea-pig and rat vas deferens, cat nictitating membrane)
which had been loaded with ^3H-adenine or ^3H-adenosine, release ^3H-purines
in response to sympathetic stimulation. This release was reduced by α-adrenoceptor antagonists, indicating the post-synaptic origin of the released nucleotides. From these data it seems very likely that adenosine or a related
analogue could be released from both pre- and post-synaptic sites. When the
relative proportion of nerves to smooth muscle of the rat vas deferens is
markedly increased following castration, the time course of labelled purine
release is different from that of labelled NA (Fredholm *et al.*, 1982). Blockade
of alpha-adrenoceptors by phentolamine increased the release of noradrenaline but decreased that of purines. This shows that the major release of
purines from the castrated rat vas deferens by nerve stimulation is not concomitant with noradrenaline release. In the presence of dipyridamole, stimulated ^3H-NA overflow was unaltered while stimulated purine release was
essentially abolished. This finding also argues against a substantial release of
ATP at a co-transmitter. These results suggest a predominantly post-

junctional site of origin of these purines. L-adrenaline produced purine release from the smooth muscle of arterial segments even after denervation (Katsuragi and Su, 1980).

Direct evidence for the presence of ATP in superfusates from electrically stimulated taenia coli has been reported utilizing the luciferin–luciferase assay method: this ATP release was tetrodotoxin-sensitive (Burnstock *et al.*, 1978), indicating its nervous origin. However, tetrodotoxin-resistant release of ATP from the guinea-pig intestine has been demonstrated when high-voltage, high-frequency pulses were applied (White *et al.*, 1981; Kuchii *et al.*, 1973).

In 1976, Israel *et al.* described a method for directly and continuously detecting the release of ATP from the electric organ of *Torpedo marmorata* following stimulation of its motor nerve. They superfused the organ with a medium containing firefly luciferin–luciferase and directly detected the release of ATP by monitoring the light produced from the ATP luciferin–luciferase reaction with a photomultiplier. Using a modification of this technique, the depolarization-induced release of ATP from isolated pre-synaptic nerve terminals (synaptosomes) has been directly and continuously monitored (White, 1977, 1978).

6.2 PROSTAGLANDINS

The discovery of 'prostaglandins' was based mainly on the finding that extracts of seminal fluid caused a marked and long-lasting fall in the arterial blood pressure of anaesthetized animals of several species (Euler, 1936). The mechnanism of this phenomenon remained obscure until Hedqvist and Brundin (1969) for PGE_1 and Hedqvist (1970a,b) for PGE_2 showed that PGEs act pre-junctionally by depressing the secretion of NA from sympathetic autonomic nerves (cf. Hedqvist, 1977).

6.2.1. Effect on noradrenaline release

The autonomic nervous system

The pre-synaptic component of the effect of PGEs was first demonstrated by Hedqvist and Brundin (1969) and Hedqvist (1969) in the isolated perfused spleen of the cat. It has been shown that both PGE_2 and PGE_1 reduce the overflow of NA evoked by sympathetic nerve stimulation. Evidence for a role of prostaglandins in the regulation of the per pulse release of NA has now been obtained in many organs. Pre-synaptic influence on autonomic neurotransmission has been reviewed by Hedqvist (1973a,b,c), Horton (1973), Stjärne (1975a), and Starke (1977).

The pre-junctional inhibitory effect of PGE_2 also occurs in human vasoconstrictor nerves (Stjärne and Gripe, 1973). Stjärne and Brundin (1976) showed that addition of PGE_2 at concentrations ranging from 10^{-10} to 10^{-6} M resulted in dose-dependent and parallel depression of the secretion of ^{3}H-NA and of the contractile response of the human omental vein (1 Hz, 300 shocks). PGE_2 (10^{-8} M) reduced both the secretion of ^{3}H-NA and the contraction by about 60%. PGEs do not change the neuronal (Blakeley *et al.*, 1969) or extraneuronal (Salt, 1972) uptake of NA, nor the activities of MAO or COMT in tissue homogenates (Bhagat *et al.*, 1972), therefore their inhibitory action is mainly due to an effect on transmitter release.

PGE_1 and PGE_2 reduce not only release evoked by electrical stimulation but also that induced by high extracellular potassium concentrations or by nicotine (Stjärne, 1973c,d; Westfall and Brasted, 1974).

The central nervous system

Since in some tissues the release of prostaglandins of the E series is associated with sympathetic nerve stimulation it was suggested that endogenously released prostaglandins were involved in a trans-synaptic negative feedback mechanism that regulates noradrenaline release (Hedqvist, 1969, 1973a).

Prostaglandin E_2 (Bergström *et al.*, 1973) and PGE_1 (Taube *et al.*, 1977) reduced the release of ^{3}H-noradrenaline caused by electrical stimulation from rat brain slices. The inhibition induced by PGE_1 does not involve α-adrenoceptors or opiate receptors, since these effects are not antagonized by phentolamine or naloxone (Taube *et al.*, 1977). The inhibition of PG synthesis by indomethacin resulted in an increase of ^{3}H-NA release from rat brain slices, a finding made by Bergström *et al.* (1973). However, a positive feedback action has also been observed (Roberts and Hillier, 1976).

6.2.2 Effect on ACh release

The autonomic nervous system

Wennmalm and Hedqvist (1971) have suggested, on the basis of indirect evidence, that PGEs depress cholinergic transmitter release from vagal nerve endings in the pacemaker region of the rabbit heart. This conclusion was later questioned by Hadházy *et al.* (1973), who measured ACh release directly. Moreover, it was reported that PGEs did not affect the outflow of acetylcholine (ACh) from parasympathetic nerve-endings of the guinea-pig ileum (Illés *et al.*, 1974). PGEs have been shown to depress the MEPP frequency at the neuromuscular junction of the frog sartorius muscle (Illés and Vyskocil, 1978) and to produce intermittent failures in the generation of EPP's of the rat diaphragm, indicating a reduced release of the transmitter.

In contrast, Ehrenpreis *et al.* (1973) suggested that in the guinea-pig ileum PGEs couple 'cholinergic nerve terminals excitation with ACh release'. This conclusion was reached because low concentrations of PGE_1 or E_2 reversed the inhibition of electrically induced neurogenic contractions by various substances (morphine, PG antagonists, PG synthesis inhibitors) in isolated preparations. Indomethacin, a PGE synthesis inhibitor, at a concentration of 45 μg/ml, blocked field-stimulated longitudinal muscle contractions of the

154

guinea-pig ileum (Ehrenpreis *et al.*, 1973). It was concluded that the release of ACh should be inhibited in the absence of PGE's. These proposals were not based on measurements of ACh output but, rather, on the end-organ effect of either exogenous PGE or PGE synthesis inhibitors. Kadlec *et al.* (1974) reached the same conclusions, as in their experiments the neurogenic contractions were selectively inhibited by a small dose of indomethacin (1 μM) and restored by the administration of PGE_2 (6 νM). This was partly confirmed by Bennett *et al.* (1975), who found that 40 μg/ml of indomethacin suppressed contractions induced by electrical stimulation at low frequencies by acting not only on nerves because responses to ACh and histamine, which directly stimulate muscle, were also reduced. Thus a selective inhibition of nerve-mediated contractions could not be obtained with the low drug-concentration used by Kadlec *et al.* (1974). The difference seems due to the type of bathing fluid employed by Kadlec *et al.* (1974), since with the 'modified Krebs solution', concentration of Ca_0 may have decreased in the organ bath since the Krebs solution was bubbled with air instead of a CO_2 and oxygen mixture. Bennett *et al.* (1975) also obtained a selective reduction of electrically induced contractions with 0.36 μg/ml indomethacin. Hadházy *et al.* (1973) and Illés *et al.* (1974) provided direct evidence for the release of ACh from parasympathetic nerve terminals of guinea-pig ileum and PGE_1 had no effect on ACh release induced by low-frequency stimulation and under resting conditions. Supporting these data, Hazra (1975) showed that, in the longitudinal muscle strip with attached Auerbach's plexus, neither spontaneously released ACh nor that released by field stimulation is affected by indomethacin (15–45 μg/ml), although PGE_2-like material was released during stimulation of guinea-pig isolated ileum. At variance with these observations, Kadlec *et al.* (1978) have found under similar conditions that the output of ACh was increased by 6 nM PGE_2. This increase was inversely related to the initial level of ACh output, thus it was marked only in the preparations with a low initial output of ACh.

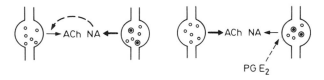

Fig. 48. Indirect effect of PGE_2 on acetylcholine (ACh) release. PGE_2 disinhibits the release of ACh by inhibiting the release of noradrenaline (NA)

Since the stimulated outflow of NA at a frequency of 5 Hz was increased by indomethacin (50 μM) and decreased by PGE_2 (30 nM), it seems likely that the effect of PGE_2 on ACh release is indirect (Fig. 48) and is due to the removal of pre-synaptic inhibitory control. This suggestion seems to be supported by the data of Hársing *et al.* (1979b) in the central nervous system.

The central nervous system

Cerebral ventricular perfusion with prostaglandin E_1 (PGE_1, 1.4×10^{-8} M) reduced the release of acetylcholine (ACh) into the ventriculo-cisternal perfusate of the unaesthetized cat brain (Hársing *et al.*, 1979b). Neither a halving (1.07×10^{-3} M) nor a threefold rise (6.45×10^{-3}M) in the calcium concentration if the perfusate had any effect on the spontaneous release of the cholinergic transmitter. The action of PGE_1 was counteracted by decreasing and potentiated by increasing the extracellular calcium concentration. Polyphloretin phosphate (PPP 10^{-4} g/ml) by itself did not significantly affect the ACh outflow, but it readily antagonized the depressant effect of PGE_1. Ouabain (2.5×10^{-5} M), added to the perfusion fluid, increased the ACh output from the cat brain and PGE_1 prevented this effect. A single intraventricular injection of 6-hydroxydopamine (1 mg) counteracted the inhibitory effect of PGE_1 on spontaneous ACh release (Hársing *et al.*, 1979b). A perfusion with the potent PG synthesis inhibitor suprofen (3.8×10^{-7} M) did not influence the change in the mean output of the transmitter. This data indicates that PGE_1 reduces ACh release in the central nervous system via a pre- or post-synaptic modification of monoaminergic neurotransmission. Monoamines are known to modify the release of ACh from both peripheral and central cholinergic nerve-endings (cf. Vizi, 1979). Inhibition by PGE_1 of cholinergic transmitter outflow may therefore be related to the well-known interference of prostaglandins with the release (Taube *et al.*, 1977), and with the action of monoamines in the central nervous system. This assumption is supported by the finding that 6-hydroxydopamine pre-treatment abolished the effect of PGE_1 on ACh release from the central nervous system of the cat.

Prostaglandin E_2 produces a slight inhibition of the electrically induced release of dopamine from the rat striatum (Bergström *et al.*, 1973). Westfall and Kitay (1977) have shown inhibition of ³H-dopamine release by PGE_1 and PGE_2 but not by $PGF_{2\alpha}$. In agreement with this inhibitory role of prostaglandins, Westfall and Kitay (1977) reported facilitation of ³H-dopamine release in animals pre-treated with indomethacin to inhibit the synthesis of prostaglandins.

6.2.3 Effect on neuroeffector transmission

Noradrenergic transmission

Studies carried out provide indirect evidence that PGE_1 and/or PGE_2 inhibits NA release in several organs—cat spleen, rabbit jejunum, guinea-pig vas deferens—(Baum and Shropshire, 1971; Hedqvist, 1972a, b; Hedqvist and Euler, 1972; Illés *et al.*, 1973; Ambache *et al.*, 1975; Hedqvist and Persson, 1975). No or minimal inhibition of effector responses to nerve stimulation were found in the spleen of the dog (Davies and Withrington, 1968), the

156

nictitating membrane of the cat (Brody and Kadowitz, 1974; Illés *et al.*, 1974), the vas deferens of the rat (Ambache *et al.*, 1972; Hedqvist and Euler, 1972; Illés *et al.*, 1973), or in the isolated rabbit ear artery (Hadházy *et al.*, 1976). Figure 49 shows the effect of PGE₁ on the vasoconstrictor responses of the central ear artery of the rabbit in response to nerve stimulation and noradrenaline. In bovine splenic nerves, PGE₂ does not have any effect on impulse conduction (Hedqvist, 1970b). This finding suggests that its effect is on the axon terminals.

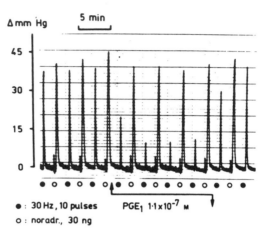

Fig. 49. Effect of PGE₁ on the vasoconstrictor responses of the central ear artery to nerve stimulation and to noradrenaline PGE₁ was added to the organ bath, noradrenaline was injected into the tube near the perfusion cannula. The basic pressure in this vessel was 13 mm Hg. Note that the effect of nerve stimulation was markedly depressed by PGE₁ while the constrictor responses to noradrenaline was not significantly affected
(from Hadházy *et al.*, 1976)

The pre-synaptic inhibitory effect of PGEs is inversely related to the stimulation frequency: the post-synaptic responses were inhibited only in those cases where a low frequency of stimulation was applied (Baum and Shropshire, 1971; Illés *et al.*, 1973; Hedqvist and Persson, 1975; Hadházy *et al.*, 1976).

The effect of PGEs also depends on the number of shocks applied (Illés *et al.* 1973). In the majority of studies, the frequency-dependence of inhibition of an electrically induced response is tested over fixed time, so therefore the pulse number varies.

This can be objected to since in this case pulse number is also varied by altering the rate of stimulation. As a result, a steep frequency-response curve is obtained since more pulses release higher amounts of transmitter. A constant number of 10 pulses at 1, 3, 10 and 30 Hz was used by Hadházy *et al.* (1976) to establish the inhibitory action of PGE₁ as a function of stimulation frequency. It was found that the PGE₁-induced reduction of vasoconstrictor responses to nerve stimulation is dependent on the train length of stimuli.

Cholinergic transmission

It has been shown that prostaglandins stimulate the longitudinal gastrointestinal muscle by acting solely on muscle receptors (Bennett and Fleshler, 1970). PGEs can potentiate the effect of acetylcholine (ACh) and electrical stimulation on the guinea-pig ileum (Baum and Shropshire, 1971; Illés et al., 1973; Kadlec et al., 1974).

The inhibitor of PG synthesis, indomethacin decreases intestinal muscle tone (Ferreira et al., 1972) and inhibits electrically induced contractions of guinea-pig ileum (Ehrenpreis et al., 1973; Kadlec et al., 1974; Bennett et al., 1975; Zséli et al., 1979). However, it was found that pre-treatment of guinea-pig ileum with guanethidine or α-methyl-p-tyrosine prevented the inhibitory effect of indomethacin on electrically induced contractions (Kadlec et al., 1974). Ehrenpreis et al. (1973) had already suggested that PGE_2 enhanced the release of ACh from Auerbach's plexus and Kadlec et al. (1974) had reached a very similar conclusion. On an isolated longitudinal muscle strip of guinea-pig ileum evidence had been obtained showing that the effect of PGEs on cholinergic neuro-effector transmission was post-synaptic (Zséli et al., 1979).

Relatively high doses of PGE_1 contracted the longitudinal muscle; this agrees with previous reports (Horton and Main, 1963; Bennett et al., 1968a, b; Kottegoda, 1969; Bennett and Fleshler, 1970). Lower concentrations of PGE_1, which did not affect muscle tone, enhanced field-stimulated or CCK-induced contractions. PGE or F compounds inhibited sympathetic neuro-transmission in several tissues of different species (Hedqvist, 1970a, 1973d; Illés et al., 1973). If this also occurs in the longitudinal muscle of the guinea-pig ileum, the potentiating effect of PGE_1 may be explained by an inhibition of the noradrenergic depression of ACh release. However, this seems unlikely since reserpine pre-treatment and phentolamine administration did not affect the PGE_1 potentiation of these contractions. A direct PGE_1 potentiation of cholinergic activity is more likely. The site of action of PGE_1 seems post-junctional, since PGE_1 does not affect the overflow of ACh from guinea-pig ileum at rest or during 0.1-Hz electrical stimulation (Hadházy et al., 1973), while responses to exogenous ACh were increased by PGE_1 (Zséli et al., 1979). However, in the presence of atropine, PGE_1 increased spike activity in guinea-pig taenia caecum preparations and depolarized the muscle in a calcium-dependent manner (Vizi et al., 1976; Zséli et al., 1979). PGE_1 may therefore increase the response to ACh by non-selectively sensitizing the muscle membrane to agonists (Bennett et al., 1968a, b). A similar observation was made with non-steroidal anti-inflammatory drugs on histamine-induced contractions of guinea-pig ileum segments (Famaey et al., 1977).

Pre-treatment of the ileal longitudinal muscle with indomethacin significantly reduced both the indirectly and directly induced contractions. The relatively low concentration of indomethacin used and the need for a long contact time both suggested that indomethacin may be inhibiting these contractions by inhibiting PG synthesis. Since use of the adrenergic α-adrenoceptor blocking agent phentolamine or the catecholamine-depleter reserpine failed to significantly influence the inhibitory effect of indomethacin on indirectly induced contractions, it seemed unlikely that there was a link between indomethacin and adrenergic neuroeffector transmission in the longitudinal muscle preparation, as suggested by Kadlec et al., (1974). PGE_1 reversed not only the inhibitory effect of indomethacin on contractions of the longitudinal muscle layer but also that of morphine and noradrenaline or clonidine (Zséli et al., 1979). These results support those of Chong and Downing (1974) and refute the suggestion of Ehrenpreis et al. (1973) that the reversal effect of PGE_1 may be via a non-selective sensitization of the muscle cell membrane to ACh.

Since cholecystokinin-induced contractions are mediated via the release of ACh from cholinergic nerves of the gastrointestinal tract (Vizi et al., 1972, 1973a), and since these contractions are potentiated by small doses of PGE_1 and antagonized by indomethacin, it is possible that endogenous PGs may be involved in the CCK-induced contractions of guinea-pig ileum.

Using the sucrose-gap technique, PGE_1 depolarized the muscle in a concentration-dependent manner and this was accompanied by an increased number of spike discharges and an elevation of muscle tone. Atropine did not affect this action (Zséli et al., 1979).

6.2.4 Source of prostaglandins

As first proposed by Hedqvist (1970a), prostaglandins are formed during noradrenergic transmission, probably in post-synaptic effector cells, and they depress further secretion of NA (trans-synaptic feedback). Although it has been suggested (Stjärne, 1972, 1973d) that prostaglandins formed upon nerve stimulation are neural in origin, the major source of the prostaglandins are probably not the nerve fibres but the extraneuronal, mainly effector, cells (Fig. 48). This is shown by the finding that: (1) not only nerve impulses but also exogenous catecholamines induce prostaglandin biosynthesis (Gilmore et al., 1968; Wennmalm, 1975); (2) α-adrenolytic drugs prevent the formation of prostaglandins caused by either nerve stimulation or exogenous catecholamines (Davies et al., 1968); (3) after noradrenergic nerves have been destroyed by surgical denervation (Gilmore et al., 1968) or 6-OH-DA (Junstad and Wennmalm, 1973) the release of prostaglandins induced by exogenous NA is not reduced.

6.3 NORADRENALINE

Catecholamines are able to depress and facilitate ganglionic transmission apparently mediated via α_2 and β_2 adrenoceptors (cf. Brown and Caulfield, 1981). Brown and Medgett (1982) found that adrenaline ($10^{-5} - 10^{-4}$ M) hyperpolarized the tissue and inhibited the compound action potential evoked by supramaximal pre-ganglionic stimulation (0.2 Hz) of the desheathed rat superior cervical ganglia.

Phentolamine (1 μM) itself significantly enhanced the height of the action potential by $11\pm2\%$ (SEM, $n=13$). It was therefore suggested that with continuous stimulation NA is released from the post-synaptic site, from the dendrites of the post-ganglionic neurones (Martinez and Adler-Grashinsky, 1980) or from other sites (Noon *et al.*, 1975), and that it acts trans-synaptically, inhibiting the release of ACh from axon terminals via α_2-adrenoceptor stimulation.

Chapter 7

Physiological, pharmacological, and clinical implications

Pre-synaptic inhibition is a very important form of interaction between two neurons and plays a critical role in the modulation of transmission. This type of inhibition of chemical transmission results from a decrease of transmitter/modulator release. There are two types of pre-synaptic modulation:

(1) Synaptic, when an inhibitory input makes synaptic contact with an axon terminal or the initial segment of the cell and the inhibitory modulator released inhibits the release of the principal transmitter;

(2) Non-synaptic, when a substance (modulator) is released from an axon terminal and/or varicosity devoid of synaptic characteristics, but it is able to reach the target cell.

The modulators exert their actions via different receptors. If there is no direct contact between two or more neurons, transmitters might reach their target cell by diffusion. This remote control of neuronal activity can be monitored only by those cells that are equipped with receptors sensitive to modulator (discrimination).

It seems unlikely that intercellular communication in both the peripheral and the central nervous system is limited to synaptic transmission mediated by transmitters. Modulators released (1) from non-synaptic axon terminals, (2) from axon terminals making synaptic contact but acting both on synaptic and non-synaptic target cells, or (3) from regions of the nerve cell other than the pre-synaptic axon terminal (dendrites, axon, soma) should play a physiological role in the modulation of neurochemical transmission.

It has been concluded (cf. Vizi, 1979, 1980b, 1981a,b, 1983) that the non-conventional interactions between neurons might be of physiological importance. The idea that a modulator could be involved in chemical neurotransmission in systems that do not have clearly defined anatomical synapses was discussed by Nicoll and Alger (1979). Pickles and Simmonds (1976) found that in the mammalian lateral olfactory tract pre-synaptic fibres may have GABA receptors, even though they do not receive axo-axonic synapses. Electrical stimulation of the pre-synaptic fibres can produce activation of these receptors, presumably due to the diffusion of GABA from remote axon terminals. These findings raised the possibility that pre-synaptic inhibition might occur in regions which do not possess axo-axonic synapses. Subsequently, much morphological (see sections 3.1 and 3.2) and neurochemical (Chapters 2 and 4) evidence has been provided showing that there are free varicose axon terminals which do not make synaptic contact and which are able to transmit signals (Fig. 50).

The trichotomy of neurotransmitter versus neuromodulator versus neuro-

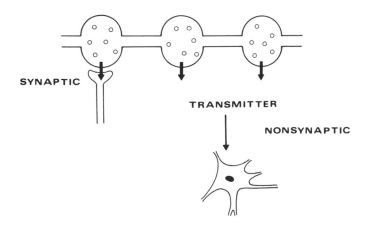

Fig. 50. Two possibilities for transmitting signals: (1) synaptic, (2) non-synaptic

hormone mainly depends on the function of the substance. For example, NA can be regarded as a synaptic and as a non-synaptic modulator (e.g. in rabbit pulmonary artery, where the cleft between axon terminal and effector cell is greater than 1000 nm) and possibly as a neurohormone, which acts after transport via the cerebrospinal fluid.

There are a large number of peptides that possess effects both on the central nervous system and the autonomic nervous system. Peptides can act locally on target cells by (1) release from nerve terminals, (2) release from paracrin cells, and (3) release from endocrin cells (Fig. 51). Therefore neuropeptides can serve an endocrine or paracrine role, they can function as a transmitter and as a modulator, they can exert an inhibitory effect on an axon terminal (e.g. enkephalin) but can stimulate cell bodies, thus enhancing transmitter release and having a non-synaptic transmitter role (e.g. cholecystokinin).

There are widely used drugs in clinical practice which have been developed or applied on the concept of pre-synaptic modulation of neurochemical transmission. Receptors located pre-synaptically or pre-junctionally are the targets of different drugs having either an agonist or antagonist activity. There is no direct evidence for this so far, but it is highly probable that the sensitivity of those receptors, is much higher which are not located in a narrow synaptic gap. They are, therefore, much more accessible to drugs and a much lower concentration of drug is needed to affect them.

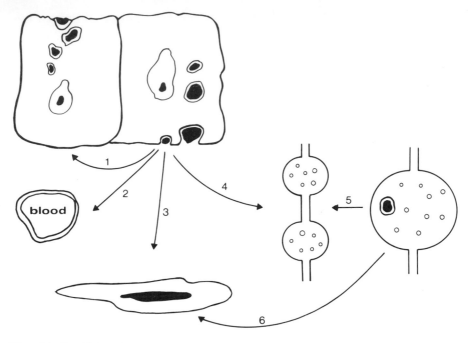

Fig. 51. Possible release and target sites of peptides. (1) Paracrin function; (2) hormonal effect; (3) paracrin function; (4) effect on nerve cell; (5) released from axon terminal and effect on neuron: (6) released from axon terminal and effect on e.g. smooth muscle

Table 15. Classification of receptors which play a role in chemical neurotransmission and drug action

Classification of receptors	Types of receptors		Criteria
Anatomical	(1) Pre-synaptic (pre-junctional)	Intrasynaptic Extrasynaptic	Based on localization
	(2) Post-synaptic (post-junctional)		
Pharmacological	α_1, α_2, β_1, β_2, D_1, D_2, P_1, P_2, H_1, H_2, etc.		Based on kinetic study with relatively selective drugs having agonist or antagonist activity
Functional	(1) Of physiological importance		Target of endogenous substances
	(2) Of physiological and pharmacological importance		Receptors are targets of both endogenous and exogenous substances
	(3) Of pharmacological importance (usually silent receptors)		Accessible only for drugs

Receptors involved in chemical neurotransmission and drug action can be classified (Table 15) (1) morphologically (anatomically), (2) pharmacologically, and (3) functionally.

Many studies have been published on the binding property of tissues in which the authors claim that they are dealing with receptor sites. However, binding sites can be equated with receptor sites only when the former can be shown to be associated with a physiological event. This is important to emphasize, since a number of ligand-binding assays have led to the detection of binding sites which may have no physiological relevance. Therefore the classification cannot be based simply on binding studies.

There are intrasynaptic receptors which are difficult for drugs to reach and, being relatively insensitive, drugs are able to affect them only in toxic concentrations. These receptors, are not target cells for remote (non-synaptic) modulation. However, there are receptors (rudimentary) which are not used by endogenous substances (transmitters, modulators, etc.) but which might be affected by relatively low concentrations of drugs. An example is the pre-synaptic (pre-junctional) dopamine receptor on noradrenergic nerve-endings. It has been proposed that noradrenergic nerve-endings are endowed with sites resembling dopamine receptors. Stimulation of these receptors by exogenous dopamine-receptor agonists depresses stimulation-induced noradrenaline release and the responses of the effector cells. Both effects are unaffected by phentolamine but are reversed by dopamine-receptor antagonists, which differentiates them from pre-junctional α_2-adrenoceptors. According to the pre-junctional receptor theory, facilitation of transmitter release by blockade of the receptor is an essential consequence of the interruption of a functioning inhibitory system. Several studies have shown, however, that dopamine-receptor antagonists do not facilitate noradrenaline efflux during electrical stimulation, neither do they increase the size of the post-junctional response. This implies that dopamine is not involved in the regulation of noradrenaline release under physiological conditions (Enero and Langer, 1975b; Hope et al., 1977; Dubocovich and Langer, 1980), but that dopamine receptors located pre-junctionally can be used as drug-receptors.

7.1 PHYSIOLOGICAL IMPLICATIONS

The non-conventional interaction between neurons, and the non-synaptic modulatory role of different amines and peptides (NA, DA, 5-HT, enkephalins, etc.) in neurochemical transmission appears to be feasible: anatomical, neurophysiological, neurochemical, and pharmacological evidence supports this suggestion. The non-synaptic control of chemical neurotransmission by different modulators released from axonal varicosities lacking junctions might play a physiological role both in the central nervous system and in the neurovegetative system in shaping emotion, behavior, or learning processes, or in controlling the balance between the parasympathetic and sympathetic nervous systems.

Since the release from varicosities devoid of typical synaptic specialization has long been known (e.g. the peripheral nervous system) the question arises as to whether NA, DA, and 5-HT released from non-synaptic varicosities can actually reach their target cells and supppress neuronal firing, and inhibit the release of genuine transmitters. All available neurochemical and electrophysiological evidence seems to favour this hypothesis. First, the surgical or neurochemical removal of the inhibitory pathway results in an enhanced release of the principal transmitter (see Chapters 3 and 4). Second, the stimulation of an inhibitory pathway results in a reduced release of the principal transmitter. Third, administration of the modulator leads to a reduction of the genuine transmitter release and of neuronal firing. In the cortex, for example, endogenous NA can probably be released from all varicosities of noradrenergic neurons, not only from the small proportion which make synaptic contact. It is thus likely that NA does not exert its effect solely on a 'one-to-one' basis, on restricted areas of post-synaptic specializations, but can diffuse to relatively distant targets. In this respect there is a striking resemblance between the autonomic transmitter and central modulators (NA, 5-HT, neuropeptides, etc.). The long-distance message can be monitored by cells equipped with receptors sensitive to modulators (discrimination).

7.1.1 Noradrenaline

Analysis of the actions of NA on the myenteric plexus suggests that the primary site of action of the amine is not on ganglion cells. Usually these show no change in membrane potential in response to the application of NA (Holman *et al.*, 1972). Instead, it reduces the size of the cholinergically mediated fast EPSP (Nishi and North, 1973a). In addition, it has been shown (Hirst and McKirdy, 1974) that sympathetic stimulation suppresses the EPSP without affecting the electrical properties of the post-ganglionic neurons.

Gillespie and Khoyi (1977) have shown that lumbar sympathetic stimulation inhibits the responses of rabbit isolated colon to pelvic parasympathetic stimulation. Phentolamine, but not propranolol, prevented the inhibitory effect of sympathetic stimulation on contractions induced by pelvic stimulation. In addition, the findings that in vagus–oesophagus and vagus–stomach (Vizi, 1974a) preparations the contractions resulting from lower frequencies of stimulation (0.1–5 Hz) were inhibited by low concentrations (10^{-8} to 5×10^{-6} M) of NA without influencing the effect of ACh added to the bath. This indicates that in the whole gastrointestinal tract, including the oesophagus, the sympathetic transmitter, NA, controls the cholinergic nerve-effector transmission by reducing the ACh release pre-synaptically.

The interaction between noradrenergic (sympathetic) and cholinergic enteric transmission in the gastro-intestinal tract seems to be non-synaptic (cf. section 3.1) since the projecting noradrenergic fibres exert an inhibitory modulatory effect on ACh release without making any close synaptic contact. The release of transmitters from nerve-endings devoid of synaptic membrane specialization in the peripheral autonomic nervous system has long been known. Vizi, 1974d suggested that NA released from varicosities reached its target cells by diffusion: '. . . There is morphological and histological evidence . . . that the adrenergic nerve terminals terminate not far from the cholinergic ones. This enables the released noradrenaline to reach the cholinergic nerve ending by diffusion . . .'

The cholinergic transmission of the autonomic nervous system is modulated in the following ways (see Fig. 52):

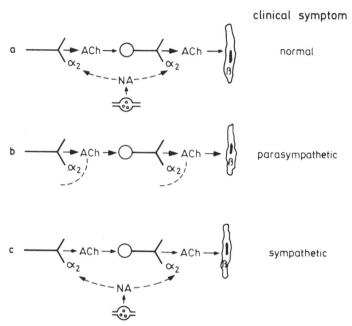

Fig. 52. Non-synaptic interaction between cholinergic and noradrenergic nerves in the gastrointestinal tract. NA released from the sympathetic nerves continuously controls the release of ACh from Auerbach's plexus: (a) when the noradrenergic control is suspended, cholinergic transmission is enhanced resulting in a parasympathetic tone (diarrhoea); (b) however, when the noradrenergic outflow is increased (shock, stress, etc.) the release of ACh is inhibited, producing symptoms of sympathetic dominance (obstipation, paralytic ileus, etc.). (c) (Vizi, 1979). (Reproduced by permission of Pergamon Press Ltd)

(1) Partly opposed by NA released from the adrenergic axon terminals. In addition, circulating catecholamines might also be effective. This happens under normal conditions.

(2) Unopposed by NA. The cholinergic transmission is increased when the 'brake' is off. The partial removal of the adrenergic restraint (e.g. depletion of NA by reserpine, by 6-OH-dopamine, or by preventing the inhibitory effect of NA by α-blocking agents on the α_2-receptors of nerve terminals) leads to an enhanced ACh release, producing typical parasym-

pathetic symptoms: diarrhoea, salivation, bradycardia, increase of gastric secretion, etc.

(3) Opposed by increased adrenergic activity. When the transmission in the cholinergic autonomic nervous system is reduced, a lack of parasympathetic tone appears, resulting in sympathetic symptoms. In addition, the direct effect of catecholamines on the effector cells (e.g. in the heart) also plays an important role in producing sympathetic tone.

When the noradrenergic outflow is increased (shock, stress, etc.) the cholinergic transmission is partly or completely inhibited, producing the symptoms of sympathetic dominance (e.g. constipation, paralytic ileus). Although evidence for the pre-synaptic interaction between nerves has been obtained recently, the notion that the sympathetic inhibitory nerves act on cholinergic neurons is a nineteenth century idea. In 1858 Lister claimed that 'the inhibitory influence does not operate directly on the muscular tissue, but on the nervous apparatus, by which its contractions are, under ordinary circumstances, elicited'.

Sympathetic control of ACh output can be viewed as a kind of pre-synaptic inhibition and, when compared with an antagonism at the effector level, offers the physiological advantage of economy in transmitter release.

A similar interaction has been observed in the cerebral cortex; evidence has been obtained showing that the coerulocortical noradrenergic fibres have a tonic inhibitory effect on cortical ACh release: the removal of noradrenergic input by using phentolamine, 6-hydroxy-dopamine pre-treatment or by an ipsilateral locus coeruleus lesion, results in an increase of ACh release (Vizi, 1974a, 1981). The electro-physiological findings that the administration of iontophoretic NA or the stimulation of the locus coeruleus (Phillis and Kostopoulos, 1969) inhibits the neuronal firing in the cortex are consistent with neurochemical observations. In the cortex, where 95% of the noradrenergic varicosities are devoid of synaptic contact (Beaudet and Descarries, 1979), this type of modulation of ACh release is highly probable. Since it has been stated that a relative paucity of synaptic junctions may be a feature common to all types of central monoaminergic terminals (Beaudet and Descarries, 1978), it seems likely that noradrenaline, in general, may act as a 'non-synaptic' modulator, rather than as a phasic transmitter in the brain.

It seems very likely that, due to non-synaptic inter-neuronal communication between axon terminals and varicosities, the NA concentration at the vicinity of pre-synaptic α_2-adrenoceptors may reach levels considerably higher than expected, solely on the basis of release from a single varicosity (Baumann and Koella, 1980).

7.1.2 Dopamine

The question arises as to whether specific pre-junctional dopamine receptors have any physiological significance. Rand et al. (1980) suggested that 'If . . . dopamine is released and activates an inhibitory feedback mechanism it may have an important physiological role to play in transmitter conservation, especially during periods of high sympathetic nerve activity'. This idea is supported by the finding that noradrenergic nerves in the rabbit ear artery can be converted into dopaminergic nerves (Hope *et al.*, 1979).

Clark and Menninger (1980) have presented evidence indicating a dopaminergic component in the sympathetic innervation of mesenteric blood vessels in the dog. They showed that electrical stimulation of the greater splanchnic or peri-arterial mesenteric nerves in animals pre-treated with phenoxybenzamine or guanethidine resulted in frequency-dependent increases in blood flow in the cranial mesenteric artery. The increases in flow were resistant to atropine and propranolol but were attenuated or abolished by haloperidol. The transmitter mediating the vasodilator responses was therefore considered to be dopamine.

One of the best-documented peripheral actions of dopamine is its influence on transmission of nerve impulses through sympathetic ganglia. Histological evidence has demonstrated the presence of small, intensely fluorescent cells (SIF cells) within sympathetic ganglia which give rise to processes that impinge on the post-ganglionic cells. These small interneurons contain dopamine which, released by pre-ganglionic muscarinic excitation of the SIF cells, acts on receptors present in the ganglion.

Electrophysiological and neurochemical evidence indicates that dopamine (DA) released from nigro-striatal axon terminals controls the axonal firing and the release of ACh from cholinergic interneurons of the striatum (see section 4.3.1). The removal of the dopaminergic input leads to enhancement of ACh release. In addition, it has been shown (Vizi and Volbekas, 1980a,b) that DA reduces the release of various neurohypophysial hormones from the intermediate posterior lobe of the rat pituitary and removal of dopaminergic input enhances their release, indicating a continuous control of hormone release.

7.1.3 Acetylcholine

The inhibitory muscarinic receptors can no longer be viewed as being only of pharmacological interest as targets for drugs. The stimulation of the vagus nerve decreases the amount of NA released from the isolated perfused rabbit atria during sympathetic nerve stimulation. ACh is usually rapidly destroyed by cholinesterases after its release. Because of the very low acetyl- and butyryl-cholinesterase activity of the heart muscle, it is conceivable that some of the ACh released by vagal stimulation will survive long enough to exert a modulatory function. Even so, there is very little chance that following release ACh would survive long enough to diffuse any distance without being destroyed, so any modulatory function would require an axo-axonic contact. In this respect, acetylcholine is not an ideal modulator. However, muscarinic receptor stimulation results in a reduction of NA release from the heart, the perfused cat spleen, the rabbit ear artery and the dog cutaneous vein. The data available on the cortex is rather controversial. Westfall (1977) showed that dopaminergic axon terminals are also equipped with inhibitory muscarinic receptors.

Evidence has been obtained showing that, even in the absence of cholinesterase inhibition, muscarinic blocking agents are able to enhance ACh release, an observation which strongly suggests that the released ACh can affect the release of ACh (see section 5.3). This phenomenon might also have physiological importance. It must be conceded that the exact localization of muscarinic autoreceptors has not been directly proven. Nevertheless, it should be mentioned that enhanced ACh release, in response to muscarinic receptor blockade by atropine or hyoscine, could be explained by inhibition of post-synaptic muscarinic receptors whose activation results in a release of a substance with a 'trans-synaptic' (see section 6.1) inhibitory effect on ACh release from axonal varicosities.

7.2 SITE AND MODE OF ACTION

7.2.1 Site of action

It is well known that there are axons both in the central nervous system and in the autonomic nervous system which form varicose (*boutons-en-passant*) branches. It is highly probable that varicose axon terminals are the target of pre-synaptic modulation (Fig. 53).

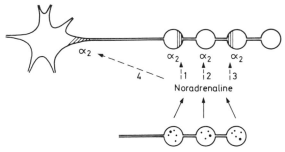

Fig. 53. Possible site of action of a modulator (noradrenaline) on a neuron (different hypotheses): (1) inhibits action-potential generation in the varicosity hillock (Alberts *et al.*, 1981; Stjärne, 1981); (2) inhibits secretion-coupling (cf. Starke, 1977; Langer, 1981); (3) inhibits conduction by sudden increase of diameter (E.S. Vizi, this book); and (4) axon hillock (Vizi, 1981a,b)

A substance released in or diffused to the vicinity of the axon terminal can modulate the release of the principal transmitter or of another modulator provided that the axon is equipped with sensitive receptors.

Mainly isolated strip- or slice-preparations have been used for studying transmitter release and its modulation. In preparations where only the axon and axon terminals are present (e.g. the isolated pulmonary artery, the vas deferens, etc.) it is likely that the site of action is on the axon terminals. However, in preparations in which the cell body is also present (e.g. isolated longitudinal muscle strip with Auerbach's plexus) it is very difficult to assess the site of action of a modulator. In those studies where the release was induced by field stimulation the site of action of an inhibitory modulator might be on the axon terminals, since using this stimulation technique not only the cell body but also the axon is stimulated.

Effect on axon hillocks

When a modulator is released from a remote non-synaptic varicosity it can reach the target cell by diffusion and exert its inhibitory effect on the axon hillock (Fig. 53). A modulator released in or diffused to the vicinity of the axon hillock might prevent initiation of firing in two ways: (1) hyperpolarization of the membrane, e.g. by stimulating (Na^+, K^+)ATPase, which leads to hyperpolarization, driving the potential away from the threshold, or (2) increasing conductance, thereby reducing excitability and initiation of action potentials. The best strategic locus for the modulation of transmitter release is the axon hillock (Vizi, 1979, 1981a,b) where action-potential generation takes place.

Stimulation of the sodium pump reduces the probability of action-potential generation and consequent transmitter release from post-ganglionic terminals without changing the sensitivity of receptors to transmitters. This might be why the release of transmitter in response to ganglionic or to pre-ganglionic stimulation is more readily blocked by exogenous NA than release induced by field stimulation of the axon terminals. Thus it is easy to explain why NA is able to inhibit cholinergic transmission on the vagus stomach preparation when a high frequency of stimulation is applied pre-ganglionically (Vizi, 1974a). In this preparation, cholinergic transmission is interrupted at least once before the impulses arrive at the target smooth muscle cell. Thus NA can act on the nerve terminals, and at the site of action-potential generation on the cell body of the post-ganglionic neuron.

It is possible that when an inhibitory modulator acts on the axon terminals its effect is frequency-dependent, but when it acts on the initial segment of the axon the effect becomes frequency-independent.

Effect on varicosities

The varicose axon is the most probable site of modulation, since axonal conduction cannot be modulated by transmitters or modulators. There is no convincing evidence that modulators (NA, 5-HT, etc.) inhibit the release of labelled transmitters from synaptosomal preparations. The inhibition by modulators of potassium-induced transmitter release would be a strong evidence of an axon terminal effect, since all varicosities arc depolarized and transmitter is released from each varicosity. Nevertheless, it is true that transmitter release evoked by high potassium concentrations cannot be inhibited.

7.2.2 Mode of action

The essential mechanism of transmitter release (Brown *et al.* 1982) is as follows:

(1) Nerve terminal depolarization opens membrane Ca-channels within about 700 μs.

(2) The inward Ca-current (I_{Ca}) increases internal Ca^{2+} by a factor $> 10^2$ within about 200 μs. The increase in Ca-conductance (G_{Ca}) produced by the membrane depolarization follows fifth-order kinetics. The relationship between pre-synaptic current and post-synaptic voltage is linear, and there is a logarithmic relationship between pre-synaptic depolarization and post-synaptic response. This means that a relatively small decrease in the pre-synaptic action potential may lead to a substantial reduction in I_{Ca} and hence in the amount of transmitter release.

(3) Internal Ca^{2+} accelerates spontaneous quantal release by a factor of about 10^5. It has been shown that the inward Ca-current is subject to some inactivation by intracellular Ca^{2+}. There may also be an element of voltage-dependent inactivation. Such effects provide mechanisms whereby long-lasting pre-synaptic depolarization or a sustained elevation of Ca^{2+}-influence can depress I_{Ca} and so impair transmitter release.

(4) Transmitter released from the axon terminal diffuses rapidly across the synaptic cleft, within about 1 μsec for a 20-nm cleft (i.e. with the speed of about 7.2 km/h).

Taking into account the above considerations, the release of the transmitter may be modulated or inhibited by the following mechanisms:

Reduction of pre-synaptic depolarization

Reduction of action potential amplitude

Failure of terminal invasion in those terminals into which action potentials propagate

Reduced electrotonic spread in terminals where release is mediated by passive invasion

Impairment of I_{Ca} by Ca^{2+}, by Ca^{2+} channel blockade or modification, or by Ca^{2+} inactivation.

There are discussions in the literature on the site of action which, in fact, are very relevant to the problem of the mode of action. Many authors support the idea that pre-synaptic modulation is a question of secretion-coupling. However, others (Stjärne, 1978, 1981; Vizi, 1981a,b) suggest that mechanisms related to failure of varicosity invasion are responsible for pre-synaptic inhibition.

Effect on secretion-coupling

There is the suggestion, mainly based on studies with synaptosomal preparations and potassium-induced release, that the site of action is on the axon terminals. Therefore the authors propose that secretion-coupling is affected by the modulators (Starke, 1977; Langer, 1981a). According to Langer (1981a), the '. . . presynaptic inhibition of noradrenergic neurotransmission through α-adrenoceptors may be mediated through a depression of the electrosecretory coupling rather than by restricting the entry of nerve impulses into the more distal parts of the nerve terminals (i.e. depressing the recruitment of varicosities)'.

Reduction of Ca-influx

There is an hypothesis, advanced on the basis of neurochemical evidence, that upon stimulation the release of transmitters from axonal varicosities may be inhibited, causing an inhibition of Ca^{2+} availability for the stimulus secretion-coupling process. This hypothesis is largely based on indirect evidence indicating, for instance, that the efficiency of the pre-synaptic modulator is inversely related to the external Ca^{2+} concentration (De Langen and Mulder, 1980; Göthert and Wehking, 1980). Ribeiro *et al.* (1979) have shown that adenosine (0.1 to 2 mM) and ATP (0.2 mM) decreased the uptake of ^{45}Ca by potassium-depolarized synaptosomes prepared from rat cerebral cortex. Although these concentrations are high, the results support the view that adenine inhibition is mediated by a decrease in calcium entry into nerve-endings. However, the inhibitory effect of the adenine nucleotides was not reduced by a calcium excess (5 mM) (Vizi and Knoll, 1976).

The modulatory effects of morphine, noradrenaline, and phentolamine on ^3H-noradrenaline release were diminished in the presence of 4-aminopyridine. One possible explanation of this is that pre-synaptic modulation of ^3H-noradrenaline release is brought about by an alteration of the Ca^{2+} influx through the nerve terminal membrane. Another possibility is that pre-synaptic modulation involves a change in Ca^{2+} utilization by the stimulus secretion-coupling process subsequent to Ca^{2+} entry. The latter possibility is supported by the findings of Schoffelmeer and Mulder (1982).

Although Stjärne (1978) and Stjärne *et al.* (1979) gave little support to the notion of secretion-coupling inhibition by α_2-adrenoceptor stimulation, Alberts *et al.* (1981) suggested that its '. . . main action is to restrict invasion, but it appears also to depress depolarization–secretion coupling although to a smaller extent'. The role of Ca_o has now been re-evaluated by Stjärne (Alberts *et al.*, 1981) who suggests that a rise in external Ca apparently results in 'recruitment' of varicosities.

Two mechanisms are critical in the modulator-induced inhibition of secretion-coupling:
(1) The action on the Ca-channel, directly inhibiting the Ca-influx, or
(2) Actions on voltage-dependent Ca-conductance.

Effect on the Ca-channel

Modulators (NA, 5-HT, adenosine, etc.) might inhibit the entry of Ca^{2+} but the evidence available is very indirect simply because the voltage-dependent calcium current (I_{Ca}) cannot be directly measured in the autonomic nerve terminals, where the effects of modulator are most pronounced.

It has been shown that the α_2-type of sympathetic terminal receptors are also present on sympathetic neuron somata (Brown and Caulfield, 1979). Brown and his colleagues, used electrophysiological techniques, to analyse the effect of NA more directly (Brown and Medgett, 1982). It has been shown that there are two effects of NA on mammalian neuron somata (Brown and Caulfield, 1981; Horn and McAfee, 1980; Ivanov and Skok, 1980): a very small (< 5 mV) membrane hyperpolarization and a change in spike configuration, suggesting a reduction in I_{Ca}.

This effect is independent of voltage, i.e. the number of available Ca-channels changed rather than their voltage-sensitivity. If (and this is still only speculation) it is occurring at terminals, this would clearly explain the inhibition of transmitter release.

Using microelectrode techniques Galvan and Adams (1982) showed that noradrenaline acts by reducing the number of available calcium channels in rat sympathetic neurons.

They had previously shown (Adams and Galvan 1981) that NA produced a reversible depression of I_{Ca} in isolated rat sympathetic neurons by direct measurement using a voltage clamp.

Effects on voltage-sensitive Ca conductance

Recently, Ca-dependent potentials have been characterized in the mammalian nervous system. The Ca-dependent potential is a regenerative spike that cannot be produced in the absence of Ca_o, but can be evoked in the absence of Na_o and in the presence of TTX. Noradrenaline reversibly inhibited both the rate of rise (dv/dt) and the amplitude of the Ca-spike (MaAfee *et al.*, 1981). These data are consistent with the hypothesis that NA inhibits an inward Ca-current that generates the Ca action potential. The order of increasing potency is: isoprenaline $<$ dopamine $<$ 1-noradrenaline $<$ adrenaline. The EC_{50} value of adrenaline is 100 nM. An α-adrenoceptor antagonist, phentolamine, inhibited the effect of adrenaline (McAfee *et al.*, 1981). These findings indicate a possible role of the α_2-adrenoceptor in the inhibition of Ca-conductance.

Fischbach and co-workers (Dunlap and Fischbach, 1978; Dunlap and Fischbach, 1981; Horn and McAfee, 1980) have obtained evidence showing that modulators such as NA, GABA, and enkephalin can also reduce voltage-dependent calcium current in vertebrate neurons. Both the hyperpolarization and the effects on action-potential configuration are mediated by α_2-receptors (Brown and Caulfield, 1979; McAfee *et al.*, 1981). It will be interesting to see whether the pharmacological profile of the calcium current inhibition (Galvan and Adams, 1981) will turn out to be also that of α_2-receptors.

Involvement of cyclic nucleotides: inhibition of transmitter release

The presence of an α-adrenoceptor-mediated calcium-dependent pre-synaptic mechanism for the generation of cyclic GMP has been described in the rat pineal gland (O'Dea and Zatz, 1976). The effect of α-adrenoceptor agonists on cyclic GMP levels is lost after sympathetic denervation in the rat pineal gland. There is also evidence for a pre-synaptic α-adrenoceptor-mediated regulation of the potassium-evoked release of noradrenaline (Pelayo *et al.*, 1977). These authors reported that exposure to dibutyryl cyclic GMP reduced the potassium-induced release of ^{3}H-noradrenaline from the rat pineal gland, but failed to modify the release of the tritiated transmitter induced by tyramine. In addition, a decrease in the stimulation-evoked release of ^{3}H-noradrenaline was observed during exposure to an inhibitor of the cyclic GMP specific phosphodiesterase. These results suggest that in the rat pineal gland cyclic GMP may be a link in the chain of events following activation of pre-synaptic α-adrenoceptors, leading to a reduction in the release of noradrenaline. It was suggested that inhibition of neurotransmission through the activation of pre-synaptic inhibitory α_2-adrenoceptors may be linked to a decrease in cyclic AMP levels (cf. Wikberg, 1979).

Muscarinic agonists acting at autoreceptors inhibit release of acetylcholine from hippocampal cholinergic nerve-endings in the rat. This effect could be mimicked by permeant cGMP analogues, 8-Br-cGMP and dibutyryl-cGMP, or by incubation with nitrosoamines which activate the guanylate cyclase (Bártfai, 1980, 1982; Nordström and Bártfai, 1981).

Facilitation of transmitter release

Catecholamines have previously been reported to produce an α-mediated hyperpolarization (Brown and Caulfield, 1979) and a β-mediated elevation of cyclic AMP (Lindl and Cramer, 1975) in rat superior cervical ganglia. Changes in extracellular DC potential were recorded from desheathed freshly isolated rat superior cervical ganglia. Catecholamines produced small (0.4 mV) reversible, concentration-dependent changes in potential. Adrenaline and noradrenaline (1–100 μM) produced a hyperpolarization. Isoprenaline (1–100 nM, mean EC_{50} 3.4 nM) and salbutamol (0.01–1 μM, EC_{50} 50 nM) produced only depolarizations, which were noticeably more delayed and prolonged than the α-hyperpolarization. Isoprenaline-induced depolarizations were antagonized by (−)propranolol (pA_2 8.9), (+)practolol (5.1), and (+)butoxamine (7.4), but not at all by (+) propranolol at 0.1 μM, suggesting that β_2-adrenoceptors were involved. Isoprenaline (0.01–1 μM, mean EC_{50} 28 nM) produced a dose-dependent elevation of cAMP (up to thirty times after 10 min incubation). This effect was antagonized by (+)practolol and (+)butoxamine, at concentrations comparable to those required to antagonize the β-depolarizations (Brown and Caulfield, 1979).

In the guinea-pig vas deferens 8-Br-cAMP enhances the electrical stimulation-induced release of noradrenaline (Stjärne *et al.*, 1979). This action mimics the effect of β-receptor agonists. The facilitatory action of β₂-agonist adrenoceptor stimulation on transmitter release is as yet little understood. Krnjevic and Miledi (1958) provided evidence that adrenaline facilitated high-frequency transmission at the neuromuscular junction by reducing the incidence of 'branch-point failure' and so increasing the number of nerve impulses that invade the terminal bouton.

In sympathetic neurons, β-agonists produce a membrane depolarization in contrast to the hyperpolarization produced by α-agonists (Haefely, 1969). This may be related to the effect described by Krnjevic and Miledi (1958). Isoprenaline also depolarizes pre-ganglionic nerve-endings in sympathetic ganglia and facilitates cholinergic pre-ganglionic transmission at this site. This effect is mimicked by dibutyryl cAMP (Brown *et al.*, 1982). It can therefore be concluded that β-agonists produce a cAMP-mediated increase in I_{Ca}.

Mechanisms related to invasion of varicosities

Another mechanism of pre-synaptic inhibition is the reduction of the safety factor for terminal varicosity invasion by the axonal nerve impulse (Stjärne, 1981). The release of NA from sympathetic nerve terminals by the invading nerve impulse is a very 'chancy' process, with a high proportion of 'failures' at any individual varicosity (Blakely and Cunnane, 1979), so that any procedure marginally reducing the chance of invasion may have profound effects.

Direct evidence for invasion control is as yet lacking. Nevertheless, indirect support for this notion has been obtained, and is applicable to both noradrenergic (Holman and Surprenant, 1980) and non-noradrenergic (cf. Levy, 1980) release. It has been proposed that control of invasion may be a major mechanism in α-adrenoceptor-mediated autoinhibition of noradrenaline secretion in the guinea-pig vas deferens (Stjärne, 1978; Stjärne, 1981; Alberts *et al.*, 1981) and in ACh release from the Auerbach's plexus (Vizi, 1981a,b). The individual varicosity in the guinea-pig vas deferens is mostly 'turned off' by α-autoinhibition, by analogy with the finding that varicosities releasing NA are mostly 'silenced' via activation of α-adrenoceptors (Blakeley and Cunnane, 1979). The enhancing effect of an α-blockade on electrically induced secretion is thus assumed to be due in part to 'turning on' previously 'silent' varicosities. The electrically induced secretion was more strongly depressed by α-autoinhibition than by high (80 μM) potassium (threshold concentration, 60 μM; cf. Stjärne, 1978, 1980, 1981; Alberts *et al.*, 1981).

Using electrophysiological techniques, Blakely *et al.* (1982) succeeded in showing the release of transmitter from single release sites in the sympathetic neuro-effector junction of the rodent vas deferens: α-adrenoceptor agonists and antagonists produced, respectively, left and right shifts in the amplitude distributions of discrete events, with no change in the amplitudes. There is, however, no evidence of an inhibitory relationship between either discrete events induced by successive stimuli, or early and late discrete events following a given stimulus. Blakeley *et al.* (1982) failed to find evidence of any local feedback inhibition of transmitter release.

Failure of action potential generation at varicosity hillocks

Alberts *et al.* (1981) suggested that varicosity hillocks could control the excitability of the varicosity and the invasion of the more distal parts of the branch. The wave of depolarization arriving at the 'varicosity hillock' reaches the firing level and generates a propagating impulse in the next intervaricose section. The effect of NA via α_2-adrenoceptors, is to lower the safety factor for generation of an action potential (Fig. 54).

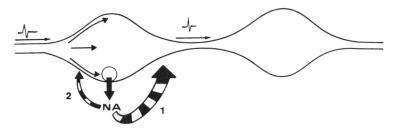

1. EFFECT ON RECRUITMENT

2. EFFECT ON SECRETION—

   —COUPLING OF VARICOSITIES

Fig. 54. Site of action of negative feedback control of noradrenaline (NA) secretion (redrawn from the paper of Alberts *et al.*, **1981**, Fig. 12). (1). Noradrenaline acts on the varicosity hillock, lowering excitability to generate action-potential thereby inhibiting recruitment of varicosities: (2) noradrenaline inhibits secretion-coupling

Shunting of current at boutons

It has been suggested that a possible microanatomical site for invasion control at varicosities would be the varicosity hillock (Vizi, 1980c; Alberts *et al.*, 1981). However, this site of action seems very unlikely since it is based on the acceptance that the method of action potential conduction in a varicose terminal is similar to that of a neuron, and this is not the case. In a neuron the axon hillock is the site where the action potential generation takes place, which is not so at the varicosity. The action potential attempts to invade the whole branch without alternating depolarization with action-potential generation and intervaricosital axonal conduction. The bouton is small (Fig. 55), therefore the action potential arriving at the first and consecutive boutons tries simply to depolarize and pass them. Although methods for analysis of the mode of impulse conduction in *boutons-en-passage* terminals are not available, it is tempting to speculate that the difference in size between the cross-section of the intervaricosital axonal part and the bouton (Fig. 55) makes it difficult for impulses to invade the whole branch. The sudden change in size of the varicose branch at the bouton might produce changes for example in the length constant.

A decrease in the length constant might limit the spread of electrotonic

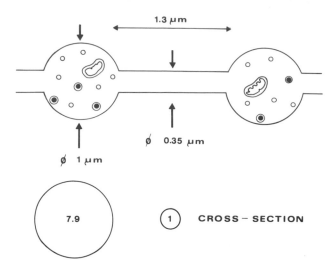

Fig. 55. Schematized representation of a varicose axon terminal. The drawing is based on the quantitative data of Beaudet and Descarries (1978). Note that at the bouton there is a sudden, almost eight-fold, increase in the calibre. At the intervaricosital axon: $0.185 \times 0.175 \times 3.14 = 0.1$ μm^2 and at the bouton: $0.5 \times 0.5 \times 3.14 = 0.79$ μm^2

currents, thereby reducing transmitter release. The receptors might be located near the oral part of the bouton (Fig. 56) or at the axonal bifurcation, and if their stimulation results in a conduction failure or hyperpolarization, the invasion of the branch is stopped at that bouton where the current is shunted. Khodorov *et al.* (1969) made a theoretical analysis of the conduction of action potentials through a portion of axon with increased diameter, and reached the conclusion that any inhomogeneity in an axon could cause a failure of action-potential conduction. Therefore it is suggested that a shunting by hyperpolarization of currents at the proximal part of the bouton is sufficient to produce a complete conduction failure in the varicosity branch. In fact, Dudel (1965) was the first to consider a blocking of conduction as the main effect of pre-synaptic inhibition. Grossman *et al.* (1973) provided electrophysiological evidence that the axonal bifurcation can be an area with a low safety for spike-propagation.

Both hypotheses (failure of action-potential generation at the varicosity hillock and a current shunting at the boutons) agree in that there is a conduction failure in the varicose branch and that the amount of transmitter/modulator released depends on the number of varicosities invaded (Fig. 57), i.e. it has nothing to do with secretion-coupling. A frequency-dependent intermittency of conduction and secretion of transmitter was suggested (Cunnane and Stjärne, 1982). The question arises now as to what kind of mechanism can stop invasion of the varicose branch. *Hyperpolarization* could be one of the basic mechanisms in conduction failure. Membrane potential can be increased by (1) promotion of K-permeability, or (2) stimulation of NaK-ATPase (Vizi, 1978, 1979).

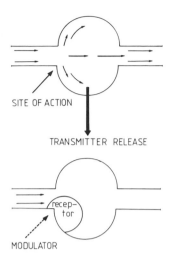

Fig. 56. Site of action of a modulator: the oral part of the bouton where there is a sudden increase in diameter (see Fig. 55). It is likely that the receptors are located near here

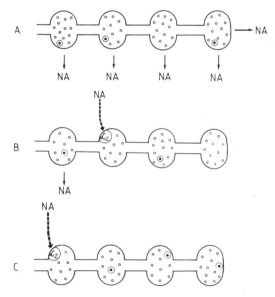

Fig. 57. The amount of transmitter released depends on the number of varicosities invaded, which, in fact, depends on the location of the modulatory receptor. (A) No modulatory receptor in the varicosity branch, therefore the whole branch is invaded; (B) there is no release distal to the modulatory receptor (e.g. α_2); (C) the modulatory receptor (e.g. α is located immediately on the oral part of the first bottom, therefore there is no bouton. The release from a varicosity is all-or-none (since the secretion-coupling is not affected), but the number of invaded varicosities depends on the location of the modulatory receptor. Any change in volley output is, in fact, dependent on the number of varicosities invaded

Promotion of K-permeability

In different synapses, pre-synaptic inhibition has been attributed to a depolarization of the synaptic terminals (Eccles, 1964). However, a number of chemical synapses have now been described in which depolarization enhances rather than depresses transmitter release (Shimahara and Tauc, 1975; Nicholls and Wallace, 1978; Shapiro *et al.*, 1980). As has been shown (Shapiro *et al.*, 1980) pre-synaptic depolarization enhances transmitter release by (1) activation of a steady-state Ca-current, and (2) decreasing the outward potassium-current and consequently increasing spike duration. Therefore, other mechanisms are now suggested as the mode of action of pre-synaptic modulators.

Many neurons possess a potassium conductance that is Ca-dependent but voltage-independent, and its stimulation (by increased Ca-influx, noradrenaline, clonidine) results in hyperpolarization. Morita and North (1981) have provided evidence that clonidine and adrenaline activated potassium-conductance in myenteric neurons, and that low potassium enhanced and high potassium reduced the hyperpolarizing action of clonidine. The effect was increased by lowering Ca_o and reduced by increasing extracellular Ca ion concentration. It was suggested that clonidine and adrenaline activated potassium-conductance by elevating free intracellular Ca concentration. Their effects were antagonized by phentolamine.

Stimulation of NaK-ATPase

The NaK-ATPase is the membrane-embedded enzyme responsible for the active transport of Na and K ions. It plays an important role in the finely tuned control of the ionic environment which underlies nerve activity, it is an electrogenic pump, transporting 3 Na ions for 2 K ions and its activity can hyperpolarize the nerve cell, modulating the firing threshold for nerve impulse initiation. It has been suggested that it plays a trigger role in transmitter release (cf. Vizi, 1978).

It was also suggested that release is terminated by stimulation of the membrane ATPase (Vizi, 1972, 1977b). It has been shown that K-deprivation enhances the volley output of ACh (Vizi, 1977b) and the release of NA from the isolated rat iris (Vizi, 1977b). A similar observation was made by others (Bonaccorsi *et al.*, 1977; Reading and Isbir, 1980; Wakade, 1981; Palaty, 1981). It has been reported by different authors (Rang and Ritchie, 1968; Bolton, 1973; Vizi, 1977b) that the removal of potassium followed by its reintroduction produces a stimulation of the sodium pump and a hyperpolarization. Under these conditions the transmitter release, in response to stimulation, is completely inhibited, even in response to high frequency stimulation. Both ACh and NA release from stimulated Auerbach's plexus and rabbit pulmonary artery were completely inhibited when potassium was reintroduced (Vizi, 1972, 1977; Vizi *et al.*, 1982, 1984).

Much evidence has been presented to show that NA enhances NaK-

ATPase activity (Schaefer *et al.*, 1972; cf. Vizi, 1978; cf. Powis, 1981). Since it has been suggested that the inhibitory effect of NA on transmitter release is related to its stimulatory effect on NaK-activated ATPase activity (cf. Vizi, 1978) it is very important to know whether there is any link between α_2-receptor-mediated pre-synaptic inhibition of transmitter release and membrane ATPase stimulation. Convincing data is available concerning the involvement of membrane ATPase in the hyperpolarizing action of NA. Additional experimental evidence has been adduced to support this hypothesis (cf. Powis, 1981).

The minimum requirement which has to be fulfilled to accept the notion that NA, acting via α-adrenoceptors, stimulates neuronal membrane ATPase, and thereby exerts its pre-synaptic inhibitory action on transmitter release, is to show that the α_2-adrenoceptor stimulants enhance enzyme activity. But this was not the case in the experiments of Adam-Vizi and Seregi (1982), in which there was no correlation between the degree of stimulation of the enzyme by α-agonists and their inhibitory effect on transmitter release. Before it can be concluded that the link between the action of NA on membrane ATPase and transmitter release is a mere arterfact or coincidence, it seems important to show that (1) homogenization of the tissue does not destroy the linkage between adrenoceptor and enzyme and (2) the preparation used does not contain heterogenous ATPase.

7.3 PHARMACOLOGICAL AND CLINICAL IMPLICATIONS: PRE- AND POST-SYNAPTIC RECEPTORS

Almost a century ago, study of the interaction between atropine and pilocarpine on salivary flow in cats led Langley (1878) to suggest that '. . . there is some substance or substances in the nerve endings or gland cells, with which both atropine and pilocarpine are capable of forming compounds . . . according to some low of which their relative mass and chemical affinity for the substance are factors.' This was probably the first expression of the 'receptor concept'. Later Langley (1905) concluded that '. . . there is evidence that the majority of substances which are ordinarily supposed to act upon nerve-endings (as nicotine, curari, atropine, pilocarpine, strychnine) act upon the receptive substances of the cells', i.e. upon the receptors. He might have not been aware, but he suggested that there is some substance or substances (Ehrlich (1913) named them receptors) in the nerve endings and gland, i.e. pre- and post-synaptically.

The impact of the receptor theory of Langley and Ehrlich on pharmacology and drug research was very limited in their lifetimes. After 1960 receptor theory became a major area of research interest in pharmacology.

The search for receptor and organ selective drugs which either mimic the action of endogenous modulators/transmitters or prevent their effect is very promising, and this field of research should make great progress in the near future.

7.3.1 α-adrenoceptors

Ahlquist (1948) first classified adrenergic receptors into α and β. The concept of pre-synaptic control of neurotransmitter release via α-adrenoceptors was suggested by Paton and Vizi (1969) when they found that noradrenaline and adrenaline inhibited acetylcholine release from Auerbach's plexus and phentolamine prevented their action. The need for sub-classification of the α-adrenoceptors was first realized after the discovery of α-adrenoceptor modulation of noradrenaline release from noradrenergic neurons (cf. Starke, 1981; Langer, 1981a). Starke et al. (1972) found that in the rabbit heart the pre-junctional α-adrenoceptors could be differentiated from the receptors located post-junctionally with respect to the relative potencies of agonists and antagonists. In the rabbit pulmonary artery, agonists (Starke et al., 1975b,c) and antagonists (Borowski et al., 1977) showed significant selectivity of α-adrenoceptors located pre- and post-synaptically. The sub-classification of α-adrenoceptors into α_1, which mediates excitatory responses, and α_2, which mediates transmitter release, was suggested by Langer (1974).

At first the terms α_1- and α_2-adrenoceptors were suggested as synonyms for pre- and post-synaptic α-adrenoceptors, i.e. it was based on anatomical localization. Later, the evidence of extrasynaptic localization of α_2-adrenoceptors, even at post-junctional sites, forced pharmacologists to use the terms α_1 and α_2 based on the pharmacological characterization of the receptors, irrespective of their localization or function (Berthelsen and Pettinger, 1977; Starke and Langer, 1979; Timmermans and van Zwieten, 1982). The α_2-adrenoceptors located on axon terminals might be a target for different drugs. Table 16 shows the relative selectivity of different α-adrenoceptor agonists and antagonists. Recently Carbon et al. (1982) succeeded in producing a very selective α_2-antagonist. The N-ethyl-imidazol-13(1-ethyl-2-1(1,4-benzadioxan-2-yl) methyl) imidazole selectivity was more than eleven times that of rauwolscine.

Table 16. Relative orders of selectivity of agonists and antagonist for α_1- and α_2-adrenoceptors

---

Agonists
Xylazine > tramazoline > clonidine > α-methylnoradrenaline > oxymetazoline α_2
  Naphazoline > adrenaline > noradrenaline $\alpha_2 = \alpha_1$
Phenylephrine > methoxamine > (-)-amidephrine[a] α_1

Antagonists
Rauwolscine > yohimbine > piperoxan, Imidazole-13[b]
  > dihydroergocryptine > tolazoline α_2
Mianserine > phentolamine $\alpha_2 = \alpha_1$
Phenoxybenzamine > WB 4101[c] > E-643[d]> labetolol= prazosin α_1

---

[a]3 (2-methylamino-1-hydroxyethyl)methanosulphonalide (Vizi and Ludvig, 1983).
[b]ethyl-2-(1,4-benzodioxan-2-yl)methyl)-imidazole (Caroon et al., 1982).
[c]4-amino, 6,7-dimethoxy-2-4 (5-methyl-thio-1,3,4-oxadiazole-2-carbonyl) piperazin-1-yl guinazoline (Graham et al., 1977).
[d]2–4–(n-butyryl)-homopiperazine-1-yl-4-amino-6, dimethoxy-guinazoline (Shoji et al., 1980).

Evidence has been obtained showing that the pre-junctional α-adrenoceptors located on the enteric nerve terminals were the α_2-sub-type, pharmacologically identical to those on the sympathetic nerve terminals (Drew, 1977; Cambridge and Davey, 1980). A similar correlation of α_2-adrenoceptors was observed in the cerebral cortex (Vizi, 1979). No difference between α_2-adrenoceptors located on cholinergic and noradrenergic axon terminals was found (Vizi, 1982b). It would be of clinical importance to find α_2-adrenoceptor agonists or antagonists with a selective action either on cholinergic or on noradrenergic axon terminals. Some drugs seem to be able to distinguish between cardiac and vascular post-synaptic α_2-adrenoceptors (De Jonge *et al.*, 1981). BE-2254 was shown to have a different potency on pre- and post-synaptic α_2-adrenoceptors (Hicks, 1981).

Table 17. Potency of adrenergic agonists and antagonists at ^{3}H-clonidine binding sites and for modulating cholinergic transmission in guinea-pig ileum

	Guinea-pig ileum inhibition of ^{3}H-clonidine binding[a]	Rat cerebral cortex inhibition of ^{3}H-clonidine binding[b]	
		High K_1(nM)	Low K_1 (nM)
Agonists			
1-noradrenaline	58	2.5	15
d-noradrenaline	–	57	351
α-methylnoradrenaline	18	2.8	3.9
Phenylephrine	2300	50	160
Clonidine	5.7	0.62	4.3
Tramazoline	6.9	1.2	2.6
Antagonists			
Phentolamine	12	2.5	1.9
WB-4101	230	106	92
Prazosin	5400	6400	2600
Yohimbine	82	73	40

[a]Data from Starke *et al.* (1980).
[b]Data from U'Prichard and Snyder (1979).

The binding of the preferentially pre-synaptic agonists and antagonists is shown in Table 17. ^{3}H-Clonidine binding to membranes from the guinea-pig ileum was also studied (Starke *et al.*, 1980). The specific binding of ^{3}H-clonidine, i.e. the binding that could be inhibited by a high concentration of noradrenaline or unlabelled clonidine, was of high affinity: $K_D=3$ nM (Starke *et al.*, 1980). The number of binding sites was 30 fmoles/mg protein.

In the central nervous system there are two types of α-adrenoceptors, and two distinct α-adrenoceptor binding sites specifically labelled by $(^3$H)-clonidine and ^{3}H-WB-4101 have been shown (U'Prichard *et al.*, 1977, 1978) in the rat cerebral cortex. Prazosin was highly effective in displacing ^{3}H-WB-4101 bound to membranes and, by contrast, it was 1000 and 10 000 times less effective in displacing ^{3}H-clonidine bound to cerebral cortex membranes. Phentolamine was approximately equi-effective. This finding indicates that

α_2-adrenoceptors are present in the cerebral cortex. So far there is no evidence of a difference between the α_2-adrenoceptors located on cholinergic and noradrenergic and/or adrenergic axon terminals so it is therefore very difficult to say where the binding sites are located. Nevertheless, both are present in the cerebral cortex. Greenberg *et al.* (1976) failed to observe any difference in the binding of ³H-clonidine following intraventricular 6-hydroxydopamine treatment.

In the study of U'Prichard and Snyder (1979) ³H-clonidine binds to two α-adrenoceptor sites, with K_D values of 0.5 nM and 3 nM. It has been shown that α-adrenoceptor agonists are more potent at the high-affinity site, while α-adrenoceptor antagonists are not. 6-hydroxydopamine pre-treatment doubles the number of high-affinity sites, but fails to influence the number of low-affinity sites. There was a significant difference (fourteen-fold) between them.

Recent radioligand binding studies have revealed the heterogenicity of α-adrenoceptors in various tissue preparations (Wood *et al.*, 1979). It should be kept in mind that the preferential affinity of a labelled drug to a receptor sub-type, as determined in binding studies, gives no evidence for exact binding sites, since a similar receptor may be present on cholinergic or adrenergic neurons or on other neurons. Nevertheless, a dissociation of pre-synaptic α_2-adrenoceptors was observed on rat vas deferens following prazosin administration (Vizi *et al.*, 1983a). In the presence of prazosin yohimbine failed to affect the pre-synaptic α_2-adrenoceptor stimulatory effect of l-NA while the effect of xylazine was almost completely abolished.

A frontal post-mortem human brain cortex revealed two binding sites for ³H-clonidine with K_D-values of approximately 1 and 8 nM as indicated by the biphasic Scatchard plot, the biphasic pattern of dissociation kinetics, and the binding of ³H-clonidine. The high-affinity binding sites for ³H-WB 4101 were also detected in the human frontal cortex, with K_D-values of about 0.09 and 1.5 nM.

Three other brain regions revealed very similar patterns exhibited by the frontal cortex, except that the density of the ³H-clonidine sites of either high or low affinity was highest in the hypothalamus, whereas the density of ³H-WB 4101 sites was highest in the hippocampus (Weinreich and Seeman, 1981). The rank order of potencies of a series of catecholamine agonists and antagonists in inhibiting the binding of the high-affinity ³H-clonidine sites was: phentolamine > noradrenaline > adrenaline > yohimbine > clonidine > LSD > WB 4101 > prazosin. The relative potencies of various drugs in competing for the high affinity ³H-WB 4101 binding sites were: prazosin > phentolamine > WB 4101 > phenoxybenzamine > LSD > adrenaline > yohimbine > clonidine > noradrenaline.

The relative proportions of α_1- and α_2-adrenoceptors in several tissues were estimated by Hoffman *et al.* (1979). Recent autoradiographic studies with ³H-WB 4101 and ³H-clonidine revealed that α_1- and α_2-adrenoceptors are distributed differently in the rat brain (Young and Kuhar, 1979). While high densities of α_2-adrenoceptors are found in the limbic system and in the nucleus tractus solitarius, α_1-adrenoceptors are concentrated predominantly in the olfactory bulb and in the dentate gyrus of the hippocampus. Two weeks after chemical sympathectomy with 6-hydroxydopamine there was a significant decrease in the maximal binding of ^{3}H-DHE in the rat heart, indicating that some of the α_2-adrenoceptors labelled by ^{3}H-DHE are associated with noradrenergic nerve-endings (Story *et al.*, 1979). However, chemical denervation with 6-hydroxydopamine does not reduce the binding of ^{3}H WB 4101 and ^{3}H-clonidine-labelled α_1- and α_2-adrenoceptors localized postsynaptically in this central nervous system.

In agreement with Langer (1981a), further and extensive studies are required to clarify the significance of changes in the binding parameters of α-adrenoceptor ligands after surgical or chemical denervation.

There are widely used drugs in clinical practice that are believed to act via the pre-synaptic modulation of neurochemical transmission. There are different clinical symptoms which are related to modulation of chemical neurotransmission via α-adrenoceptors.

The paralytic ileus

Involvement of the sympathetic nervous system in the pathomechanism of the so-called 'paralytic' ileus was recognized by Wangensteen (1955), who termed it the 'inhibition' ileus. More recent data (Vizi and Knoll, 1971; cf. Vizi, 1979, 1982a) derived from isolated organ experiments demonstrate that there is an interneuronal interaction between adrenergic and cholinergic neurons. Noradrenaline reduces the acetylcholine release from cholinergic nerve terminals via pre-synaptic α-adrenoceptors thereby modulating gastro-intestinal motility. Drugs which act on pre-synaptic α-adrenoceptors decrease both the noradrenaline and the acetylcholine release. However, pre-synaptic β-receptor excitation increases the noradrenaline outflow. According to this theory, the release of acetylcholine under physiological or pathological conditions is controlled by sympathetic nerves. However, the release of noradrenaline by itself is controlled by pre-synaptic α-receptors. The released noradrenaline decreases the acetylcholine release from cholinergic nerve-endings and results in diminished intestinal contractions.

The fact that drugs with an α-receptor blocking effect are beneficial in paralytic ileus provides clinical evidence that cholinergic transmission in the gut is continuously controlled by α-receptor stimulation produced by noradrenaline or by circulating adrenaline, and the removal of their inhibitory effects leads to an increase in intestinal movement.

Diarrhoea, observed as a side-effect following catecholamine depletion by reserpine or guanethidine treatment, also suggest a local pre-synaptic control of ACh release in the gut.

184

Clinical experiments show that the paralytic ileus is due to the inhibition of the peristaltic reflex by catecholamines. Petri (1971) observed that α- or β-adrenoceptor blocking agents, or their combination, are able to restore the failing peristalsis. In the presence of practolol the pre-synaptic β-receptors are blocked and the positive feedback mechanism is extinguished. Therefore the release of noradrenaline is decreased and the gut contractions are augmented. There is clinical and *in vitro* (Szenochradszky *et al.*, 1979) experimental evidence showing that practolol may suspend the inhibitory effect of increased endogenous sympathetic stimulation on intestinal motor activity. This could explain its beneficial effect in the treatment of paralytic ileus.

Hypertension

The administration of either α₂-adrenoceptor agonists, (e.g. clonidine) or α₁-adrenoceptor antagonists (e.g. prazosin) results in a decrease of blood pressure. Both reduce the sympathetic neuro-effector transmission of the vessels, thereby resulting in vasodilation (Fig. 58). Nevertheless, the antihypertensive action of α₂-adrenoceptor agonists (clonidine, xylazine) is related to their action on the central nervous system, leading to a decrease in sympathetic and an increase in vagal tone (Kobinger, 1974).

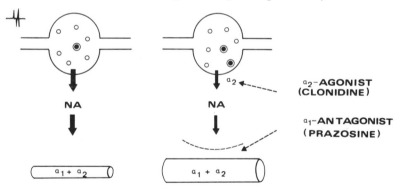

Fig. 58. Site of action of α₂-agonist (clonidine) and α₁-antagonist (prazosin) in noradrenergic neuroeffectorical transmission. Note that through reduction of noradrenaline (NA) release or prevention of its action on the α₁-adrenoceptor, the vessel is dilated

The action of clonidine on animals depleted of central noradrenaline led van Zwieten (1980) to propose that the central receptors stimulated by clonidine were post-synatically located. Since the hypotensive effects of clonidine were inhibited by compounds such as yohimbine, piperoxan, and rauwolscine (selective antagonists for peripheral pre-synaptic α₂-adrenoceptors), van Zwieten (1980) concluded that the central receptors involved were α₂ in nature.

α-Adrenolytic agents, such as phenoxybenzamine and phentolamine, augment the induced release of adrenergic neurotransmitters in the blood vessel wall. This is masked by their α-adrenolytic properties on the vascular smooth muscle cells, while in the heart the greater release of noradrenaline results in a β-adrenergically-mediated increase in heart rate. This is not seen with α_1-blocking drugs like prazosin (Lefevre-Borg *et al.*, 1979).

During episodes of vasovagal syncope both the heart and the cholinergically innervated blood vessels escape very rapidly from sympathetic adrenergic control because the released acetylcholine decreases the release of adrenaline from axon terminals. Conversely, when muscarinic blocking drugs are administered on a background of cholinergic nerve activity, the resulting increase in release of adrenergic transmitter explains the cardiovascular acceleration and the increases in plasma catecholamine concentration they sometimes cause. These clinical symptoms are seen especially with drugs such as atropine, amitryptyline, gallamine, and pancuronium bromide (Van Hee and Vanhoutte, 1978; Collis and Shepherd, 1980; Vercruysse *et al.*, 1979).

Differences in the distribution of sub-types (α_1 and α_2) of peripheral α-adrenergic receptors explains the variations in sensitivity to the hypotensive effects of drugs such as clonidine and prazosin among hypertensions of different origin (Shepherd and Vanhoutte, 1981). The adrenergic sensitivity of vascular smooth muscle cells, and the way they are modulated by local temperature changes, may well explain the vasospastic episodes seen in primary Raynaud's disease on exposure of the peripheral vessels to cold.

The relative lack of β-receptor sensitivity, whether ideopathic, due to hypothyroidism, or caused by the administration of β-adrenergic blocking drugs, exaggerates the vasoconstriction due to the increased affinity of the post-junctional α-adrenergic receptors caused by local cold, and thus precipitates the vasospasm (Shepherd and Vanhoutte, 1979; Vanhoutte and Janssens, 1980).

It has been reported that naloxone inhibits the hypotensive effect of the central receptor agonists clonidine and α-methyldopa in spontaneously hypertensive rats (Kunos *et al.*, 1981). This functional antagonism could be due to the release of an endogenous opiate by activation of α_2-receptors that participate in the central control of blood pressure. Evidence was provided by Kunos *et al.* (1981) who showed that clonidine and α-methylnoradrenaline increased the release of a substance with β-endorphin immunoreactivity from slices of brainstem of spontaneously hypertensive rats but not from normotensive rats. These findings indicate that release of β-endorphin by central β_2-adrenoceptor agonists may contribute to the antihypertensive action of these drugs (Farsang *et al.*, 1982).

The lack of effective antihypertensive activity of classical α-adrenoceptor antagonists is most likely due to blockade of pre-junctional α_2-adrenoceptors, thus removing the inhibitory feedback control and thereby increasing the release of noradrenaline. This noradrenaline will compete with the α_1-adrenoceptor antagonist for the post-junctional receptors, overcoming the blockade. The tachycardia produced by most α-adrenoceptor antagonists is due to blockade of pre-junctional α-adrenoceptors of cardiac sympathetic nerves, resulting in an increased release of noradrenaline and a greater effect on the post-junctional β-adrenoceptors.

Bradycardia and tachycardia

Phentolamine potentiated the tachycardia due to nerve stimulation of the guinea-pig atria (Langer *et al.*, 1977), in pithed rats (Doxey, 1977), and in anaesthetized dogs (Lokhandwala and Buckely, 1976), suggesting that an α-inhibitory feedback process does modulate noradrenergic transmission. However, other investigators failed to demonstrate a potentiating effect of phentolamine in the guinea-pig atria (Vizi *et al.*, 1973), in the rat isolated atria (Idowu and Zar, 1977), or in anaesthetized dogs (Antonaccio *et al.*, 1974; Cavero *et al.*, 1977).

In vivo it is possible that the bradycardic effects of clonidine are particularly related to the activation of the pre-synaptic inhibitory α$_2$-adrenoceptors in the accelerans nerve of the heart due to the high affinity of this drug for the pre-synaptic α$_2$-adrenoceptors.

Inhibition of pre-junctional α$_2$-adrenergic receptors is also seen with anti-depressant drugs such as dexamisole and amitryptyline (Vanhoutte, 1977; Collis and Shepherd, 1980). In the case of amitryptyline, this may help to explain the tachycardia and arrhythmias that it can cause.

Depression and sedation

Evidence has been provided showing that depression is linked to a decrease in the release of monoamines in the brain. Thus drugs such as reserpine and tetrabenzine are able to deplete catecholamines, thereby reducing the release of noradrenaline and causing depression while others such as amphetamines, that increase the release of catecholamines, can alleviate it. Schildkraut (1965) proposed that 'some, if not all, depressions are associated with an absolute or relative deficiency of catecholamines, particularly norepine-phrine, at functionally important adrenergic receptor sites in the brain'. According to this way of thinking, drugs able to inhibit the release of NA will produce sedation. Drew *et al.* (1979) and Cavero and Roach (1978) found that sedation is mediated by control α$_2$-adrenoceptors. Anti-depressants and α$_2$-adrenoceptor blocking agents are able to block clonidine-induced sedation (cf. Timmermans and van Zwieten, 1982).

In veterinary medicine, xylazine (2-2,6-dimethylphenyl-amino 4H-5,6-dihy-dro-1,3-thiazine) is widely used as a sedative analgesic restraining agent.

In human beings the sedative, hypotensive effects of the xylazine-related drug clonidine are antagonized by frequent dosages of the α-adrenoceptor antagonist tolazoline (Houston, 1981). Xylazine sedation was antagonized by yohimbine in dogs (Hatch *et al.*, 1982). Xylazine and clonidine are thought to cause sedation and analgesia mainly by stimulating central pre-synaptic α$_2$-adrenoceptors (Timmermans *et al.*, 1981).

Anti-depressive actions

Drugs able to block the inhibitory pre-synaptic α-receptors increase the synaptic concentration of noradrenaline and, according to the catecholamine hypothesis, should have an anti-depressant effect.

The first indication that anti-depressant drugs might block pre-synaptic α_2-adrenoceptors was provided by studies of the drug mianserin (Bauman and Maitre, 1977). In the presence of desmethylimipramine, when neuronal uptake was inhibited, mianserin was still able to cause an increase in the stimulated overflow of transmitters in isolated brain slices that had been incubated with ³H-noradrenaline. However, when pre-synaptic α-receptors were also blocked by phentolamine, mianserin had no effect. The reversal by mianserin of the inhibitory effects of the pre-synaptic α_2-adrenoceptor agonist clonidine further supported the conclusion that the anti-depressant drug was blocking α_2-receptors (Collis and Shepherd, 1980).

Blockade of pre-synaptic α_2-receptors by mianserin has been confirmed in the rat vas deferens (Brown et al., 1980). Other anti-depressant drugs such as trazodone, amitryptyline, and nortryptyline also appeared to inhibit pre-synaptic receptors. It has been found in ten patients that long-term desipramine treatment produced a α_2-adrenoceptor subsensitivity. However, it is also known that other anti-depressants do not share the property of desipramine in reducing the sensitivity of α_2-adrenoceptors.

Symptoms of opiate withdrawal

Selective α_2-adrenoceptor agonists like clonidine are capable of suppressing opiate withdrawal symptoms (Gold et al., 1978).

Glaucoma

α_2-adrenoceptor stimulants have been found efficient in decreasing intra-ocular pressure (Innemee et al., 1981), therefore selective α_2-adrenoceptor agonists might be useful against glaucoma. No clinical data are available.

Epilepsy

The experimental data (Vizi and Pásztor, 1981) obtained in the isolated human cerebral cortex shows that ouabain-induced ACh release can be reduced by NA or by phenytoin. Since there is a relationship between EEG conversive activity and ACh release (Gardner and Webster, 1977), it is suggested that cortical epileptogenic activity could result at least partly from the excessive release of ACh (Spehlman et al., 1971). This suggestion is supported by the findings that convulsants are able to decrease ACh concentrations and enhance ACh release (Beleslin et al., 1965; Celesia and Jasper, 1966; Hemsworth and Neal, 1968). There is considerable evidence to support the idea that central catecholamines also play a role in the convulsive activity of the cortex (Papanicolaou *et al.*, 1982), as α_2-agonists of clonidine-type had anti-convulsant activity against pentetrazol-induced convulsions.

7.3.2 Dopamine receptors

Several sub-types of dopamine receptor have been proposed. Recent evidence allows a distinction to be made between sub-types of dopamine receptor. These include D_1 (linked to adenylate cyclase) and D_2 (not linked to adenylate cyclase) (Kebabian and Calne, 1979), DA_e (excitation-mediating), DA_i (inhibition-mediating) (Cools and Van Rossum, 1980), and D_1, D_2, D_3, and D_4, according to different affinities to radio-ligands (Seeman, 1980).

The D_1 receptor is coupled to an adenylate cyclase and is responsive to such agonists as apomorphine at rather high concentrations, while the D_2 receptor is responsive to the ergot drugs at very low concentrations (Kebabian and Calne, 1979). Behavioural and neurochemical correlates of activation of these two receptors are not well defined.

In the central nervous system it was reported that dopamine did not modify the stimulation-induced release of noradrenaline from rat cortical slices when cocaine (Taube *et al.*, 1977) or desipramine were used to inhibit neuronal uptake of noradrenaline. When similar experiments were carried out in the absence of cocaine in slices of the rabbit hypothalamus, apomorphine and pergolide, two potent dopaminergic agonists, reduced the stimulation-induced release of ^{3}H-noradrenaline (Galzin and Langer, 1982). These inhibitory dopamine receptors are blocked by sulpiride while they remain unaffected in the presence of phentolamine or yohimbine, indicating that DA-receptors are present and that central pre-synaptic inhibitory dopamine receptors can only be demonstrated in the absence of inhibitors of neuronal uptake of noradrenaline.

Since the inhibition of the stimulated release of noradrenaline is obtained by apomorphine and pergolide, but not by dopamine itself, the physiological role of these receptors remains to be clarified.

Available pharmacological data (cf. Cavero *et al.*, 1982) clearly favours the existence of two distinct sub-types of DA-receptors (Fig. 59), DA_2-receptors, which are located pre-junctionally, and DA_1-receptors, found post-junctionally.

The vascular dopamine receptor has been termed DA_1 (Langer 1981a), and have only been located post-synaptically. The second peripheral dopamine receptor described has been termed DA_2. This is located on the pre-synaptic post-ganglionic sympathetic nerve and functions to inhibit noradrenaline release (Goldberg, 1978; Langer, 1973; cf. Cavero *et al.*, 1982). This classification is similar to the subdivision of α-adrenergic receptors.

DA_1-DA_2-receptors differ in the following ways: N,N-di-n-propyldopamine has a much higher affinity for the DA_2- than for the DA_1-receptor. A rela-

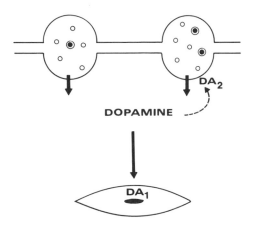

Fig. 59. Pre- (DA₂) and post-synaptic (junctional) (DA₁) dopamine receptors. Dopamine can inhibit its own release

tively weak activity of N,N-di-n-propyldopamine on DA₁-receptors has been shown. Bulbocapnine was found to be a selective antagonist at the DA₁-receptor (Massingham *et al.*, 1980).

Pre-synaptic inhibitory dopamine receptors are present on peripheral nora-drenergic neurons (Enero and Langer, 1975b; Lockhandwala *et al.*, 1979; Dubocovich and Langer, 1980; Massingham *et al.*, 1980). Their stimulation by different agonists reduces the stimulation-evoked release of noradrenaline, and this effect is accompanied by a decrease in the end-organ responses to sympathetic nerve stimulation. This is not affected by α-adrenoceptor blocking agents while it is selectively antagonized by dopamine receptor blocking agents such as sulpiride.

There are pharmacological differences between pre-synaptic inhibitory dopamine receptors on peripheral noradrenergic nerve-endings and post-synaptic dopamine receptors that mediate vasodilatation in some vascular beds.

It seems that pre-synaptic inhibitory dopamine receptors on peripheral noradrenergic nerves appear to be involved in the hypotensive and brady-cardic effects of dopamine receptor agonists (Lokhandwala *et al.*, 1979; Massingham *et al.*, 1980).

DA₁-receptors could be regarded as those receptors which, upon stimu-lation by agonists (6,7 − ADTN > epinine > dopamine > DPDA), produce responses which are preferentially blocked by bulbocapnine (bulbocapnine > α-flupenthixol > haloperidol > metodopramid > sulpiride). DA₂-receptors could be called those receptors which, upon stimulation by agonists (dopa-mine, DPI, 3-PPP, TL-99, apomorphine), produce responses which are preferentially blocked by sulpiride.

At present, it is not possible to give a precise rank order of selectivity of substances acting on DA-receptors. In general, ergoline and aminotetraline derivatives, as well as piribedil, seem to exhibit a certain selectivity for the DA_2-receptors. The phenylethylamine derivatives stimulate both DA_1 and DA_2-receptors. Table 18 shows the selective agonists for DA-autoreceptors.

Table 18. Selective agonists of dopamine autoreceptors

	References
3-PPP (3-(3-hydroxyphenyl)-N-n-propyldiperidine)	Nilsson and Carlsson (1982)
DPI (3,4-dihydroxyphenyl-amino-2-imidazoline)	Vizi et al. (1976)
TL-99 (6,7-dihydroxy-2-(dimethylamino)-tetralin	Gooddale et al. (1980)

Important evidence for the existence of DA-autoreceptors is the so-called 'paradoxical' response to apomorphine (Carlsson, 1975). At least one order of magnitude below that required for producing stereotyped behaviour in the rat can inhibit the DA synthesis in the striatum and in the limbic system in a dose-dependent manner. These low doses also cause an inhibition of locomotor activity, compatible with a reduction in the release of DA. In man, low doses of apomorphine have been shown to exert anti-psychotic and anti-dyskinetic effects (Tamminga et al., 1978; Smith et al., 1977), whereas in higher doses it stimulates locomotor activity probably via post-synaptic dopamine receptors. A DA analogue 3-(3-hydroxy-phenyl)-N-n-propyldiperidine (3-PPP) produces all the actions observed after low doses of apomorphine (Nilsson and Carlsson, 1982). 3-PPP seems to be one of the first apparently selective dopaminergic autoreceptor agonists (Nilsson and Carlsson, 1982). Hopefully, it will be possible to obtain an anti-psychotic agent without extrapyramidal side-effects. The most striking behavioural action of 3-PPP in rats and mice is a marked inhibition of exploratory activity. No stereotyped behaviour due to stimulation occurs at any dosage, and the action of reserpine is not antagonized. Vizi et al. (1976) suggested that DPI is a selective stimulant for DA autoreceptors of the nigro-striatal pathway. Gooddale et al. (1980) reported 6,7-dihydroxy-2-(dimethylamino)-tetralin (TL-99) to be a selective agonist for DA autoreceptors.

DA is not present uniformly throughout the brain. DA-rich nerve terminals are found in the striatum and the limbic system, so it is considered that these areas are under dopaminergic control. Anti-psychotic drugs (neuroleptics) with central DA receptor blocking effects are able to increase the synthesis and release of striatal DA and the firing of dopaminergic neurons in the substantia nigra. These effects have been postulated to be secondary to the drug-induced blockade of striatal post-synaptic DA receptors, and mediated by a negative feedback loop impinging in nigral DA neurons (Carlsson and Lindqvist, 1963). Alternatively, it has been suggested that neuroleptics stimulate dopaminergic firing by blocking nigral DA receptors (autoreceptors), thus relieving DA neurons from the inhibitory action of DA released from dopaminergic neurons.

There is a growing body of evidence suggesting that dopamine receptors at the sympathetic nerve terminals may differ from those in vascular smooth muscle (Goldberg *et al.*, 1978; Cavero *et al.*, 1982), i.e. there are pharmacological differences between DA-receptors located pre- (D_2) and post-synaptically (D_1).

The lowering of blood pressure by bromocriptine, observed in hypertensive patients and in experimental animals, results from effects exerted on both vascular and pre-junctional dopamine receptors. Since there appear to be differences between vascular and pre-junctional receptors, it should be possible to synthesize compounds with selective effects on one of these two receptor types. Progress has already been made in this direction (Kyncl *et al.*, 1979; cf. Brodde, 1982).

Dopamine receptor stimulants are theoretically of potential therapeutic value, not only in hypertension but also in congestive cardiac failure and renal failure. Since DA receptor agonists can produce natriuresis and inhibit aldosterone-release these findings constitute the basis for considering DA-receptors as potential targets for a new class of drugs useful in treating a number of cardiovascular diseases.

Hypertension

The hypotensive effect of levodopa in patients with Parkinson's disease could be the result of uptake of the levodopa into sympathetic nerves, and thus the release of dopamine would activate a negative feedback loop, causing a decrease in transmitter release and hypotension. The hypotensive effect of monoamine oxidase inhibitors may also be explained in terms of a build-up of dopamine in sympathetic neurons and activation of a negative feedback loop when it is released as a transmitter.

Parkinson's disease

The diminished dopaminergic influence, together with the enhanced cholinergic activity and ACh release induced by treatment with neuroleptics, is believed to result in a relative imbalance between the cholinergic and dopaminergic systems in the striatum. This might be related to parkinsonism observed in patients undergoing treatment with these drugs. This notion is consistent with the observation that bromocriptine, a DA agonist, is beneficial for patients with Parkinson's disease.

The results obtained (Vizi *et al.*, 1976; Stoof *et al.*, 1979) indicate that the relative sensitivity and affinity of the two pre-synaptic DA receptors (one is located on the dopaminergic axon terminals, the other is on the cholinergic interneurons) are different for some drugs, and suggest that these differences might be exploited in the development of more specific stimulators and blocking agents which might ultimately have some therapeutic advantages over drugs now used in medical practice.

Clozapine is an anti-psychotic drug that inhibits DA receptors (Miller and Hiley, 1974) but it does not induce Parkinson's syndrome (DeMaio, 1972; Matz *et al.*, 1974). However, clozapine is also a potent anti-muscarinic agent (Miller and Hiley, 1974). It is well known that anti-muscarinic agents are useful in the treatment of Parkinson's disease and it has been suggested that the lack of Parkinson's syndrome in patients treated with clozapine may be related to its ability to block muscarinic receptors (Miller and Hiley, 1974; Andén and Stock, 1973). In addition to blocking muscarinic receptors on the post-synaptic site, clozapine also reduces cholinergic activity pre-synaptically by reducing the amount of ACh available for synaptic transmission. The increase in dopaminergic output in the striatum should improve Parkinson's syndrome but in the limbic system it should produce psychosis. This would be an explanation of the psychosis observed after L-DOPA treatment.

As regards the anti-psychotic effect of neuroleptics, the blockade of DA receptor activity as well as the receptor-mediated acceleration of DA synthesis could be of major importance for the therapeutic treatment of psychosis. Since in schizophrenic patients a synergistic effect was found with α-methyl-p-tyrosine, which blocks catecholamine synthesis, and thioridazine, which blocks DA receptors, it would appear that the direct DA receptor blocking effect of neuroleptics is best related to anti-psychotic efficacy (Carlsson *et al.*, 1972). Following this line of thinking, it could be proposed that a selective DA_2-agonist which also has anatomical selectivity in that it acts only on dopaminergic autoreceptors would be a potential candidate as a drug against hyperkinetic diseases and psychosis. In neurochemical studies (Vizi *et al.*, 1976) DPI has been found to act in this way.

7.3.3 Muscarinic receptors

In addition to post-junctional muscarinic receptors, pre-synaptic receptors have been localized (see section 4.2). Pre-synaptic or pre-junctional muscarinic receptors are involved in the modulation of transmitter release.

Pre-synaptic receptors are now being considered as potential targets for the development of new drugs and as being involved in the side-effects of certain drugs. According to Muscholl (1980), there is no difference between pre- and post-junctional muscarinic receptors (Fig. 60a and b) in the rabbit heart. Despite the range of affinities of 5 log units among the muscarinic antagonists, their pA_2 values measured on atrial tension (post-synaptic effect) and NA release (pre-synaptic effect) did not differ (Muscholl, 1980).

Although Bowen and Marek (1982) showed no indication of a pharmacological difference between the central pre-synaptic muscarinic autoreceptor and the central and peripheral post-synaptic muscarinic receptors, the experiments made with antagonists suggest that there is tissue selectivity in pharmacological antagonism (Barlow *et al.*, 1976). Hammer *et al.* (1980) provided evidence with pirenzepine, a new anti-muscarinic agent, that there is hetero-

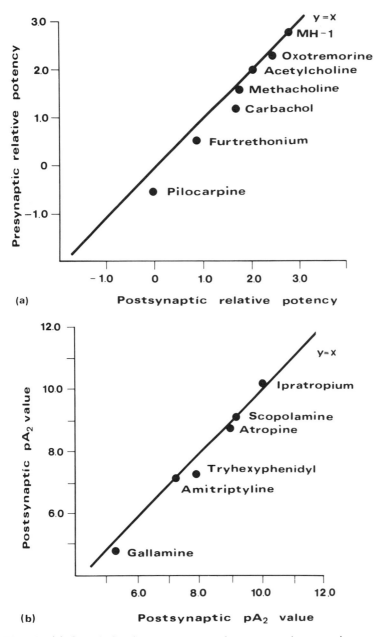

Fig. 60. (a) Correlation between pre- and post-synaptic muscarine agonistic activities of different substances (taken from Muscholl, 1980): rabbit heart. (b) Correlation between pre- and post-synaptic muscarine-receptor inhibitory effects of different antagonists. pA2-values (taken from Muscholl, 1980). Inhibition of noradrenaline release is taken as a presynaptic effect and a decrease of atrial systolic tension by drugs as a postsynaptic response

geneity in binding which correlates well with the pharmacological activity. Whilst its effect on gastric secretion is seen at very low doses, very much higher doses are needed to affect the smooth muscle. Indeed, in clinical practice (cf. Hammer *et al.*, 1980) it proved to be a drug with beneficial effects on gastric secretion.

7.3.4 Purine receptors

The modulatory role of adenosine (see section 6.1) is generally accepted (cf. Phillis and Wu, 1981). Burnstock (1978b) proposed terms for classification or purine receptors (Table 19): a P_1 receptor for adenosine, for which the methylxanthines, caffeine, and theophylline are antagonists, and the P_2 receptor for adenine nucleotides. Theophylline is a competitive antagonist of adenosine receptors on axon terminals (Vizi and Knoll, 1976) and on hippocampal neurons (Dunwiddie and Hoffer, 1980). Theophylline shifted dose response curves to the right without altering the maximal response. Nevertheless, Table 20 shows the classification of purine receptors made by others (cf. Stone, 1982). The drugs used in everyday clinical practice are shown in Table 21.

Table 19. Sub-types of purinoceptors (see Burnstock, 1981)

Classification of receptor	Antagonist	Agonist potencies
P_1	Methylxanthines (theophylline, caffeine, aminophylline)	$AD \geqslant AMP > ADP \geqslant ATP$
P_2	Quinidine 2-substituted imidazolines 2'2 pyridilisatogen	$ATP \geqslant ADP > AMP \geqslant D$

AD: Adenosine.

Table 20. Classifications of purine receptors

Receptor	Agonist	Antagonist	References
P_1	Adenosine $>$ ATP	Methylxanthines	Burnstock (1978b)
P_2	ATP $>$ adenosine	Quinidine	
A_1	PIA[a] $>$ adenosine	Methylxanthines	Van Calker *et al.* (1979)
A_2	Adenosine $>$ PIA	Methylxanthines	
P	2'.5'-dideoxyadenosine	methylxanthine-resistant	
R	R_a ECA[a] $>$ adenosine $>$ Methylxanthines PIA		
	R_i PIA $>$ adenosine $>$ Methylxanthines ECA		Londos *et al.* (1980)

[a]Abbreviations: PIA, N^6-phenylisopropyladenosine; ECA, 5-N-ethylcarboxamide adenosine.

Table 21. Drugs involved in the effect of adenosine on chemical neurotransmission

Drugs	Effect	Modification of adenosine action
Methylxanthines		Inhibits[a]
Theophylline	Inhibition of P_1 receptor	
Theobromine	(competitive)[a,b]	
Caffeine		
Dipyridamole	Uptake inhibition[c]	Potentiates[c,d,e]
Dilazep	Uptake inhibition[f]	Potentiates[d]

References:
[a]Vizi and Knoll (1976).
[b]Vizi et al. (1982).
[c]Stachell et al. (1972).
[d]Gustafsson et al. (1980).
[e]Moritoki et al. (1978).
[f]Pohl and Brock (1974).

It has been demonstrated that adenosine concentration in brain tissue rises rapidly during periods of brain ischemia (Kleihues et al., 1974) and under such conditions may reach 1 mM in cerebrospinal fluid (Berne et al., 1974). Pull and McIlwain (1971) found adenosine in the central nervous system in concentrations of 0.15-0.4 mM following nerve stimulation.

Both ATP and adenosine produce dilatation of the cerebral blood vessels (Wahl and Kuchinsky, 1976). Perivascularly applied ATP has a dilatory effect on pial arterioles. Threshold concentrations for adenosine were about two orders of magnitude higher than those for ATP (10^{-11} M). The dilatory effect of purines on cerebral arteries is antagonized by the P_1-antagonist, theophylline. As the concentrations of ATP and adenosine required to cause vascular dilatation are well within the range of those occurring in the extracellular space (Berne et al., 1974) it is reasonable to assume that neuronally released adenosine and adenine nucleotides are an important part of the homeostatic mechanism which controls the vascular supply to the brain and chemical neurotransmission. Indeed, Schubert and Kreutzberg (1976) concluded that release of adenosine from nerve cell soma, dendrites, and terminals can lead to a highly localized and effective increase in the extraneuronal level of free adenosine, and thus cause vasodilatation. In addition, Forrester et al. (1979) have postulated that circulating purines, released from actively metabolizing peripheral tissues, may influence cerebral blood flow, and thus mediate a form of metabolic communication in the body.

Caffeine-withdrawal headaches may result from an adenosine-induced vasodilatation following the unblocking of adenosine receptors in cerebral vascular smooth muscle as caffeine levels are reduced by excretion. Caffeinism may be associated with symptoms of anxiety, depression, and schizophrenia (Greden et al., 1978; Lutz, 1978). Inherited disorders of purine metabolism in the brain have been related to symptoms of psychomotor retardation, athetosis, and self-mutilation (Lesch–Nyhan syndrome) (Berman *et al.*, 1969; Seegmiller, 1979) in which there is a deficiency of the enzyme hypoxanthine–guanine phosphoribosyltranferase. Adenine therapy has been used for this condition (Schulman *et al.*, 1971). Adenosine triphosphate has been used to prevent or ameliorate the extrapyramidal syndromes associated with the administration of neuroleptic drugs to man and animals. Its effect might be related to pre-synaptic inhibition of ACh and/or DA release in the extrapyramidal system.

It is clear that substitution in the 3-position of alkylxanthine is sufficient for potent bronchodilatory activity, whereas substitution in the 1-position seems to be essential for adenosine receptor antagonism. The 3-propyl derivate (enprofylline) exemplifies these relationships (cf. Persson, 1982). Enprofylline is about five times more potent than theophylline *in vitro* and *in vivo* in animals (Persson and Kjellin, 1981).

The most frequently used psychotropic agent is a methylxanthine. We consume enormous amounts of caffeine-containing beverages which increase fitness and alertness. As far as stimulant effects on the central nervous system are concerned (Andersson and Persson, 1980), theophylline is even more potent than caffeine. In asthma therapy these effects can be regarded as side-effects and massive CNS-stimulation by theophylline may produce life-threatening seizures.

Antagonism of adenosine has been proposed to explain the CNS-stimulant behavioural effects of caffeine and theophylline (Phillis *et al.*, 1979).

The involvement of adenosine antagonism in both slight and massive CNS-stimulation was supported by lack of stimulant behavioural effects, including lack of theophylline-like seizure activity in different animals species receiving enprofylline (Persson, 1982).

7.3.5 Opiate receptors

The habit of pharmacologists to classify types and sub-types of receptors has proved to be very useful: now we have different pre- and post-synaptic receptors and selective drugs. When the problem of multiple receptors arises the first question that has to be answered is the selectivity of agonist and antagonists used for studies. Table 22 shows the 'selective' agonists and antagonists of different opiate receptors (μ, κ, σ, δ, ϵ). Of the different pharmacological models, the guinea-pig ileum has mainly μ-receptors and the mouse vas deferens mainly σ-receptors. This interpretation was strongly supported by the finding that, in the mouse vas deferens, the opioid peptides are difficult to antagonize by naloxone in that they require a concentration about ten times higher than that needed for morphine.

Knoll (1977) suggested a completely different classification of opiate receptors: A and B. Opiate A receptors are located on the cholinergic axon terminals and B receptors on the catecholaminergic axon terminals.

Table 22. Agonists and antagonists of opiate receptors

Opioid receptor sub-type	Agonist	Antagonist	Reference
μ	Morphine	Naloxone[a]	Martin et al. (1976)
		Naltrexone[a]	Lord et al. (1977), Kosterlitz et al. (1980)
	FK – 33 824		Kosterlitz et al. (1980), Gillan and Kosterlitz (1982)
	Morphiceptin		Zhang et al. (1981)
κ	Ketocyclazocine	Naloxone[b]	Martin et al. (1976)
	Ethylketazocine	Naltrexone[b]	Lord et al. (1977), Kosterlitz et al. (1980). Gillan and Kosterlitz, 1982
	Bremazocine	MR 2266[a]	Römer et al. (1980)
	Dynorphin		Huidoboro-Toro et al. (1981)
σ	SKF 10 467	Naloxone[b]	Martin et al. (1976)
		Naltrexone	
	Phencyclidine		Zukin and Zukin, 1981
δ	Leu-enkephalin	Naloxone[b]	Lord et al. (1977)
	D-Ala-D-Leu-Enk.	Naltrexone[b]	Kosterlitz et al. (1980)
	Tyr-D-Ser-Gly-Phe-Leu-Thr		Gacel et al. (1980)
		ICI 154129[a]	Morley (1983)
ε	β-endorphin	Naloxone[b]	Wüster et al. (1979)
		Naltrexone[b]	Rónai et al. (1982)

[a]Preferential antagonism.
[b]Non-selective antagonism.

Clinical aspects of opioid peptides

Besides the known pharmacological spectra of morphine congeners the regional distribution of opioid peptides and their receptors serve as guidelines to establish the possible fields of therapeutic application of opioid peptides as well as to disclose their involvement in different pathophysiological processes.

Pain

In the central nervous system prominent opioid peptide receptor levels are present in areas related to pain sensation (pain, perception, motivation), in the limbic system, in the striatum, in the mediobasal hypothalamus and pituitary, and also in certain brainstem sites involved in central vasoregulation (Atweh and Kuhar, 1977a,b,c; Hökfelt et al., 1980c). At the periphery, the opiate receptors are the most abundant in certain neurons of the gastrointestinal tract, in some autonomic ganglia, in the adrenal medulla and in the carotic body. Some probably non-neural locations have also been described

(Tang *et al.*, 1982). As pain-killers when administered systemically the natural opioid peptides and the hitherto synthetized analogues failed to fulfil preliminary expectations. However, in administration by a intrathecal route—which is a well-established method in anesthesiology—the results are more promising (Ogama *et al.*, 1980). It is now widely accepted that some classical or impaired analgesic manipulations, like acupuncture, electro-acupuncture, certain kinds of trans-cutaneous nerve stimulation, and the stimulation of properly chosen brain sites through implanted electrodes produced analgesia which is mediated, at least in part, by endogenous opioid systems (cf. Terenius, 1978; Watkins and Mayer, 1982). It was suggested that in certain stress-related phenomena, endogenous opioids are also involved (Watkins and Mayer, 1982). The involvement of opioids in the pathomechanism of some states which altered pain sensation could also be assumed (Terenius, 1978).

Schizophrenia

Theoretically, it seems possible to alleviate schizophrenia by manipulating endogenous opioid systems. However, even the theories are rather controversial. Bloom *et al.* (1976) attributed a causative role to the assumed overproduction of endogenous opioid in schizophrenia, whilst Jacquet and Marks (1976) claimed that endorphin-related opioids might serve as endogenous anti-psychotic drugs. De Wied (1978) related the symptomatology of several psychiatric disorders to the hypothetized dysfunction of the processing of opioid peptides (cf. Praag and Verhoveen, 1980; Berger *et al.*, 1980). Supporting data have been reported for each theory, although clinically they have not been proved (Terenius *et al.*, 1976; Watson *et al.*, 1978; Kline and Lehmann, 1979; Verhoeven *et al.*, 1979; Praag and Verhoeven, 1980; Berger *et al.*, 1980). The most convincing evidence is the alleviation produced by the opiate antagonist naloxone in well-defined sub-populations of schizophrenics (Berger *et al.*, 1980).

Neuroendocrinology

It has long been known that the opioid peptides, even in a sub-analgesic dose range, are capable of influencing the secretion of some hypophyseal hormones, mostly by acting at hypothalamic regulatory sites (Meites *et al.*, 1979). The findings in laboratory animals have been corroborated by clinical experience in humans (Graffenried *et al.*, 1978) both for the Sandoz peptide FK 33824 (Roemer *et al.*, 1977) and for D-Met^2,Pro^5-enkephalinamide (Bajusz *et al.*, 1977). The possibilities of therapeutic or diagnostic utilization of this pharmacological effect of opioid peptides appear to be very modest.

Hyptertension

Some centrally acting anti-hypertensive drugs have been shown to act via central opioid systems (Kunos *et al.*, 1981; Farsang *et al.*, 1982). The contribution of endogenous opioids to the haemodynamic changes during various shock states has also been demonstrated (Holaday and Faden, 1981a,b).

Chapter 8

Summary

While the tremendous complexity of synaptic interactions of the nervous system is known and appreciated, some unconventional patterns of inter-neuronal communication begin to emerge, and have to be taken into account.

All available neurochemical, electrophysiological, pharmacological, and morphological evidence favours the hypothesis that substances (neuromodulators, neurotransmitters, and neurohormones) are not only released from varicosities devoid of typical synaptic specialization but are also able to reach remote target cells and to modulate signal processing and transmitter/modulator release or the effect of transmitters. This kind of modulation works as a tuning system, and their effectiveness depends on the distribution (presence and density) of the receptor which is able to recognize the long-distance message. This hypothesis is given credence by the findings that first, the surgical or neurochemical removal of the inhibitory pathway results in an enhanced release of the transmitter or the modulated substance, and second, administration of the modulator or the stimulation of the inhibitory axon leads to a reduction of the genuine modulator/transmitter release and the neuronal firing. The signal processing and the evoked activity pattern is not only the result of turning on or off given neuronal circuits, but the communication between neurons or neuron and target cell(s) can be modulated pre- and post-synaptically and pre- and post-junctionally.

In the cortex endogenous noradrenaline, for example, can probably be released from all varicosities and not only from the small proportion which make synaptic contact. It is thus likely that noradrenaline does not exert its effect solely on a one-to-one basis, on restricted areas of specializations, but can diffuse to reach distant targets. In this respect there is a striking resemblance between autonomic transmitters and central modulators.

It seems very likely that intercellular communication in both the peripheral and the central nervous systems is mediated by substance released from (1) non-synaptic axon terminals, (2) from axon terminals making synaptic contact but acting both on synaptic and non-synaptic target cells, or (3) from regions of the nerve cell other than the pre-synaptic axon terminal (dendrites, axon, soma). All or a combination of these procedures should play a physiological role in the modulation of neurochemical transmission and even in signal transmission. Neuropeptides, such as cholecystokinin, are stored and released from neurons and are able to excite remote neurones (cell body, dendrite), resulting in an increased firing and transmitter release.

Both in the central and peripheral nervous systems there is a rich diversity of chemical signals, together with a unique distribution of receptors for these signals, for modulator/transmitters, on some target cells but not on others. This would allow for chemical interaction between neurons without the need for precise synaptic connections. The modulator-containing neurons are varicose, highly branched and divergent and may involve a diffuse non-synaptic modulation of signal transmission. This is a novel mechanism for tonic modulation of excitability in large populations of neurons.

It seems very likely that different receptors located non-synaptically (extrasynaptically) might be the target of different drugs with either agonist or antagonist properties and stimulating or inhibiting the receptor, thereby mimicking or removing the effect of endogenous substances.

References

Abood, L. G., Kokitsu, K., and Miyamoto, S. (1962). Outflow of various phosphates during membrane depolarization of excitable tissues. *Am. J. Physiol.*, **202**, 469–74.

Abrahams, V. C., Koelle, G. B., and Smart, P. (1957). Histochemical demonstration of cholin esterases in the hypothalamus of the dog. *J. Physiol. (Lond.)*, **139**, 137–44.

Adams, P. R., and Galvan, M. (1981). Noradrenaline reduces calcium current in rat sympathetic neurones. *J. Physiol. (Lond.)*, **318**, 41–2.

Adam-Vizi, V., and Vizi, E. S. (1978). Direct evidence of acetylcholine releasing effect of serotonin in the Auerbach plexus. *J. Neural Transm.*, **42**, 127–38.

Adam-Vizi, V., and Seregi, A. (1982). Receptor independent stimulatory effect of noradrenaline on Na,K-ATPase in rat brain homogenate: role of lipid peroxidation. *Biochem. Pharmacol.*, **31**, 2231–6.

Adler-Graschinsky, E., and Langer, S. Z. (1975). Possible role of a β-adrenoceptor in the regulation of noradrenaline release by nerve stimulation through a positive feedback mechanism. *Br. J. Pharmacol.*, **53**, 43–50.

Aghajanian, G. K. (1972). Influence of drugs on the firing of serotonin-containing neurons in brain. *Fedn Proc.*, **31**, 91–6.

Aghajanian, G. K., and Bunney, B. S. (1973). In *Frontiers in Catecholamine Research* (Eds E. Usdin and S. Synder), Pergamon Press, New York, pp. 643–8.

Aghajanian, G. K., and Bunney, B. S. (1977). Dopamine 'autoreceptors': pharmacological characterization by micro-iontophoretic cell recording studies. *Naunyn–Schmiedeberg's Arch. Pharmacal.*, **297**, 1–7.

Ahlquist, R. P. (1948). A study of the adrenotropic receptors. *Amer. J. Physiol.*, **153**, 586–600.

Ajika, K., and Hökfelt, T. (1973). Ultrastructural identification of catecholamine neurons in the hypothalamic preventricular arcuate nucleus–median eminence complex with special reference to quantitative aspects. *Brain Res.*, **57**, 97–117.

Akasu, T., Hirai, K., and Koketsu, K. (1981). Increase of acetylcholine-receptor sensitivity by adenosine triphosphatase: a novel action of ATP on ACh-sensitivity. *Br. J. Pharmacol.*, **74**, 505–7.

Alberts, P., and Stjärne, L. (1982). Facilitation, and muscarinic and α-adrenergic inhibition of the secretion of ³H-acetylcholine and ³H-noradrenaline from guinea-pig ileum myenteric nerve terminals, *Acta Physiol. Scand.*, **116**, 83–92.

Alberts, P., Bártfai, T., and Stjärne, L. (1981). Site(s) and ionic basis of α-autoinhibition and facilitation of ³H-noradrenaline secretion in guinea-pig vas deferens, *J. Physiol. (Lond.)*, **312**, 297–334.

Ally, I., and Nakatsu, K. (1976). Adenosine inhibition of isolated rabbit ileum and antagonism by theophylline, *J. Pharmac. Exp. Ther.*, **199**, 208–15.

Ambache, N., and Zar, A. M. (1971). Evidence against adrenergic motor transmission in the in guinea-pig vas deferens. *J. Physiol.*, **216**, 359–89.

Ambache, N., Dunk, L. P., Verney, J., and Zar, M. A. (1972). Inhibition of post-ganglionic motor transmission in the vas deferens by indirectly acting sympathomimetic drugs. *J. Physiol. (Lond.)*, **227**, 433–56.

Ambache, N., Killick, S. W., Srinivasan, V., and Zar, M. A. (1975). Effects of lysergic acid diethylamide on autonomic post-ganglionic transmission. *J. Physiol. (Lond.)*, **246**, 571–93.

Anden, N. E., and Grabowska, M. (1977). FLA-136: Selective agonist at central α-adrenoceptors mediating changes in the turnover of noradrenaline. *Naunyn-Schmiedeberg's Arch. Pharmacol.*, **298**, 239–43.

202

Anden, N. E., and Stock, G. (1973). Effect of clozapine on the turnover of dopamine in the corpus striatum and in the limbic system. *J. Pharm. Pharmacol.*, **25**, 346–8.

Anderson, P., and Curtis, D. R. (1964). The pharmacology of synaptic and acetylcholine induced excitation of ventrobasal thalamic neurones. *Acta Physiol. Scand.*, **61**, 100–20.

Andersson, K. E., and Persson, C. G. A. (1980). Extrapulmonary effects of theophylline. *Eur. J. Respir. Dis. Suppl.*, **61**, 17–28.

Andréjak, M., Pommier, Y., Mouillé, P., and Schmitt, H. (1980). Effects of some α-adrenoceptor agonists and antagonists on the guinea-pig ileum. *Naunyn-Schmiedeberg's Arch. Pharmacol.*, **314**, 83–7.

Andersen, P., Eccles, J. C., and Voorhoeve, P. E. (1963). Inhibitory synapses on somas of Purkinje cells in the cerebellum. *Nature*, **199**, 655–6.

Angus, J. A., and Korner, P. I. (1980). Evidence against pre-synaptic α-adrenoceptor modulation of cardiac sympathetic transmission. *Nature*, **286**, 288–91.

Antonaccio, M. J., Halley, J., and Kerwin, L. (1974). Functional significance of α-stimulation and α-blockade on responses to cardiac nerve stimulation in anaesthetized dogs. *Life Sci.*, **15**, 765–77.

Arbilla, S., and Langer, S. Z. (1978a). Effects of γ-aminobutyric acid on the potassium and tyramine induced release of ^{3}H-noradrenaline from rat occipital cortex slices. *Br. J. Pharmacol.*, **63**, 389–90P.

Arbilla, S., and Langer, S. Z. (1978b). Morphine and β-endorphin inhibit release of noradrenaline from cerebral cortex but not of dopamine from rat striatum. *Nature*, **271**, 559–61.

Arbilla, S., Kamal, L., and Langer, S. Z. (1979). Pre-synaptic GABA autoreceptors on gabaergic nerve endings of the rat substantia nigra. *Eur. J. Pharmacol.*, **57**, 211–17.

Argiolas, A., Melis, M. S., Fadda, F., and Gessa, G. L. (1982). Evidence for dopamine autoreceptors controlling dopamine synthesis in the substantia nigra. *Brain Res.*, **234**, 177–81.

Arluison, M., and DeLa Manche, I. S. (1980). High resolution radioautographic study of the serotonin innervation of the rat corpus striatum after intraventricular administration of ^{3}H-5-hydroxytryptamine. *Neurosci.*, **5**, 229–40.

Armstrong, J. M., and Boura, A. L. A. (1973). Effects of clonidine and guanethidine on peripheral sympathetic nerve function in the pithead rat. *Br. J. Pharmacol.*, **47**, 850–2.

Arrang, J.-M., Garbarg, M., and Schwartz, J.-C. (1983). Auto-inhibition of brain histamine release mediated by a novel class (H₃) of histamine receptor. *Nature*, **302**, 832–7.

Atweh, S. F., and Kuhar, M. J. (1977a). Autoradiographic localization of opiate receptors in rat brain. I. Spinal cord and lower medulla. *Brain Res.*, **124**, 53–67.

Atweh, S. F., and Kuhar, M. J. (1977b). Autoradiographic localization of opiate receptors in rat brain. II. The brainstem. *Brain Res.*, **129**, 1–15.

Atweh, S. F., and Kuhar, M. J. (1977c). Autoradiographic localization of opiate receptors in rat brain. III. The telencephalon. *Brain Res.*, **134**, 393–403.

Atweh, S., Simon, J. R., and Kuhar, M. J. (1975). Utilization of sodium-dependent uptake *in vitro* as a measure of the activity of cholinergic neurons *in vivo*. *Life Sci.*, **17**, 1535–44.

Babad, H., Ben Zri, R., Bdolah, A., and Schram, M. (1967). The mechanism of enzyme secretion by cell. *Eur. J. Biochem.*, **1**, 96–104.

Bacopoulos, N. G., Spokes, E. G., Bird, E. D., and Roth, R. H. (1979). Antipsychotic drug action of schizophrenic patients: Effect on cortical dopamine metabolism after long-term treatment. *Science*, **205**, 1405–7.

Baer, H. P., and Drummond, G. I. (1979). *Physiological and Regulatory Functions of Adenosine and Adenine Nucleotides*, Raven Press, New York.

Bajusz, S., Rónai, A. Z., Székely, J. I., Gráf, L., Dunai-Kovács, Zs., and Berzétei, I. (1977). A superactive antinociceptive pentapeptide D-Met²-Pro⁵-enkephalinamide. *Febs. Lett.*, **76**, 91–2.

Baker, S., and Marshall, I. (1982). The effect of R781094. A selective α₂-adrenoceptor antagonist on ^{3}H-noradrenaline release in the mouse vas deferens. *Br. J. Pharmacol.*, **76**, 212P.

Baker, P. F., Hodgkin, A. L., and Ridgway, E. B. (1971). Depolarization and calcium entry in squid giant axons. *J. Physiol. (Lond.)*, **218**, 709–55.

Baldessarini, R. J., and Kopin, I. J. (1967). The effect of drugs on the release of norepinephrine-H from central nervous system tissue by electrical stimulation *in vitro*. *J. Pharmac. Exp. Ther.*, **156**, 31–8.

Baldessarini, R. J., and Tarsy, D. (1976). Mechanisms underlying tardive dyskinesia, in *The Basal Ganglia* (Ed. M. D. Yahr), Raven Press, New York, pp. 433–46.

Balfour, D. J. K., and Clasper, P. (1982). Evidence for a pre-synaptic dopamine receptor in rat hippocampus. *Br. J. Pharmacol.*, **76**, 292P.

Barchas, J. D., Akil, H., Elliott, G. R., Holman, R. B., and Watson, S. J. (1978). Behavioral neurochemistry: neuroregulators and behavioral states. *Science*, **200**, 964–73.

Barker, J. L. (1976). Peptides: roles in neuronal excitability. *Physiol. Rev.*, **56**, 435–48.

Barker, J. L., and Nicoll, R. A. (1973). The pharmacology and ionic dependency of amino acid responses in the frog spinal cord. *J. Physiol. (Lond.)*, **228**, 259–77.

Barlow, R. B., Berry, K. J., Glenton, P. A. M., Nicolson, N. M., and Sik, K. S. (1976). A comparison of affinity constants for muscarine-sensitive acetylcholine receptors of guinea-pig atrial pacemaker cells at 29°C and in ileum at 29°C and 37°C. *Br. J. Pharmacol.*, **58**, 613–20.

Bártfai, T. (1980). Cyclic nucleotides in the central nervous system. *Current Topics in Cellular Regulation* **16**, 225–69.

Bártfai, T. (1982). Cyclic nucleotides and the nervous system. *Trends in Pharm. Sciences*, **8**, 338–9.

Bartholini, G., Stadler, H., and Gadea-Ciria, M. (1977). Interaction of dopaminergic and cholinergic neurons in the extrapyramidal and limbic systems. *Adv. Biochem. Psychopharmacol.*, **16**, 391–5.

Bartholini, A., and Pepeu, G. (1967). Investigations into the acetylcholine cutput from the cerebral cortex of the cat in the presence of hyoscine. *Br. J. Pharmacol.*, **31**, 66–73.

Baum, R., and Shropshire, A. T. (1971). Influence of prostaglandins on autonomic responses. *Am. J. Physiol.*, **221**, 1470–5.

Baumann, P. A., and Koella, N. P. (1980). Feedback control of noradrenaline release as a function of noradrenaline concentration in the synaptic cleft in cortical slices of the rat. *Brain Res.*, **189**, 437–48.

Baumann, P. A., and Maitre, L. (1976). Is drug inhibition of dopamine uptake a misinterpretation of *in vitro* experiments? *Nature*, **264**, 789–90.

Baumgarten, H. G., Holstein, A. F., and Owmann, C. (1970). Auerbach's plexus of mammals and man: electron microscopic identification of three different types of neuronal processes in myenteric ganglia of the large intestine from Rhesus monkeys, guinea-pigs and man. *Z. Zellforsch. Mikros. Anat.*, **106**, 376–97.

Baumgarten, H. G., Björklund, A., Holstein, A. F., and Notin, A. (1972). Organization and ultrastructural identification of the catecholamine nerve terminals in the neural lobe and pars intermedia of the rat pituitary. *Z. Zellforsch. Mikrosk. Anat.*, **126**, 483–517.

Bayliss, W. M., and Starling, E. H. (1899). The movements and innervation of the small intestine. *J. Physiol. (Lond.)*, **24**, 99–143.

Baylor, D. A., and Nicholls, J. G. (1969). After-effects of nerve impulses on signalling in the central nervous system of the leech. *J. Physiol. (Lond.)*, **203**, 571–89.

Beani, L., and Bianchi, C. (1973). Effect of amantadine of cerebral acetylcholine release and content in the guinea-pig. *Neuropharmac.*, **12**, 283–9.

Beani, L., Bianchi, C., Santinoceto, L., and Crema, A. (1968). The cerebral acetylcholine release in conscious rabbits with semi-permanently implanted epidural cups. *Int. J. Neuropharmac.*, **7**, 469–81.

Beani, L., Bianchi, C., and Crema, A. (1969). The effect of catecholamines and sympathetic stimulation on the release of acetylcholine from the guinea-pig colon. *Br. J. Pharmacol.*, **36**, 1–17.

Beani, L., Bianchi, C., Gracomelli, A., and Tamberi, F. (1978). Noradrenaline inhibition of acetylcholine release from guinea-pig brain. *Eur. J. Pharmacol.*, **48**, 179–83.

Beani, L., Bianchi, C., and Siniscalchi, A. (1982). The effect of naloxone on opioid-induced inhibition and facilitation of acetylcholine release in brain slices. *Br. J. Pharmacol.*, **76**, 393–401.

Beaudet, A., and Descarries, L. (1976). Quantitative data on serotonin nerve terminals in adult rat neocortex. *Brain Res.*, **111**, 301–9.

Beaudet, A., and Descarries, L. (1978). The monoamine innervation of rat cerebral cortex: synaptic and non-synaptic axon terminals, *Neurosci.*, **3**, 851–60.

Beleslin, D., and Polak, R. L. (1965). Depression by morphine and chloralose of acetylcholine release from the rat's brain. *J. Physiol. (Lond.)*, **177**, 411–19.

Beleslin, D., Polak, R. L. and Sprouli, D. H. (1965). The effect of leptazol and strychnine on the acetylcholine release from the rat brain. *J. Physiol. (Lond.)*, **181**, 308–20.

Bell, C. (1980). Effects of dopamine on adrenergic neuromuscular transmission in the guinea-pig vas deferens. *Br. J. Pharmacol.*, **68**, 505–12.

Bennett, A. (1969). Studies on the avian gizzard: histochemical analysis of the extrinsic and intrinsic innervation. *Z. Zellforsch.*, **98**, 188–99.

Bennett, A., and Fleshler, B. (1970). Prostglandins and gastrointestinal tract. *Gastroenterology*, **59**, 790–8.

Bennett, A., Eley, K. G., and Scholes, G. B. (1968a). Effect of prostaglandins E_1 and E_2 on intestinal motility in the guinea-pig and rat. *Brit. J. Pharmacol.*, **34**, 639–47.

Bennett, A., Eley, K. G., and Scholes, G. B. (1968b). Effects of prostaglandins E_1 and E_2 on human guinea-pig and rat isolated small intestine. *Brit. J. Pharmacol.*, **34**, 630–8.

Bennett, M. R., Eley, K. G., and Stockley, H. L. (1975). Modulation by prostaglandins of contractions in guinea-pig ileum. *Prostaglandins*, **9**, 377–84.

Beny, J. L., and Baertschi, A. J. (1981). Release of vasopressin from the microdissected median eminence of the rat. *Brain Res.*, **206**, 469–73.

Berger, B., Tassin, J. P., Baluc, G., Moyne, M. A., and Thierry, A. M. (1974). Histochemical confirmation for dopaminergic innervation of the rat cerebral cortex after destruction of the noradrenergic ascending pathways. *Brain Res.*, **81**, 332–7.

Berger, P. A., Watson, S. J., Akil, H., Barchas, J. D., and Li, C. H. (1980). Clinical studies with naloxone and β-endorphin in chronic schizophrenia, in *Enzymes and Neurotransmitters in Mental Disease* (Eds E. Usdin, T. L. Sourkes, and M. B. H. Youdim), Plenum Press, New York, pp. 45–64.

Bergström, S., Farnebo, L. O., and Fuxe, K. (1973). Effect of prostaglandin E_2 on central and peripheral catecholamine neurons. *Eur. J. Pharmacol.*, **21**, 362–8.

Berman, P. H., Balis, M. E., and Dancis, J. (1969). Congenital hyperuricemia, an inborn error of purine metabolism associated with psychomotor retardation, athetosis and self-mutilation. *Arch. Neurol.*, **20**, 44–53.

Bernard, C. (1858). De l'influence de deux ordres de nerfs qui determinent les variations de couleur de sang veineux dans les organes glandulaires. *C.R. Acad. Sci.*, **47**, 245–53.

Berne, R. M., Rubio, R., and Curnish, R. R. (1974). Release of adenosine from ischemic brain. Effect of cerebral vascular resistance and incorporation into cerebral adenine nucleotides. *Circulat. Res.*, **35**, 262–71.

Bertaccini, G. (1982). Peptides: gastrointestinal hormones, in *Handbook of Experimental Pharmacology*, Vol. 59 (Ed. G. Bertaccini), Springer-Verlag, Hamburg, pp. 11–83.

Berthelsen, S., and Pettinger, W. A. (1977). A functional basis for classification of α-adrenergic receptors. *Life Sci.*, **21**, 596–606.

Bertler, A., and Rosengreen, E. (1959). Occurrence and distribution of dopamine in brain and other tissues. *Experientia*, **15**, 10–11.

Bevan, J. A., and Su, C. (1974). Variation of intra- and perisynaptic adrenergic transmitter concentrations with width of synaptic cleft in vascular tissues. *J. Pharmacol. Exp. Ther.*, **190**, 30–8.

Bevan, P., Bradshaw, C. M., and Szabadi, E. (1975). Effect of iprindole on responses of single cortical and caudate neurones to monoamines and acetylcholine. *Br. J. Pharmacol.*, **55**, 17–25.

Bevan, P., Bradshaw, C. M., and Szabadi, E. (1977). Comparison of the effects of dopamine and noradrenaline on single cortical neurones. *Br. J. Pharmacol.*, **59**, 468–9P.

Bhagat, B., Dhalla, N. S. Ginn, D., Montagne, A. E., and Montier, A. D. (1972). Modification by prostaglandin E_2 (PGE_2) of the response of guinea-pig isolated vasa deferentia and atria to adrenergic stimuli. *Br. J. Pharmacol.*, **44**, 689–98.

Bianchi, C., Tanganalle, S., and Beani, L. (1979). Dopamine modulation of acetylcholine release from the guinea-pig brain. *Eur. J. Pharmacol.*, **58**, 235–46.

Biscoe, T. J., and Straughan, D. W. (1966). Micro-electrophoretic studies of neurones in the rat hippocampus. *J. Physiol. (Lond.)*, **183**, 341–59.

Biswas, B., and Carlsson, A. (1977). The effect of intracerebroventricularly administered GABA on brain monoamine metabolism. *Naunyn–Schmiedeberg's Arch. Pharmacol.*, **299**, 41–56.

Björklund, A., and Lindvall, O. (1975). Dopamine in dendrites of substantia nigra neurons: suggestions for a role in dendritic terminal. *Brain Res.*, **83**, 531–7.

Björklund, A., Cegrell, L., Falck, B., Ritzén, M., and Rosengren, E. (1970). Dopamine containing cells in sympathetic ganglia. *Acta Physiol. Scand.*, **78**, 334–8.

Blakeley, A. G. H., and Cunnane, T. C. (1979). The packeted release of transmitter from the sympathetic nerves of the guinea-pig vas deferens: an electrophysiological study. *J. Physiol. (Lond.)*, **296**, 85–96.

Blakeley, A. G. H., Brown, G. L., Dearnaley, D. P., and Woods, R. I. (1969). Perfusion of the spleen with blood containing preostaglandin E_1: transmitter liberation and uptake. *Proc. R. Soc. B.*, **174**, 281–92.

206

Blakeley, A. G. H., Cunnane, T.C., and Peterson, S. A. (1982). Local regulation of transmitter release from rodent sympathetic nerve terminals. *J. Physiol. (Lond.)*, **325**, 93–109.

Blaschko, H. (1957). Metabolism and storage of biogenic amines. *Experimentia*, **13**, 9–13.

Blaustein, M. P., Johnson, E. M., and Needleman, P. (1972). Calcium-dependent norepinephrine release from pre-synaptic nerve endings *in vitro*. *Proc. Nat. Acad. Sci. (Wash.)*, **69**, 2237–40.

Bloom, S. R., and Edwards, A. V. (1980). Vasoactive intestinal peptide in relation to atropine resistant vasodilatation in the submaxillary gland of the rat. *J. Physiol. (Lond.)*, **300**, 41–53.

Bloom, F. E., and Iversen, L. L. (1971). Localizing ^{3}H-GABA in nerve terminals of rat cerebral cortex by electron-microscopic autoradiography. *Nature*, **229**, 628–30.

Bloom, F. E., Costa, E., and Salmoiraghi, G. C. (1965). Anesthesia and the responsiveness of individual neurons of the caudate nucleus of the cat to acetylcholine, norepinephrine and dopamine administered by microelectrophoresis. *J. Pharmacol. Exp. Ther.*, **150**, 244–52.

Bloom, F. E., Oliver, A. P., and Salmoiraghi, G. C. (1963). The responsiveness of individual hypothalamic neurones to microelecetrophoretically administered endogenous amines. *Int. J. Neuropharmac.*, **2**, 181–93.

Bloom, F. E., Segal, D., Ling, N., and Guillemin, R. (1976). Endorphins: profound behavioral effects in rats suggest new etiological factors in mental illness. *Science*, **194**, 630–2.

Bogdansky, D. F., Weissbach, H., and Udenfriend, S. (1956). The distribution of serotonin, 5-hydroxytryptophan decarboxylase and monoamine oxydase in the brain. *J. Neurochem.*, **1**, 272–8.

Bolton, B. (1973). Effects of electrogenic sodium pumping on the membrane potential of longitudinal smooth muscle from terminal ileum of guinea-pig. *J. Physiol. (Lond.)*, **228**, 693–712.

Bonaccorsi, A., Hermsmeyer, K., Smith, C. G., and Bohr, D. F. (1977). Norepinephrine release in isolated arteries induced by K-free solution. *Am. J. Physiol.*, **232**, 140–5.

Borowski, E., Starke, K., Ehrlich, H., and Endo, T. (1977). A comparison of pre- and post-synaptic effects of α-adrenolytic drugs in the pulmonary artery of the rabbit. *Neuroscience*, **2**, 285–96.

Bosler, O. (1978). Radioautographic identification of serotonin axon terminals in the rat organum vasculosum laminae terminals, *Brain Res.*, **150**, 177–81.

Bourdois, P. S., Mitchell, J. F., Somogyi, G. T., and Szerb, J. C. (1974). The output per stimulus of acetylcholine from cerebral cortical slices in the presence or the absence of cholinesterase inhibition. *Br. J. Pharmacol.*, **52**, 509–17.

Bourgoin, S. F., Artaud, A., Enjialbert, F., Héry, J., Glowinski, J., and Hamon, M. (1977). Acute changes in central serotonin metabolism induced by the blockade or stimulation of serotoninergic receptors during ontogenesis in the rat. *J. Pharmacol. Exp. Ther.*, **202**, 519–31.

Bower, A., Hadley, M. E., and Hruby, V. I. (1974). Biogenic amines and control of melanophore stimulating hormone release. *Science*, **184**, 70–2.

Bowern, D. M., and Marek, K. L. (1982). Evidence for the pharmacological similarity between the central pre-synaptic muscarinic autoreceptor and post-synaptic muscarinic receptors. *Br. J. Pharmacol.*, **75**, 367–72.

Bowery, N. G., Hill, D. R., Hudson, A. L., Doble, A., Middlemiss, D. N., Shaw, J., and Turnbull, M. (1980). (−)Baclofen decreases neurotransmitter release in the mammalian CNS by an action at a novel GABA receptor. *Nature*, **283**, 92–4.

Bowman, W. C., and Raper, C. (1976). Adrenotropic receptors in skeletal muscle. *Ann. NY. Acad. Sci.*, **139**, 741–50.

Bowman, W. C., and Webb, S. N. (1976). Tetanic fade during partial transmission failure produced by non-depolarizing neuromuscular blocking drugs in the cat. *Clin. Exp. Pharmacol, and Physiol.*, **3**, 545–55.

Bradford, H. F. (1970). Metabolic response of synaptosomes to electrical stimulation: release of amino acids. *Brain Res.*, **19**, 239–47.

Branisteanu, D. D., Haulica, I. D., Proca, B., and Nhue, B. G. (1979). Adenosine effects upon transmitter release parameters in the Mg^{2+}-paralyzed neuro-muscular junction of frog. *Naunyn-Schmiedeberg's Arch. Pharmacol.*, **308**, 273–9.

Brase, D. A. (1980). Pre- and post-synaptic striatal dopamine receptors: differential sensitivity to apomorphine inhibition of ^{3}H-dopamine and ^{14}C GABA release *in vitro*. *J. Pharm. Pharmacol.*, **32**, 432–3.

Breckenridge, B. M., Burn, J. H., and Matschinsky, F. M. (1967). Theophylline, epinephrine and neostigmine facilitation of neuromuscular transmission. *Proc. Nat. Acad. Sci. USA*, **57**, 1893–7.

Bridges, R. E., Hillhouse, E. W., and Jones, M. T. (1979). The effect of dopamine on neurohypophysial hormones release *in vivo* and from the rat neural lobe and hypothalamus *in vitro*. *J. Physiol. (Lond.)*, **260**, 647–66.

Briggs, C. A., and Cooper, J. R. (1982). Cholinergic modulation of the release of ³H-acetylcholine from synaptosomes of the myenteric plexus. *J. Neurochem.*, **38**, 501.

Brimble, M. J., and Wallis, D. I. (1973). Histamine H₁- and H₂-receptors at a ganglionic synapse. *Nature*, **246**, 156–7.

Brodde, O. E. (1982). Vascular dopamine receptors: demonstration and characterization by *in vitro* studies of vascular DA-receptors. *Life Sciences*, **31**, 284–306.

Brodie, B. B., and Shore, P. A. (1957). A concept for a role of serotonin and epinephrine as chemical mediators in the brain. *Ann. N.Y. Acad. Sci.*, **66**, 631–42.

Brody, M. J., and Kadowitz, P. J. (1974). Prostaglandins as modulators of the autonomic nervous system. *Fedn Proc.*, **33**, 48–60.

Brown, D. A., and Caulfield, M. P. (1979). Hyperpolarizing α₂-adrenoceptors in rat sympathetic ganglia. *Br. J. Pharmacol.*, **65**, 435–45.

Brown, D. A., and Caulfield, M. P. (1981). Adrenoceptors in ganglia, in *Adrenoceptors and Catecholamine Action* (Ed. G. Kunos), John Wiley, London, pp. 265–79.

Brown, D. A., and Higgins, A. J. (1979). Pre-synaptic effects of aminobutyric acid in isolated rat superior cervical ganglia. *Br. J. Pharmacol.*, **64**, 108–9.

Brown, D. A., and Medgett, I. C. (1982). Functional role of pre-synaptic α and β adrenoceptors in rat isolated superior cervical ganglion. *Br. J. Pharmacol.*, **75**, 18P.

Brown, D. A., Dunn, P. M., and Marsh, S. (1982). Possible electrophysiological mechanisms for modulating transmitter release, in *Pre-synaptic Receptors* (Ed. J. S. De Belleroche), Ellis Horwood, London.

Brown, G. L. (1960). Release of sympathetic transmitter by nerve stimulation, in *GABA Symposium on Adrenergic Mechanisms* (Ed. J. R. Vane), Little, Brown, New York, pp. 116–24.

Brown, G. L., and Feldberg, W. (1936). The activity of potassium on the superior cervical ganglion of the cat. *J. Physiol.*, **86**, 290–305.

Brown, G. L., and Gillespie, J. S. (1956). Output of sympathin from the spleen. *Nature (Lond.)*, **178**, 980.

Brown, G. L., and Gillespie, J. S. (1957). Output of sympathetic transmitter from the spleen of the cat. *J. Physiol. (Lond.)*, **138**, 81–102.

Brown, J., Doxey, J. C., and Handley, S. (1980). Effects of α--adrenoceptor agonists and antagonists and of anti-depressant drugs on pre- and post-synaptic α–adrenoceptors. *Eur. J. Pharmacol.*, **67**, 33–40.

Brownlee, G., and Johnson, E. S. (1963). The site of the 5-hydroxytryptamine receptor on the intraneural nervous plexus of the guinea-pig isolated ileum. *Br. J. Pharmacol.*, **21**, 306–22.

Brownstein, M. J., Saavedra, J. M., Axelrod, J. M., Zeman, G. H., and Carpenter, D. O. (1974). Coexistence of several putative neurotransmitters in single identified neurons of *Aplysia*. *Proc. Natl. Acad. Sci., USA*, **71**, 4662–5.

Brücke, F. T. (1935). Über die Wirkung von Acetylcholine auf die pilomotorns. *Klin. Wschr*, **14**, 7–9.

Byrant, B. J., McCullogh, M. W., Rand, M. U., and Story, D. F. (1975). Release of ³H-(−)-noradrenaline from guinea-pig hypothalamic slices: effects of adrenoceptors agonists and antagonists. *Br. J. Pharmacol.*, **53**, 454-9.

Bülbring, E. (1944). The action of adrenaline on transmission in the superior cervical ganglion, *J. Physiol. (Lond.)*, **103**, 55–67.

Bülbring, E., and Crema, A. (1959). The action of 5-hydroxytryptamine, 5-hydroxytryptophan and reserpine on peristalsis in anaesthetized guinea-pigs. *J. Physiol.*, **146**, 29–53.

Bülbring, E., and Gershon, M. D. (1968). 5-hydroxytryptamine participation in the vagal inhibitory innervation of the stomach. *J. Physiol. (Lond.)*, **192**, 823–46.

Bülbring, E., and Tomita, T. (1969). Increase of membrane conductance by adrenaline in the smooth muscle of guinea-pig taenia coli. *Proc. R. Soc. (Lond.)*, **172**, 89–102.

Bunney, B. S., and Aghajanian, G. K. (1976). D-amphetamine-induced inhibition of central dopaminergic neurones. Mediation by a striato-nigral feedback pathway. *Science*, **192**, 391–3.

Bunney, B. S., and Aghajanian, G. K. (1979). D-amphetamine-induced depression of central dopamine neurons: evidence for mediation by both autoreceptors and a striato-nigral feedback pathway. *Naunyn-Schmiedeberg's Arch. Pharmacol.*, **304**, 255–61.

Bunney, B. S., Aghajanian, G. K., and Roth, R. H. (1973a). Comparison of effects of L-dopa, amphetamine and apomorphine on firing rate of rat dopaminergic neurones. *Nature*, **245**, 123–5.

Bunney, B. S., Aghajanian, G. K., and Roth, R. H. (1973b). Comparison of effects of L-dopa, amphetamine and apomorphine on firing rate of rat dopaminergin neurons. *Nature New Biol.*, **245**, 123–5.

Bunney, B. S., Walters, J. R., Roth, R. A., and Aghajanian, G. K. (1973c). Dopaminergic neurons: effect of antipsychotic drugs and amphetamine on single cell activity. *J. Pharmac. Exp. Ther.* **185**, 560–571.

Burn, J. H. (1976). The discovery of uptake of adrenaline in 1930–1933 and the development of the adrenergic fibre from a cholinergic fibre. *J. Pharm. Pharmac.*, **28**, 342–47.

Burn, J. H., and Rand, M. J. (1959). Sympathetic postganglionic mechanism. *Nature (Lond.)*, **184**, 163–5.

Burn, J. H., and Rand, M. J. (1965). Acetylcholine in adrenergic transmission. *Ann. Rev. Pharmac.*, **5**, 163–82.

Burnstock, G. (1970). Structure of smooth muscle and its innervation, in *Smooth Muscle* (Eds E. Bülbring, A. Brading, A. Jones, and T. Tomita), Edward Arnold, London, pp. 1–69.

Burnstock, G. (1972). Purinergic nerves. *Pharm. Rev.*, **24**, 509–81.

Burnstock, G. (1975a). Comparative studies of purinergic nerves. *J. Exp. Zoology*, **194**, 103–34.

Burnstock, G. (1975b). Innervation of vascular smooth muscle histochemistry and electron-microscopy. *Clin. Exp. Pharmacol. Physiol. Suppl.*, **2**, 7–20.

Burnstock, G. (1976). Do some nerve cells release more than one transmitter? *Neuroscience*, **1**, 239–48.

Burnstock, G. (1977). Cholinergic, adrenergic and purinergic neuromuscular transmission. *Fedn Proc.*, **36**, 2434–8.

Burnstock, G. (1978a). Cholinergic and purinergic regulation of blood vessels, in *Handbook of Physiology* (*Vascular Smooth Muscle*) (Eds D. Bohr, M. S. Somlyó, and H. V. Sparks) Williams and Wilkins, Baltimore, Md.

Burnstock, G. (1978b). A basis for distinguishing two types of purinergic receptor, in *Cell Membrane Receptors for Drugs and Hormones: A multidisciplinary Approach* (Eds R. W. Straub and L. Bolis), Raven Press, New York, pp. 107–18.

Burnstock, G. (1979a). The ultrastructure of autonomic cholinergic nerves and junctions. *Progr. Brain Res.*, **79**, 3–21.

Burnstock, G. (1979b). Post and current evidence for the purinergic nerve hypothesis, in *Physiological and Regulatory Functions of Adenosine and Adenine Nucleotides* (Eds A. P. Bocer and G. I. Drummond), Raven Press, New York, pp. 3–32.

Burnstock, G. (1981). Neurotransmitters and trophic factors in the autonomic nervous system. *J. Physiol. (Lond.)*, **313**, 1–35.

Burnstock, G., and Costa, M. (1975). *Adrenergic Neurons, their Organization, Function and Development in the Peripheral Nervous System*, Chapman and Hall, London.

Burnstock, G., Campbell, G., and Rand, M. J. (1966). The inhibitory innervation of the taenia of the guinea-pig caecum. *J. Physiol. (Lond.)*, **182**, 504–26.

Burnstock, G., Gannon, B. J., and Iwayama, T. (1970a). Sympathetic innervation of vascular smooth muscle in normal and hypertensive animals. *Circ. Res.*, **27**, Suppl. II, 5–24.

Burnstock, G., Campbell, G., Satchell, D. G., and Smythe, A. (1970b). Evidence that adenosine triphosphate or a related nucleotide is the transmitter substance released by non-adrenergic inhibitory nerves in the gut. *Br. J. Pharmacol. Chemother.*, **40**, 668–88.

Burnstock, G., Cocks, T., Kasakov, L., and Wong, H. K. (1978). Direct evidence for ATP release from nonadrenergic non-cholinergic (purinergic) nerves in the guinea-pig taenia coli and bladder. *Eur. J. Pharmacol.*, **49**, 145–9.

Butcher, L. L. (1978). Chemical communication processes involving neurons: vocabulary and syntax, in *Cholinergic–monoaminergic Interactions in the Brain* (Ed. L. L. Butcher), Academic Press, New York, pp. 3–22.

Butcher, R. W., and Sutherland, E. W. (1962). Adenosine 3,5-phosphate in biological materials. I. Purification and properties of cyclic $3',5'$-nucleotide phosphodiesterase and use of this enzyme to characterize adenosine $3',5'$-phosphate in human urine. *J. Biol. Chem.*, **237**, 1244.

Cajal, S. Ramón y (1983). Sur les ganglions et plexus de l'intestin. *C.R. Soc. Biol. (Paris)*, **5**, 217–23.

Cajal, S. Ramón y (1911). *Histologie du systeme nerveux de l'homme et des vertebres*, Vol. II, Malione, Paris.

Calas, A., Alonso, G., Arnauld, E., and Vincent, J. D. (1974). Demonstration of indolaminergic fibers in the median eminence of the duck, rat and monkey. *Nature*, **250**, 242–3.

Calas, A., Besson, M. J., Cauchy, C., Alonso, G., Glowinsky, J., and Cheramy, A. (1976). Radioautographic study of *in vivo* incorporation of ^{3}H-monoamines in the cat caudate nucleus: identification of serotoninergic fibers. *Brain Res.*, **118**, 1–13.

Cambridge, D., and Davey, M. J. (1980). Comparison of the α-adrenoceptors located on sympathetic and parasympathetic nerve terminals. *Br. J. Pharmacol.*, **69**, 345–6P.

Campbell, G. (1966). The inhibitory fibres in the vagal supply to the guinea-pig stomach. *J. Physiol. (Lond.)*, **85**, 600–12.

Cannon, J. G. (1981). Future directions in dopaminergic nervous system and dopaminergic agonists. *J. Med. Chemistry*, **24**, 1113–18.

Carlsson, A. (1975). Receptor mediated control of dopamine metabolism, in *Pre- and Post-synaptic Receptors* (Eds E. Usdin and W. E. Bunney), Marcel Dekker, New York, pp. 49–65.

Carlsson, A., and Lindqvist, M. (1963). Effect of chlorpromazine or haloperidol on formation of 3-methoxytyramine and normetanephrine in mouse brain. *Acta Pharmac. Toxic.*, **20**, 140–4.

Carlsson, A., Lindquist, M., and Magnusson, T. (1957). 3,4-Dihydroxyphenylalanine and 5-hydroxytryptophan as reserpine antagonists. *Nature*, **180**, 1200–1.

Carlsson, A., Persson, T., Roos, B. E. and Walinder, J. (1972). Potentiation of phenothiazines by α-methyltyrosine in treatment of chronic schizophrenia. *J. Neural. Trans.*, **33**, 83–90.

Caroon, J. M., Clark, R. D., Kluge, A. F., Olah, R., Repke, D. B., Unger, S. H., Michel, A. D., and Whiting, R. L. (1982). Structure–activity relationships for 2-substituted imidazoles as α_2-adrenoceptor antagonists. *J. Med. Chem.*, **25**, 666–70.

Carraway, R., and Leeman, S. E. (1976). Characterization of radioimmunoassayable neurotensin in the rat, its differential distribution in the central nervous system, small intestine and stomach. *J. Biol. Chem.*, **251**, 7045–52.

Carstens, E., Tulloc, I., Ziegelgänsberger, W., and Zimmermann, M. (1979). Presynaptic excitability changes induced by morphine in single cutaneous afferent C and A fibers. *Pflügers Arch.*, **379**, 143–7.

Cavero, I., and Roach, A. G. (1978). The effects of prazosin on the clonidine induced hypotension and bradycardia in rats and sedation in chicks. *Br. J. Pharmacol.*, **62**, 468–9.

Cavero, I., Dennis, T., Lefevre-Borg, F., Perrot, P., Roach, A. G., and Scatton, B. (1979). Effect of clonidine, prazosin and phentolamine on heart rate and coronary sinus catecholamine output during cardioaccelerator nerve stimulation in spinal dogs. *Br. J. Pharmacol.*, **67**, 283–92.

Cavero, I., Lefevre-Borg, F., and Roach, A. G. (1977). Differential effect of prazosin on the pre- and post-synaptic α-adrenoceptors in the rat and dog. *Br. J. Pharmacol.*, **61**, 469P.

Cavero, I., Massingham, K., and Lefevre-Borg, F. (1982). Peripheral dopamine receptors, potential targets for a new class of anti-hypertensive drugs. *Life Sci.*, **31**, 939–48.

Celesia, G. G., and Jasper, H. J. (1966). Acetylcholine released from cerebral cortex in relation to state of activation. *Neurology, Minneap.*, **16**, 1053–63.

Celsen, B., and Kuschinsky, K. (1974). Effect of morphine on kinetics of ^{14}C-dopamine in rat striatal slices. *Naunyn–Schmiedeberg's Arch. Pharmacol.*, **284**, 159–65.

Celuch, S. M., Dubocovich, M. L., and Langer, S. Z. (1978). Stimulation of pre-synaptic β-adrenoceptors enhances ^{3}H-noradrenaline release during nerve stimulation in the perfused cat spleen. *Br. J. Pharmacol.*, **63**, 97–109.

Cerrito, F., and Raiteri, M. (1979a). Serotonin release is modulated by pre-synaptic autoreceptors. *Eur. J. Pharmacol.*, **57**, 427–30.

Cerrito, F., and Raiteri, M. (1979b). Evidence for an autoreceptor-mediated presynaptic control of serotonin release in central nerve ending. *Br. J. Pharmacol.*, **67**, 424–5.

Chan-Palay, V. (1975). Fine structure of labelled axons in the cerebellar cortex and nuclei of rodents and primates after intraventricular infusions with tritiated serotonin. *Anat. Embryol.*, **148**, 235–65.

Chan-Palay, V. (1976). Serotonin axons in the supra- and subependymal plexuses and in the leptomeninges: their roles in local alterations of cerebrospinal fluid and vasomotor activity. *Brain Res.*, **102**, 103–30.

Chan-Palay, V. (1977). *Cerebellar Dentate Gyrus*, Springer-Verlag, Berlin, pp. 418–25.

Chan-Palay, V., and Palay, S. L. (1977). Ultrastructural identification in rat sensory ganglia and their terminals in the spinal cord by immunocytochemistry. *Proc. Nat. Acad. Sci.*, **74**, 4050–4.

Cheng, H. S., Gleason, E. M., Nathan, B. A., Lachman, P. J., and Woodward, J. K. (1981). Effects of clonidine on gastric acid secretion in the rat. *J. Pharmacol. Exp. Ther.*, **217**, 121–6.

Chéramy, A., Leviel, V., and Glowinski, J. (1981). Dendritic release of dopamine in the substantia nigra. *Nature*, **289**, 537–42.

Chéramy, A., Nieoullon, A., Michelot, R., and Glowinski, J. (1977). Effects of peripheral and local administration of picrotoxin on the release of newly synthesized ³H-dopamine in the caudate nucleus of the cat. *Naunyn-Schmiedeberg's Arch. Pharmacol.*, **297**, 31–7.

Chéramy, A., Michelot, R., Leviel, V., Nieoullon, A., Glowinsky, J., and Kerdelhué, B. (1978). Effect of the immunoneutralization of substance P in the cat substantia nigra on the release of dopamine from dendrites and terminals of dopaminergic neurons. *Brain Res.*, **155**, 404–8.

Chong, R., and Downing, K. (1974). Reversal by prostaglandin E₂ of the inhibitory effect of indomethacin on contractions of guinea pig ileum induced by angiotensin. *J. Pharm. Pharmacol.*, **26**, 729.

Christ, D. D., and Nishi, S. (1971). Site of adrenaline blockade in the superior cervical ganglion of the rabbit. *J. Physiol. (Lond.)*, **213**, 107–18.

Chubb, I. W. (1977). The release of non-transmitter substances, in *Synapses* (Eds G. A. Cottrell and P. N. R. Usherwood), Blackie, Glasgow and London, pp. 264–90.

Chubb, I. W., and Smith, A. D. (1975a). Isoenzymes of soluble and membrane-bound acetylcholin-esterase in bovine splanchnic nerve and adrenal medulla. *Proc. R. Soc. Lond. B.*, **191**, 245–61.

Chubb, I. W., and Smith, A. D. (1975b). Release of acetylcholinesterase into the perfusate from the noradrenal gland. *Proc. R. Soc. Lond. B.*, **191**, 263–71.

Chubb, I. W., Goodman, S., and Smith, A. D. (1976). Is acetylcholinesterase secreted from central neurons into the cerebrospinal fluid? *Neuroscience*, **1**, 57–62.

Cedarbaum, I. M., and Aghajanian, G. K. (1977). Catecholamines receptors on locus coeruleus neurons: pharmacological characterization. *Eur. J. Pharmacol.*, **44**, 375–85.

Clanachan, A. S., Johns, A., and Paton, D. M. (1977). Pre-synaptic inhibitory actions of adenine nucleotides and adenosine on neurotransmission in the rat vas deferens. *Neuroscience*, **2**, 597–602.

Clark, B. J., and Menninger, K. (1980). Peripheral dopamine receptors' circulation. *Brain Res.*, **46**, 1–59.

Coggershall, R. E. (1970). A cytological analysis of the bag cell control of egg laying in *Aplysia*. *J. Morphology*, **132**, 461–85.

Cohen, M. W., Gerschenfield, H. M., and Kuffler, S. W. (1968). Ionic environment of neurons and glial cells in the brain of an amphibian. *J. Physiol. (Lond.)*, **197**, 363–80.

Cohen, M. L., Wiley, K. S., and Landry, A. S. (1980). *In vitro* comparison of the pre- and postsynaptic alpha adrenergic receptor blocking properties of prazosin and tiodazosin, *Clinical and exp. Hypertension*, **2(6)**, 1067–82.

Cole, A. E., and Shinnick-Gallagher, P. (1980). Alpha-adrenoreceptor and dopamine receptor antagonists do not block the slow inhibitory post-synaptic potential in sympathetic ganglia. *Brain Res.*, **187**, 226–30.

Collis, M. G., and Shepherd, J. T. (1980). Interaction of the trycyclic anti-depressant amitriptyline with prejunctional alpha and muscarinic receptors in the dog saphenous vein. *J. Pharmacol. Exp. Ther.*, **213**, 616–22.

Consolo, S., Garattini, S., Ladinsky, H., and Thoenen, T. (1972). Effect of chemical sypathectomy on the content of acetylcholine, choline and choline acetyltransferase activity. *J. Physiol. (Lond.)*, **220**, 639–46.

Cook, R. D., and Burnstock, G. (1976). The ultrastructure of Auerbach's plexus in the guinea-pig. I. Neuronal elements. *J. Neurocytol.*, **5**, 171–94.

Cook, M. A., Hamilton, J. T., and Okwuasaba, F. K. (1978). Coenzyme A or adenosine inhibiting acetylcholine release. *Nature*, **274**, 721–2.

Cools, A. R., and Van Rossum, J. W. (1980). Multiple receptors for brain dopamine in behavior regulation: concept of dopamine-E and dopamine-I receptors. *Life Sci.*, **27**, 1237–53.

Coombs, J. S., Curtis, D. R., and Eccles, I. C. (1957). The generation of impulses in motoneurones. *J. Physiol. (Lond.)*, **139**, 232–49.

Cooper, J. R., Bloom, F. E., and Roth, R. H. (1982). *The Biochemical Basis of Neuropharmacology*, Oxford University Press, New York and Oxford.

Corrigall, W. A., and Linsemann, M. A. (1980). A specific effect of morphine on evoked activity in the rat hippocampal slice. *Brain Res.*, **192**, 227–38.

Corrodi, H., Fuxe, K., Hamberger, B., and Ljungdahl, A. (1970). A study on central and peripheral noradrenaline neurons using a new dopamine-ß-hydroxylase inhibitor. *Eur. J. Pharmacol.*, **12**, 145–55.

Costa, M., and Furness, J. B. (1971). Storage, uptake and synthesis of catecholamines in the intrinsic adrenergic neurones in the proximal colon of the guinea-pig. *Z. Zellforsch.*, **120**, 364–85.

Costa, M., and Furness, J. B. (1979a). Commentary: on the possibility that an indoleamine is a neurotransmitter in the gastrointestinal tract. *Biochem. Pharmacol.*, **28**, 565–71.

Costa, M., and Furness, J. B. (1979b). The sites of action of 5-HT in nerve muscle preparations from guinea-pig small intestine and colon. *Br. J. Pharmacol.*, **65**, 237–52.

Costa, M., and Furness, J. B. (1982). Neuronal peptides in the intestine. *Brit. Med. Bull.*, **38**, 247–52.

Costa, M., and Gabella, G. (1971). Adrenergic innervation of the alimentary canal. *Z. Zellforsch.*, **122**, 357–77.

Costa, M., Cuello, A. C., Furness, J. B., and Franco, R. (1980). Distribution of enteric neurons showing immunoreactivity for substance P in the guinea-pig ileum. *Neurosci.*, **5**, 323–31.

Costa, M., Furness, J. B., and Gabella, G. (1971). Catecholamine containing nerve cells in the mammalian myenteric plexus. *Histochemie*, **25**, 103–6.

Cottrell, G. A. (1976). Does the giant cerebral neurone of *Helix* release two transmitters: acetylcholine and serotonin? *J. Physiol. (Lond.)*, **259**, 44P.

Cottrell, G. A. (1977). Identified amine-containing neurones and their synaptic connections. *Neurosci.*, **2**, 1–18.

Cowie, A. L., Kosterlitz, H. W., and Watt, A. J. (1968). Mode of action of morphine-like drugs on autonomic neuroeffectors. *Nature*, **220**, 1040–2.

Cox, B., and Hecker, S. E. (1971). Investigation of the mechanism of action of oxotremorine on the guinea-pig isolated ileum preparation. *Br. J. Pharmacol.*, **41**, 19–25.

Cresen, I., Burt, D. R., and Snyder, H. (1976). Dopamine receptor binding predicts clinical and pharmacological potencies of anti-schizophrenic drugs. *Science*, **192**, 481–3.

Cubeddu, L. X., Barnes, E., Langer, S. Z., and Weiner, N. (1974a). Release of norepinephrine and dopamine-β-hydroxylase by nerve stimulation. I. Role of neuronal and extraneuronal uptake and of α-pre-synaptic receptors. *J. Pharmacol. Exp. Ther.*, **190**, 431–50.

Cubeddu, L. X., Barnes, E., and Weiner, N. J. (1974b). Release of norepinephrine and dopamine-β-hydroxylase by nerve stimulation. II. Effect of papaverine. *J. Pharmac. Exp. Ther.*, **191**, 444–57.

Cuello, A. C. (1978). Endogenous opioid peptides in neurons of the human brain. *Lancet*, **II**, 291–3.

Cuello, A. C. (1983). Non-classical neuronal communications. *Fedn. Proc.*, **42**, 2912–22.

Cuello, A. C., and Iversen, L. L. (1978). Interactions of dopamine with other neurotransmitters in the rat substantia nigra: a possible functional role of dendritic dopamine, in *Interactions among Putative Neurotransmitters in the Brain* (Eds S. Garattini, J. F. Pujol, and R. Samanin), Raven Press, New York, pp. 127–49.

Cuello, A. C., and Paxinos, G. (1978). Evidence for a long leu-enkephalin striopallidal pathway in rat brain. *Nature*, **271**, 178–80.

Cunnane, T. C., and Stjärne, L. (1982). Secretion of transmitter from individual varicosities of guinea-pig and mouse vas deferens: all-or-none and extremely intermittent. *Neurosci*, **7**, 2565–76.

Dahlöf, C. (1981). Studies on β-adrenoceptor mediated facilitation of sympathetic neuro-transmission. *Acta Physiol. Scand. Suppl.*, **500**, 1–147.

Dahlöf, C., Ablad, B., Borg, K. O., Ek., L., and Waldeck, B. (1975). Pre-junctional inhibition of adrenergic nervous vasomotor control due to β-receptor blockade, in *Chemical Tools in Catecholamine Research* (Eds O. Almgre, A. Carlsson, and J. Engel), Vol. 2, North-Holland, Amsterdam and Oxford, pp. 201–10.

Dahlöf, C., Ljung, B., and Ablad, B. (1978). Increased noradrenaline release in rat portal vein during sympathetic nerve stimulation due to activation of presynaptic β-adrenoceptors by noradrenaline and adrenaline. *Eur. J. Pharmacol.*, **50**, 75–8.

Dahlström, A., and Fuxe, K. (1964). Evidence for the existence of monoamine-containing neurons in the central nervous system. I. Demonstration of monoamines in the cell bodies of brain stem neurons. *Acta Physiol. Scand., Suppl.*, **232**, 1–55.

Dahlström, A., and Fuxe, K. (1965). Evidence for the existence of monoamine neurone in the central nervous system. *Acta Physiol. Scand.*, **64**, 56–66.

Dahlström, A., and Waldeck, B. (1968). The uptake and retention of ^{3}H-noradrenaline in rat sciatic nerves after ligation. *J. Pharm. Pharmacol.*, **20**, 673–7.

Dahlström, A. B., Evans, C. S., and Häggendal, C. J. (1974). Rapid transport of acetylcholine in rat sciatic nerve proximal and distal to a lesion. *J. Neural. Transm.*, **35**, 1–11.

Dale, H. H. (1934). Pharmacology and nerve endings. *Proc. R. Soc. Med.*, **28**, 319–32.

Dale, H. H. (1935). *cit.* in Dale, H. H. *Adventures in Physiology.* Pergamon Press, London, 1953.

Dale, H. H. (1937). Acetylcholine as a chemical transmitter of the effects of nerve impulses. *J. Mt Sinai Hosp.*, **4**, 401–15.

Dalsgaard, C. J., Hökfelt, T., Johansson, O., and Elde, R. (1981). Somatostatin immunoreactive cell bodies in the dorsal horn and the parasympathetic intermediolateral nucleus of the rat spinal cord. *Neurosci. Lett.*, **27**, 335–9.

Dalsgaard, C. J., Hökfelt, T., Elfvin, L. B., and Terenius, L. (1982). Enkephalin-containing sympathetic preganglionic neurones projecting to the inferior mesenteric ganglion: evidence from combined retrograde tracing and immunohistrochemistry. *Neuroscience*, **7**, 2039–50.

Daly, J. W. (1976). The nature of receptors regulating the formation of cyclic AMP in brain tissue. *Life Sci.*, **18**, 1349–58.

Das, M., Ganguly, D. K., and Vedastromoni, J. R. (1978). Enhancement by oxotremorine of acetylcholine release from the rat phrenic nerve. *Br. J. Pharmacol.*, **62**, 195–8.

Davies, B. N., and Withrington, P. G. (1968). The release of noradrenaline by the sympathetic postganglionic nerves to the spleen of the cat in response to low frequency stimulation. *Arch. Int. Pharmacology*, **171**, 85–196.

Davies, B. N., and Withrington, P. G. (1969). Actions of prostaglandin A_1, A_2, F_1 and F_2 on splenic vascular and capsular smooth muscle and their interactions with sympathetic nerve stimulation, catecholamines and angiotensin, in *Prostaglandins, Peptides and Amines*, (Ed. E. W. Horton) Academic Press, New York, pp. 53–6.

Davies, B. N., Horton, E. W., and Withrington, P. G. (1968). The occurrence of prostaglandin E_2 in splenic venous blood of the dog following splenic nerve stimulation. *Br. J. Pharmacol.*, **32**, 127–35.

Davies, J., and Tongroach, P. (1978). Neuropharmacological studies on the nigro-striatal and raphé-striatal system in the rat. *Eur. J. Pharmacol.*, **51**, 91–100.

Davies, L. P., Cook, A. F., Poonian, M., and Taylor, K. M. (1980). Displacement of ^3H diazepam binding in rat brain by dipyridamole and by 1-methylisognanosine: a marine natural product with muscle relaxant activity. *Life Sci.*, **26**, 1089–95.

Davies, G. C., Bunney, W. E. Jr, Defraites, E. G., Kleinman, J. E., Van Kammen, D. P., Post, R. M., and Wyatt, R. J. (1977). Intravenous naloxone administration in schizophrenia and affective illness. *Science*, **197**, 74–7.

Dawes, P. M., and Vizi, E. S. (1973). Acetylcholine release from the rabbit isolated superior cervical ganglion preparation. *Br. J. Pharmacol.*, **48**, 225–32.

De Belleroche, J. S., and Bradford, H. F. (1972). The stimulus induced release of acetylcholine from synaptosome and its calcium dependence. *J. Neurochem.*, **9**, 1817–19.

De Belleroche, J. S., and Bradford, H. F. (1980). Presynaptic control of the synthesis and release of dopamine from striatal synaptosomes: a comparison between the effects of 5-hydroxytryptamine, acetylcholine and glutamate. *J. Neurochem.*, **35**, 1227–34.

De Belleroche, J. S., Continho-Netto, J., and Bradford, H. F. (1982). Dopamine inhibition of the release of endogenous acetylcholine from corpus striatum and cerebral cortex in tissue slices and synaptosomes: a presynaptic response? *J. Neurochem.*, **39**, 217–22.

De Groat, W. C., and Volle, R. L. (1966). The actions of catecholamines on the transmission in the superior cervical ganglion of the cat. *J. Pharmac. Exp. Ther.*, **154**, 1–13.

De Jonge, A., Santing, P. N., Timmermans, P. B. M. W. M., and van Zwieten, P. A. (1981). A comparison of peripheral pre- and postsynaptic α_2-adrenoceptors using meta-substituted imidazolidines. *J. Auton. Pharmacol.*, **1**, 377–83.

De Groat, W. C., and Theobald, R. J. (1976). Reflex activation of sympathetic pathways to vesical smooth muscle and parasympathetic ganglia by electrical stimulation of vesical afferents. *J. Physiol. (Lond.)*, **259**, 223–37.

De Langen, C. D. J., and Mulder, A. H. (1980). On the role of calcium ions in the pre-synaptic alpha-receptor mediated inhibition ^3H-noradrenaline release from rat brain cortex synaptosomes. *Brain Res.*, **185**, 399–408.

Delbarre, B., and Schmitt, H. (1971). Sedative effects of α-sympathomimetic drugs and their antagonism by adrenergic and cholinergic blocking drugs. *Eur. J. Pharmacol.*, **13**, 356–63.

Delbarre, B., and Schmitt, H. (1974). Effects of clonidine and some α-adrenoceptor blocking agents on avoidance conditioned reflexes in rats. Their interactions and antagonism by atropine. *Psychopharmacologia (Berl.)*, **35**, 193–202.

Del Tacca, M., Lecchini, S., Frigo, C. M., Crema, A., and Benzi, C. (1968). Antagonism of atropine towards endogenous and exogenous acetylcholine before and after sympathetic system blockage in isolated distant guinea-pig colon. *Eur. J. Pharmacol.*, **4**, 188–97.

Del Tacca, M., Soldani, G., Selli, M., and Crema, A. (1970). Action of catecholamines on release of acetylcholine from human taenia coli. *Eur. J. Pharmacol.*, **9**, 80–4.

DeMaio, D. (1972). Clozapine: a novel major tranquillizer (clinical experiences and pharmacotherapeutic hypotheses). *Arzneimittel Forsch.*, **22**, 919–23.

Dennis, M. J., and Miledi, I. R. (1974). Electrically induced release of acetylcholine from denervated Schwann cells. *J. Physiol. (Lond.)*, **237**, 431–52.

De Polo, R. (1978). Calcium pump driven by ATP in squid axons. *Nature*, **269**, 557–8.

De Potter, W. P., Chubb, I. W., Put, A., and De Schaepdryver, A. F. (1971). Facilitation of the release of noradrenaline and dopamine-β-hydroxylase at low stimulation frequencies by α-blocking agents. *Arch. Int. Pharmacodyn. Ther.*, **193**, 191–7.

De Potter, W. P., De Schaepdryver, A. F., Moerman, E. J., and Smith, A. D. (1969a). Evidence for the release of vesicle porteins together with noradrenaline upon stimulation of the splenic nerve. *J. Physiol. (Lond.)*, **204**, 102–4.

De Potter, W. P., Moerman, E. J., De Schaepdryver, A. F., and Smith, A. D. (1969b). Release of noradrenaline and dopamine β-hydroxylase upon splenic nerve stimulation, in *Proc. 4th Int. Congr. Pharmac., Abstracts*, Schwab and Co., Basel, p. 146.

De Quidt, M. E., and Emson, P. C. (1983). Neurotensin potentiates the potassium-stimulated release of ³H-dopamine from rat striatal slices *in vitro*. *Br. J. Pharmacol.*, **78**, 4P.

Descarries, L., and Droz, B. (1969). Incorporation de noradrenaline ⁻³-(NA⁻³H) dans le système nerveux central du rat adulte. Etude radio-autographigue en microscopie électronique. *Acad. Sci. Paris*, **266**, 2480–2.

Descarries, L., and Léger, L., (1978). Serotonin nerve terminals in the locus coeruleus of the adult rat, in *Interactions between Putative Neurotransmitters in the Brain* (Eds S. Garattini, J. F. Pujol, and R. Samanin), Raven Press, New York, pp. 355–67.

Descarries, L., and Saucier, G. (1972). Disappearance of the locus coeruleus in the rat after intraventricular 6-hydroxydopamine. *Brain Res.*, **37**, 310–16.

Descarries, L., Beaudet, A., and Watkins, K. C. (1975). Serotonin nerve terminals in adult rat neocortex. *Brain Res.*, **100**, 563–88.

Descarries, L., Watkins, K. C., and Lapierre, Y. (1977). Noradrenergic axon terminals in the cerebral cortex of rat. III. Topometric ultrastural analysis. *Brain Res.*, **133**, 197–222.

Descartes, R. (1644). *L'Homme*, Jacoves le gras, Paris.

Dettmar, P. W., Lynn, A. G., and Tullock, I. F. (1981). Neuropharmacological evaluation of RX 781094, a new selective α₂-adrenoceptor antagonist. *Br. J. Pharmacol.*, **74**, 843–4.

DeWied, D. (1978). Psychopathology as a neuropeptide dysfunction, in *Characteristics and Function of Opioids* (Eds J. M. Van Ree and L. Terenius), Elsevier, Amsterdam, pp. 165–79.

Dillier, N., László, J., Müller, B., Koella, W. P., and Olpe, H. R. (1978). Activation of an inhibitory noradrenergic pathway projecting from the locus coeruleus to the cingulate cortex of the rat. *Brain Res.*, **154**, 61–8.

Dingledine, R., and Goldstein, A. (1975). Single neuron studies of opiate action in the guinea-pig myenteric plexus. *Life Sci.*, **17**, 59–62.

Dingledine, R., and Goldstein, A. (1976). Effect of synaptic transmission blockade on morphine action in the guinea-pig myenteric plexus. *J. Pharmacol. Exp. Ther.*, **196**, 97–106.

Dingledine, R., Goldstein, A., and Kending, J. (1974). Effects of narcotic opiates and serotonin on the electrical behaviour of neurons in the guinea-pig myenteric plexus. *Life Sci.*, **14**, 229–30.

Dismukes, K. (1977). A new look at the aminergic nervous system. *Nature*, **269**, 557–8.

Dismukes, K. (1979). New concepts of molecular communication among neurons. *The Behavioral and Brain Sciences*, **2**, 409–48.

Dismukes, K., and Mulder, A. H. (1976). Cyclic AMP and β-receptor mediated modulation of noradrenaline release from rat brain slices. *Eur. J. Pharmacol.*, **39**, 383–8.

Dismukes, K., and Mulder, A. H. (1977). Effects of neuroleptics on release of ³H-dopamine from slices of rat corpus striatum. *Naunyn–Schmiedeberg's Arch. Pharmacol.*, **297**, 23–9.

Dixon, W. E. (1906). Vagal inhibition. *Brit. Med. J.*, **2**, 1807–16.

Dockray, G. J. (1982). The physiology of cholecystokinin brain and gut. *Br. Med. Bull.*, **38**, 253–8.

214

Dockray, G. J., Gregory, R. A., Tracy, H. J., and Zhu, W. Y. (1981). Transport of cholecystokinin–octapeptide-like immunoreactivity forward the gut in afferent vagal fibres in cat and dog. *J. Physiol. (Lond.)*, **314**, 501–11.

Dodd, J., and Kelly, J. S. (1981). The action of cholecystokinin and related peptides on pyramidal neurons of the mammalian hypocampus. *Brain Res.*, **205**, 337–50.

Dodd, P. R., Edwardson, J. A., and Dockray, G. J. (1980). The depolarization-induced release of cholecystokinin C-terminal octapeptide (CCK-8) from rat synaptosomes and brain slices. *Regul. Pep.*, **1**, 17–29.

Dodge, F. A., and Rahamimoff, R. (1967). Cooperative action of calcium ions in transmitter release at the neuromuscular junction. *J. Physiol. (Lond.)*, **200**, 267–83.

Douglas, W. W., and Poisner, A. M. (1966). On the relation between ATP splitting and secretion in the adrenal chromaffin cell: extrusion of ATP (unhydrolysed) during release of catecholamines. *J. Physiol. (Lond.)*, **183**, 249–56.

Dowdall, M. J., Boyne, A. F., and Whittaker, V. P. (1974). Adenosine triphosphate: a constituent of cholinergic synaptic vesicles. *Biochem. J.*, **140**, 1–12.

Dowdle, E. B., and Maske, R. (1980). The effects of calcium concentration on the inhibition of cholinergic neurotransmission in the myenteric plexus of guinea-pig ileum by adenine nucleotides. *Br. J. Pharmacol.*, **71**, 245–52.

Doxey, J. C. (1977). Effect of clonidine on cardiac acceleration in pithead rats. *J. Pharmacol.*, **29**, 173–4.

Doxey, J. C. (1979). Pre- and postsynaptic effect of α-agonists in the anococcygeus muscle of the pithead rat. *Eur. J. Pharmacol.*, **54**, 185–9.

Doxey, J. C., and Roach, A. G. (1980). Presynaptic α-adrenoceptors, *in vitro* methods and preparations utilised in the evaluation of agonists and antagonists. *J. Auton. Pharmacol.*, **1**, 73–99.

Doxey, J. C., Smith, C. F. C., and Walker, J. M. (1977). Selectivity of blocking agents for pre- and postsynaptic α-adrenoceptors. *Br. J. Pharmacol.*, **60**, 91–6.

Dray, A. (1979). The striatum and substantia nigra: a commentary on their relationships. *Neuroscience*, **4**, 1407–39.

Dray, A., and Straughan, D. W. (1976). Synaptic mechanisms in the substantia nigra. *J. Pharm. Pharmacol.*, **28**, Suppl. 333–405.

Dray, A., Gonye, T. J., Oakley, N. R., and Tanner, T. (1976). Evidence for the existence of a raphé projection to the substantia nigra in rat. *Brain Res.*, **113**, 45–57.

Drew, G. M. (1977). Pharmacological characterisation of the presynaptic α-adrenoceptor in the rat vas deferens. *Eur. J. Pharmacol.*, **42**, 123–30.

Drew, G. M. (1979). Presynaptic α-adrenoceptors: their pharmacological characterisation and functional significance, in *Presynaptic Receptors* (Eds S. Z. Langer, K. Starke, and M. L. Dubocovich), Pergamon Press, Oxford. pp. 59–63.

Drew, G. M., and Sullivan, A. T. (1980). Effect of α-adrenoceptor agonists and antagonists on adrenergic neurotransmitter overflow from dog isolated saphenous veins. *Br. J. Pharmacol.*, **68**, 139–40P.

Drew, G. M., Gower, A. J., and Marriott, A. S. (1979). α2-adrenoceptors mediate clonidine-induced sedation in the rat. *Br. J. Pharmacol.*, **67**, 133–41.

Drury, A. N., and Szentgyörgyi, A. (1929). The physiological activity of adenine compounds with special reference to their action upon the mammalian heart. *J. Physiol. (Lond.)*, **68**, 213–37.

Dubey, M. P., Muscholl, E., and Pfeiffer, A. (1975). Muscarinic inhibition of potassium-induced noradrenaline release and its dependence on the calcium concentration. *Naunyn-Schmiedeberg's Arch., Pharmacol.*, **291**, 1–15.

Dubocovich, M. L., and Langer, S. Z. (1974). Negative feedback regulation of noradrenaline release by nerve stimulation in the perfused cat's spleen. Differences in potency of phenoxybenzamine in blocking the pre- and postsynaptic adrenergic receptors. *J. Physiol. (Lond.)*, **237**, 505–19.

Dubocovich, M. L., and Langer, S. Z. (1975). Evidence against a physiological role of prostaglandins in the regulation of noradrenaline release in the cat spleen. *J. Physiol. (Lond.)*, **251**, 737–62.

Dubocovich, M. L., and Langer, S. Z. (1980). Dopamine and alpha-adrenoceptor agonists inhibit neurotransmission in the cat spleen through different presynaptic receptors. *J. Pharmac. Ther.*, **212**, 144–52.

Du Bois-Reymond, E. (1848). *Untersuchungen über thierische Electricität*, Reimer, Berlin.

Dudar, J. D. (1977). The role of the septal nuclei in the release of acetylcholine from the rabbit cerebral cortex and dorsal hippocampus and the effect of atropine. *Brain Res.*, **129**, 237–46.

Dudar, J. D., and Szerb, J. C. (1969). The effect of topically applied atropine on resting and evoked cortical acetylcholine release. *J. Physiol. (Lond.)*, **203**, 741–62.

Dudel, J. (1965). The mechanism of presynaptic inhibition at the crayfish neuromuscular junction. *Pflügers Arch. ges. Physiol.*, **284**, 66–80.

Dudel, J., and Kuffler, S. W. (1960a). A second mechanism of inhibition at the crayfish neuromuscular junction. *Nature*, **187**, 247–8.

Dudel, J., and Kuffler, S. W. (1960b). Excitation at the crayfish neuromuscular junction with decreased membrane conductance. *Nature*, **187**, 246–7.

Dudel, J., and Kuffler, S. W. (1961a). Presynaptic inhibition at the crayfish neuromuscular junction. *J. Physiol. (Lond.)*, **155**, 543–62.

Dudel, J., and Kuffler, S. W. (1961b). Presynaptic inhibition at the crayfish muscle. *Pflügers Arch. ges. Physiol.*, **284**, 81–94.

Duggan, A. W. (1979). Morphine, enkephalin and the spinal cord, in *Advances in Pain Research and Therapy*, Vol. 23 (Eds J. J. Bonica, J. C. Liebskind, and D. G. Albe-Fessard), Raven Press, New York, pp. 449–58.

Dun, N. J., and Nishi, S. (1974). Effects of dopamine on the superior cervical ganglion of the rabbit. *J. Physiol. (Lond.)*, **239**, 155–64.

Dun, N. J., and Karczmar, A. G. (1981). Evidence for a presynaptic inhibitory action of 5-hydroxytryptamine in a mammalian sympathetic ganglion. *J. Pharm. Exp. Ther.*, **217**, 714–18.

Dunant, Y., and Walker, A. I. (1981). Presynaptic inhibition of acetylcholine release by oxotremorine and anti-cholinesterases in the electric organ of Torpedo. *J. Physiol. (Lond.)*, **318**, 45P.

Dunlap, K., and Fischbach, G. C. (1978). Neurotransmitters: action potentials. *Nature*, **176**, 837–9.

Dunlap, K., and Fischbach, G. C. (1981). Neurotransmitters decrease the calcium conductance activated by depolarization of embryonic chick sensory neurones. *J. Physiol. (Lond.)*, **371**, 519–35.

Dunwiddie, T., and Hoffer, B. (1980). Adenine nucleotides and synaptic transmission in the *in vitro* hippocampus. *Br. J. Pharmacol.*, **69**, 59–68.

Dunwiddie, C., Mueller, A., Palmer, M., Stewart, J., and Hoffer, B. (1980). Electrophysiological interactions of enkephalins with neuronal circuitry in the rat hippocampus. I. Effects on pyramidal cell activity. *Brain Res.*, **184**, 311–30.

Dzieniszewski, P., and Kilbinger, H. (1978). Muscarininc modulation of acetylcholine release evoked by dimethylphenylpiperazinium and high potassium from guinea-pig myenteric plexus. *Eur. J. Pharmacol.*, **50**, 385–91.

Eccles, J. C. (1964). *The Physiology of Synapses*, Springer-Verlag, Berlin.

Eccles, J. C. (1982). The synapse: from electrical to chemical transmission. *Ann. Rev. Neurosci.*, **5**, 325–39.

Eccles, J. C., and McGeer, P. L. (1979). Ionotropic and metabotropic neurotransmission. *TINS*, **1**, 39–41.

Eccles, J. C., Eccles, R. M., and Magni, F. (1961). Central inhibitory action attributable to pre-synaptic depolarization produced by muscle afferent volleys. *J. Physiol. (Lond.)*, **159**, 147–66.

Eccles, J. C., Katz, B., and Kuffler, S. W. (1942). Effect of eserine on neuromuscular transmission. *J. Neurophysiol.*, **5**, 211–30.

Eccles, J. C., Schmidt, R. F., and Willis, W. D. (1962). Pre-synaptic inhibition of the spinal monosynaptic reflex pathway. *J. Physiol. (Lond.)*, **161**, 500–30.

Eccles, J. C., Schmidt, R. F., and Willis, W. D. (1963). The mode of operation of the synaptic mechanism producing presynaptic inhibition. *J. Neurophysiol.*, **26**, 523–8.

Eglinton, G., Raphael, R. A., Smith, G. N., Hall, W. J., Pickles, V. A. (1963). Isolation and identification of the smooth muscle stimulants from measured fluid. *Nature*, **200**, 900–5.

Ehringer, B., Falc, B., and Sporrong, B. (1970). Possible axo–axonal synapses between peripheral adrenergic and cholinergic nerve terminals. *Z. Zellforsch.*, **107**, 508–21.

Ehrenpreis, S., Light, I., and Schonbuch, G. H. (1972). In *Drug Addition: Experimental Pharmacology* (Eds J. M. Sing, L. H. Miller, and H. Lal), Futura, Mount Kisco, New York, pp. 319–42.

Ehrenpreis, S., Greenberg, J., and Belman, S. (1973). Prostaglandin reverse inhibition of electrically induced contractions of guinea-pig ileum by morphine, indomethacin and acetylsalycylic acid. *Nature (Lond.)*, **245**, 280–2.

Ehrenpreis, S., Sato, T., Takayanagi, I., Comaty, J. E., and Takagi, K. (1976). Mechanism of morphine block of electrical activity in ganglia of Auerbach plexus. *Eur. J Pharmacol.*, **40**, 303–9.

Ehringer, B., Falk, B., Persson, H., Rosengren, A. M., and Sporrong, B. (1970). Acetylcholine in adrenergic terminals of the cat iris. *J. Physiol.*, **209**, 557–65.

Ehringer, R. H., and Hornykiewicz, O. (1960). Verteilung von Noradrenalin und Dopamin (3-Hydroxytyramin) in Gehirn des Menschen und Ihr Verhalten bei Erhrankungen des extrapyramidalen systems. *Klin. Wschr.*, **38**, 1236–9.

Ehrlich, P. (1913). Chemotherapeutics: Scientific principles, methods and results. *Lancet*, **2**, 445–6.

Eisenstadt, M., Goldman, J. E., Kandel, E. R., Koester, J., Koike, H., and Schwartz, J. H. (1973). Intrasomatic injection of radioactive precursors for studying transmitter synthesis in identified neurons of *Aplysia californica*. *Proc. Natl Acad. Sci.*, **70**, 3371–5.

Elde, R., Hökfelt, T., Johansson, O., and Terenius, L. (1976). Immunohistochemical studies using antibodies to leucine–enkephalin, initial observations on the nervous system of the rat. *Neuroscience*, **1**, 349–51.

Elliott, T. R. (1904). On the action of adrenaline. *J. Physiol. (Lond.)*, **31**, 20–1.

Emson, P. C. (1979). Peptides as neurotransmitter candidates in the mammalian CNS. *Progress in Neurobiology*, **13**, 61–166.

Emson, P. C., Hunt, S. P., Rehfeld, J. F., Golterman, N., and Fahrenkrug, J. (1980). Cholecystokinin and vasoactive intestinal polypeptide in the mammalian CNS. Distribution and possible physiological role, in *Neural Peptides and Neuronal Communication* (Eds E. Costa, and M. Trabucchi), Raven Press, New York, pp. 1–112.

Enero, M. A., and Langer, S. Z. (1975a). Pharmacological effects of histamine on the isolated cat nictitating membrane. *Br. J. Pharmacol.*, **53**, 431–2.

Enero, M. A., and Langer, S. Z. (1975b). Inhibition by dopamine of ^{3}H-noradrenaline release elicited by nerve stimulation in the isolated cat's nictitating membrane. *Naunyn–Schmiedeberg's Arch. Pharmacol.*, **289**, 179–203.

Enero, M. A., and Saidman, B. Q. (1977). Possible feed-back inhibition of noradrenaline release by purine compounds. *Naunyn-Schmiedeberg's Arch. Pharmacol.*, **297**, 39–46.

Enero, A. M., Langer, S. Z., Rothlin, R. P., and Stefano, F. J. E. (1972). Role of the α-adrenoceptor in regulating noradrenaline overflow by nerve stimulation. *Br. J. Pharmacol.*, **44**, 672–88.

Ennis, C., and Cox, B. (1981). GABA enhancement of ^{3}H-dopamine release from slices of rat striatum; dependence on slice size. *Eur. J. Pharmacol.*, **70**, 417–20.

Ennis, C., Janssen, P. A. J., Schnieden, H., and Cox, B. (1979). Characterization of receptors on postganglionic cholinergic neurons in the guinea-pig isolated ileum. *J. Pharm. Pharmacol.*, **31**, 217–21.

Ennis, C., Kemp, J. D., and Cox, B. (1981). Characterization of inhibitory 5-hydroxytryptamine receptors that modulate dopamine release in the striatum. *J. Neurochem.*, **36**, 1515–20.

Erulkar, S. D. (1983). The modulation of neurotransmitter release at synaptic junctions. *Rev. Phys. B.*, **98**, 63–175.

Erulkar, S. D., and Weight, F. F. (1974). Depression of transmitter release and accumulation of extracellular potassium at the squid giant synapse. *Society for Neuroscience, Fourth Annual Meeting Abstracts*, p. 202.

Erulkar, S. D., and Weight, F. F. (1977). Extracellular potassium and transmitter release at the giant synapse of squid. *J. Physiol. (Lond.)*, **266**, 209–18.

Esquerro, E., Cena, V., Sanchez-Garcia, P., Kirpekar, S. M., and Garcia, A. G. (1980a). Release of noradrenaline from the ligated cat hypogastric nerve. *Eur. J. Pharmacol.*, **61**, 183–6.

Esquerro, E., Schiavone, M., Garcia, A. G., Kirpekar, S. M., and Prat, J. C. (1980b). Release of endogenous noradrenaline from ligated cat hypogastric nerve by veratridine. *Eur. J. Pharmacol.*, **66**, 367–73.

Euler, U.S. von (1936). On the specific vaso-dilating and plain muscle stimulating substances from accessory genital glands in man and certain animals (prostaglandin and vesiglandin). *J. Physiol. (Lond.)*, **88**, 213–34.

Euler, U. S. von (1948). Identification of the sympatho-mimetic ergone in adrenergic nerves of cattle (Sympathin N) with laevo-noradrenaline. *Acta Physiol. Scand.*, **16**, 63–74.

Euler, U. S. von (1981). Historical perspective: growth and impact of the concept of chemical neurotransmission, in *Chemical Neurotransmission: 75 years* (Eds L. Stjärne, P. Hedqvist,

H. Lagercrantz, and A. Wennmalm), Academic Press, London, pp. 3–12.

Euler, U. S. von, and Gaddum, J. H. (1931). An unidentified depressor substance in certain tissue extracts. *J. Physiol. (Lond.)*, **72**, 74–87.

Euvrard, C., Javoy, F., Herbert, A., and Glowinsky, J. (1977). Effect of guipazine, a serotonin-like drug on striatal cholinergic interneurons. *Eur. J. Pharmacol.*, **41**, 281–9.

Ewald, D. A. (1976). Potentiation of post-junctional cholinergic sensitivity of rat diaphragm muscle by high-energy phosphate adenine nucleotides. *J. Membrane Biol.*, **29**, 47–65.

Fahn, S., and Cote, L. J. (1968). Regional distribution of gamma aminobutyric acid (GABA) in brain of the rhesus monkey. *J. Neurochem.*, **15**, 209–13.

Falck, B. (1962). Observations on the possibilities of the cellular localization of monoamines by a fluorescence method. *Acta Physiol. Scand.*, **197**, 1–25.

Famaey, J. P., Fontaine, J., and Reuse, J. (1977). The effects of non-steroidal anti-inflammatory drugs on cholinergic and histamine-induced contractions of guinea-pig isolated ileum. *Brit. J. Pharmacol.*, **60**, 165.

Farah, M. B., and Langer, S. Z. (1974). Protection by phentolamine against the effects of phenoxybenzamine on transmitter release elicited by nerve stimulation in the perfused cat heart. *Br. J. Pharmacol.*, **52**, 549–57.

Farnebo, L. O., and Hamberger, S. (1971a). Drug induced changes in the release of (^{3}H)-noradrenaline from field stimulated rat iris. *Br. J. Pharmacol.*, **43**, 97–106.

Farnebo, L. O., and Hamberger, B. (1971b). Drug induced changes in the release of ^{3}H-monoamines from field stimulated rat brain slices. *Acta Physiol. Scand. Suppl.*, **371**, 35–44.

Farnebo, L. O., and Hamberger, B. (1974a). Regulation of ^{3}H-5-hydroxytryptamine release from rat brain slices. *J. Pharm. Pharmacol.*, **26**, 642–4.

Farnebo, L. O., and Hamberger, B. (1974b). Influence of α and β adrenoceptors on the release of noradrenaline from field stimulated atria and cerebral cortex slices. *J. Pharm. Pharmacol.*, **26**, 644–6.

Farnebo, L. O., and Malmfors, T. (1971). ^{3}H-noradrenaline release and mechanical response in the field stimulated mouse vas deferens. *Acta Physiol. Scand. Suppl.*, **371**, 1–18.

Farsang, C., Kapocsi, J., Juhász, I., and Kunos, G. (1982). Possible involvement of an endogenous opioid in the antihypertensive effect of clonidine in patients with essential hypertension. *Circulation*, **66**, 1268–72.

Fatt, P., and Katz, B. (1953). The effect of inhibitory nerve impulses on a crustacean muscle fibre. *J. Physiol. (Lond.)*, **121**, 374–89.

Feden, J. S., Hogaboom, G. K., O'Donnel, J. P., Colby, J., and Westfall, D. P. (1981). Contribution by purines to the neurogenic response of the vas deferens of the guinea-pig. *Eur. J. Pharmacol.*, **69**, 41–53.

Fedulova, S. A., Kostyuk, P. G., and Veselovsky, N. S. (1981). Calcium channels in the somatic membrane of the rat dorsal root ganglion neurons, effect of cAMP. *Brain Res.*, **214**, 107–28.

Fehér, E. (1974). Effect of monoamine oxidase inhibitor on the nerve elements of the isolated cat ileum. *Acta Morphol. Acad. Sci. Hung.*, **22**, 249–63.

Fehér, E. (1975). Effects of monoamine inhibitor on the nerve elements of the isolated cat's ileum. *Verh. Anat. Ges.*, **69**, 477–82.

Feldberg, W. (1943). Synthesis of acetylcholine in sympathetic ganglia and cholinergic nerves. *J. Physiol. (Lond.)*, **101**, 432–5.

Felix, D. (1982). Angiotensin, neurohormone and neurotransmitter. *TIPS*, **5**, 208–10.

Feniuk, W., Humphrey, P. P. A. and Watts, A. D. (1978). Evidence for a presynaptic inhibitory receptor for 5-hydroxytryptamine in dog isolates saphenous vein. *Br. J. Pharmacol.*, **63**, 344–5P.

Feniuk, W., Humphrey, P. P. A., and Watts, A. D. (1979). Presynaptic inhibitory action of 5-hydroxytryptamine in dog isolated saphenous vein. *Br. J. Pharmacol.*, **67**, 247–54.

Fennessey, M. R., Heimans, R. L., and Rand, M. J. (1969). Comparison effect of morphine-like analgetics on transmurally stimulated guinea-pig ileum. *Br. J. Pharmacol.*, **37**, 436–49.

Ferreira, S. H., Herman, A. G., and Vane, J. R. (1972). Prostaglandin generation maintains the smooth muscle tone of the rabbit isolated jejunum. *Br. J. Pharmacol.*, **44**, 328–9.

Ferry, C. B. (1966). Cholinergic link hypothesis in adrenergic neuroeffector transmission. *Physiol. Rev.*, **46**, 420–56.

Fields, H. L., and Basbaum, S. I. (1978). Brain stem control of spinal pain-transmission neurons. *Ann. Rev. Physiol.*, **40**, 217–48.

Finkleman, B. (1930). On the nature of inhibition in the intestine. *J. Physiol. (Lond.)*, **70**, 145–57.

Fischer-Ferraro, O., Nahmod, V. E., Goldstein, D. J., and Finkielman, S. (1971). Angiotensin and renin in rat and dog brain. *J. Exp. Med.*, **133**, 353–61.

Fischer, I. L., and Moriarty, C. M. (1977). Control of bioactive corticotropin release from the neuro-intermediate lobe of the rat. *Endocrinology*, **100**, 1047–54.

Fletcher, P., and Forrester, T. (1975). The effect of curare on the release of acetylcholine from mammalian motor nerve terminals and an estimate of quantum content. *J. Physiol. (Lond.)*, **251**, 131–44.

Florey, E. (1967). Neurotransmitters and modulators in the animal kingdom. *Fedn Proc.*, **26**, 1164–78.

Földes, F. F., and Vizi, E. S. (1980). Modulation of presynaptic acetylcholine release at the neuromuscular junction, in *Modulation of Neurochemical Transmission* (Ed. E. S. Vizi), Pergamon Press, Oxford; Akadémiai Kiadó, Budapest, pp. 355–82.

Fonnum, F., Frizell, M., and Sjöstrand, J. (1973). Transport, turnover and distribution of choline acetyltransferase and acetylcholinesterase in the vagus and hypoglossal nerves of the rabbit. *J. Neurochem.*, **21**, 1108–20.

Fonnum, F., and Malthe-Sorenssen, D. (1973). Membrane affinities and subcellular distribution of the different molecular forms of choline acetyltransferase from rat. *J. Neurochem.*, **20**, 1351–9.

Forrester, T. (1972). An estimate of adenosine triphosphate release into the venous effluent from exercising human forearm muscle. *J. Physiol. (Lond.)*, **224**, 611–28.

Forrester, T., Harper, A. M., Mackenzie, E. T., and Thomson, E. M. (1979). Effect of adenosine triphosphate and some derivatives on cerebral blood flow and metabolism. *J. Physiol. (Lond.)*, **296**, 343–55.

Fosbraey, P., and Johnsson, E. S. (1978). Presynaptic inhibition of acetylcholine release from cholinergic neurones in the myenteric plexus of the guinea-pig ileum. *Br. J. Pharmacöl.*, **62**, 466P.

Fozard, J. R., and Mobarok, Ali, A. T. (1976). Inhibition of the stimulant effect of 5-hydroxy-tryptamine on cardiac sympathetic nerves by 5-hydroxytryptamine and related compounds. *Br. J. Pharmacol.*, **58**, 416–7.

Fozard, J. R. and Muscholl, E. (1971). Effect of several muscarinic agonists on cardiac perform-ance and the release of noradrenaline from the sympathetic nerves of the rabbit heart. *Br. J. Pharmacol.*, **43**, 454–5.

Fozard, J. R., and Mwaluko, G. M. P. (1976). Mechanism of the indirect sympathomimetic effect of 5-hydroxytryptamine on the isolated heart of the rabbit. *Br. J. Pharmacol.*, **57**, 115–25.

Frank, K., and Fuortes, M. G. F. (1951). Presynaptic and postsynaptic inhibition of non-synaptic reflexes. *Fedn Proc.*, **16**, 39–40.

Frankenhauser, B., and Hodgkin, A. L. (1956). The after effects of impulses in the giant nerve fibres of *Loligo*. *J. Physiol. (Lond.)*, **131**, 341–76.

Frankhuyzen, A. L., and Mulder, A. H. (1980). Noradrenaline inhibits depolarization-induced ^{3}H-serotonin release from slices of rat hyppocampus. *Eur. J. Pharmacol.*, **63**, 179–82.

Frazier, W. T., Kandel, E. T., Kupfermann, I., Waziri, R., and Coggeshall, R. E. (1967). Morphological and functional properties of identified neurons in the abdominal ganglion of *Aplysia californica*. *J. Neurophysiol.*, **30**, 1288–1351.

Frederickson, R. C. A. (1977). Enkephalin-pentapeptides. A review of current evidence of a physiological role in verterbrate neurotransmission. *Life Sci.*, **21**, 23–42.

Fredholm, B. B. (1974). Vascular and metabolic effects of theophylline, dibutyryl cyclic AMP and dibutyryl cyclic GMP in canine subcutaneous adipose tissue *in situ*. *Acta Physiol., Scand.*, **90**, 226–36.

Fredholm, B. B. (1976). Release of adenosine-like material from isolated perfused dog adipose tissue following sympathetic nerve stimulation and its inhibition by adrenergic alpha-receptor blockade. *Acta Physiol. Scand.*, **96**, 422–30.

Fredholm, B. B. (1981). Trans-synaptic modulation of transmitter release with special reference to adenosine, in *Chemical Transmission: 75 years* (Eds L. Stjärne, H. Lagercrantz, P. Hedqvist, and A. Wennmalm), Academic Press, London, pp. 211–33.

Fredholm, B. B., and Hedqvist, P. (1978). Release of ^{3}H-purines from ^{3}H-adenine labelled rabbit kidney following sympathetic nerves stimulation and its inhibition by alpha-adreno-ceptor blockade. *Br. J. Pharmacol.*, **64**, 239–45.

Fredholm, B. B., and Hedqvist, P. (1979). Presynaptic inhibition of transmitter release by adenosine and its antagonism by theophylline, in *Presynaptic Receptors* (Eds S. Z. Langer, K. Starke, and M. L. Dubocovich), Pergamon Press, Oxford, pp. 281–6.

219

Fredholm, B. B., and Vernet, L. (1978). Morphine increases depolarization induced purine release from rat cortical slices. *Acta Physiol. Scand.*, **104**, 502–4.

Fredholm, B. B., and Vernet, L. (1979). Release of ³H-nucleotides from ³H-adenine labelled hypothalamic synaptosomes. *Acta Physiol. Scand.*, **106**, 97–107.

Fredholm, B. B., Fried, G., and Hedqvist, P. (1982). Origin of adenosine released from rat vas deferens by nerve stimulation, *Eur. J. Pharmacol.*, **79**, 233–43.

Fuenmayer, T. L., Smith, A. D., and Vogt, M. (1976). Acetylcholinesterases in perfusates of the cat's cerebral ventricles. *J. Physiol. (Lond.)*, **263**, 165–6.

Fukuda, J., and Kameyama, M. (1980). Tetrodotoxin-sensitive and tetrodotoxin-resistant sodium channels in tissue-cultured spinal ganglion neurons from adult mammals. *Brain Res.*, **182**, 191–7.

Funk, C. J. (1911). Synthesis of dl-3,4-dihydroxyphenylalamine. *J. Chem. Soc.*, **99**, 554–7.

Furness, J. B. (1970). The origin and distribution of adrenergic fibres in the guinea-pig colon. *Histochemie*, **21**, 295, 306.

Furness, J. B., and Costa, M. (1980). Types of nerves in the enteric nervous system. *Neuroscience*, **5**, 1–20.

Fuxe, K., Hamberger, B., and Hökfelt, T. (1968). Distribution of noradrenaline nerve terminals in cortical areas of the rat. *Brain Res.*, **8**, 125–31.

Fuxe, K., Hökfelt, T., and Ungerstedt, U. (1970a). Morphological and functional aspects of central monoamine neurons. *Int. Rev. Neurobiol.*, **13**, 93–126.

Fuxe, K., Goldstein, M., Hökfelt, T.. and Joh, T. H. (1970b). Immunohistochemical localization of dopamine-ß-hydroxylase in the peripheral and central nervous system. *Res. Commun. Chem. Path. Pharmacol.*, **1**, 627–36.

Gabella, G. (1971). Synapses of adrenergic fibres. *Experientia*, **27**, 280–1.

Gabella, G. (1972). Fine structure of the myenteric plexus in the guinea-pig ileum. *J. Anat. (Lond.)*, **111**, 69–97.

Gabella, G. (1976). *Structure of the Autonomic Nervous System*, Chapman and Hall, London.

Gacel, G., Fournie-Zaluski, M. C., and Roques, B. P. (1980). Tyr–D–Ser–Gly–Phe–Leu–Thr, a highly preferential ligand for δ opiate receptors. *Febs Letters.*, **118**, 245–7.

Gaddum, J. H., and Picarelli, Z. P. (1957). Two kinds of tryptamine receptor. *Br. J. Pharmacol.*, **9**, 240–8.

Gahwiler, B. H. (1980). Excitatory action of opioid peptides and opiates on cultured hippocampal pyramidal cells. *Brain Res.*, **194**, 193–203.

Gale, L., Guidotti, A., and Costa, E. (1977). Dopamine sensitive adenylate cyclase. Location in substantia nigra. *Science*, **195**, 503–5.

Gallant, C. A., and Clement, J. G., (1981). Methylxanthines antagonize adenosine but not morphine inhibition in guinea pig ileum. *Can. J. Physiol. Pharmacol.*, **59**, 886–9.

Galzin, A. M., and Langer, S. Z. (1982). Effect of dopamine and M7 on the electrically evoked release of ³H-noradrenaline from the rabbit hypothalamus. *Br. J. Pharmacol.*, **77**, 455P.

Galzin, A. M., Langer, S. Z., and Moret, C. (1982). RX 781034 antagonizes the inhibition by α₂-agonists of (³H)-NA and (³H)-5-HT release while enhancing only the release of (³H)-NA. *Br. J. Pharmacol.*, **77**, 454P.

Galvan, M., and Adams, R. (1982). Control of calcium current rat sympathetic neurons by norepinephrine. *Brain Res.*, **244**, 135–44.

Ganguly, D. K., and Das, M. (1979). Effects of oxotremorine demonstrate presynaptic and dopaminergic receptors on motor nerve terminals. *Nature*, **278**, 645.

Garau, L., Mulas, M. L., and Pepeu, G. (1975). The influence of raphé lesions on the effect of morphine on nociception and cortical ACh output. *J. Neuropharmac.*, **14**, 259–63.

Garcia, A. G., Kirpekar, S. M., and Pascual, A. (1978). Release of noradrenaline from cat spleen slice by potassium. *Br. J. Pharmacol.*, **62**, 207–11.

Garcia–Servilla, J. A., Dobocovich, M., and Langer, S. Z. (1979). Angiotensin II facilitates the potassium-evoked release of ³H noradrenaline from rabbit hypothalamus. *Eur. J. Pharmacol.*, **56**, 173–6.

Gardner, C. R., and Webster, R. A. (1977). Convulsant–anticonvulsant interaction in seizure activity and cortical acetylcholine release. *Eur. J. Pharmacol.*, **42**, 247–56.

Geffen, L. B., and Livett, B. G. (1971). Synaptic vesicles in sympathetic neurons. *Physiol. Rev.*, **51**, 98–157.

Geffen, L. B., Jessel, T. M., Cuello, A. C., and Iversen, L. L. (1976). Release of dopamine from dendrites in rat substantia nigra. *Nature*, **260**, 258–60.

Gershon, M. D. (1967). Effects of tetrodotoxin on innervated smooth muscle preparations. *Br. J. Pharmacol.*, **29**, 259–79.

Gershon, M. D. (1981). The enteric nervous system. *Ann Rev. Neurosci.*, **4**, 227–72.

Gershon, M. D., and Jonakait, G. M. (1979). Uptake and release of 5-hydroxytryptamine by enteric serotonergic neurons: effects of fluoxetine (Lilly 110140) and chlorimipramine. *Br. J. Pharmacol.*, **66**, 7–9.

Gershon, M. D., Dreyfus, C. F., Pickel, V. M., Joh, T. H., and Reis, D. J. (1977). Serotonergic neurons in the peripheral nervous system: identification in gut by immunohistochemical localization of tryptophan hydroxylase. *Proc. Natl. Acad. Sci.*, **71**, 3086–9.

Ghose, K., and Coppen, A. (1977). Noradrenaline, depressive illness and the action of amitriptyline. *Psychopharmac.*, **54**, 57–60.

Gillan, M. G. C., and Kosterlitz, H. W. (1982). Spectrum of the μ, δ and κ binding sites in homogenates of rat brain. *Br. J. Pharmacol.*, **77**, 461–9.

Gillespie, J. S. (1960). Electrical responses of mammalian smooth muscle to the extrinsic sympathetic and parasympathetic nerves. *Physiologist*, **3**, 65–71.

Gillespie, J. S. (1980). Presynaptic receptors in the autonomic nervous system, in *Adrenergic Activators and Inhibitors* (Ed. L. Szekeres). Part I, Springer-Verlag, Berlin, pp. 353–425.

Gillespie, J. S., and Khoyi, M. A. (1977). The site and receptors responsible for the inhibition by sympathetic nerves of intestinal smooth muscle and its parasympathetic motor nerves. *J. Physiol. (Lond.)*, **267**, 767–89.

Gillespie, J. S., and McGrath, J. C. (1975). The effects of lysergic acid diethylamide on the response to field stimulation of the rat vas deferens and the rat and cat anococcygeus muscles. *Br. J. Pharmacol.*, **54**, 481–8.

Gillespie, J. S., and McKenna, B. R. (1961). The inhibitory action of the sympathetic nerves on the smooth muscle of the rabbit gut, its reversal by reserpine and restoration by catecholamines and by dopa. *J. Physiol. (Lond.)*, **156**, 17–34.

Gillespie, J. S., and Maxwell, J. D. (1971). Adrenergic innervation of sphincteric and nonsphincteric smooth muscle in the rat intestine. *J. Histochem. Cytochem.*, **19**, 676–81.

Gilmore, N., Vane, J. R., and Wyllie, J. H. (1968). Prostaglandins released by the spleen. *Nature*, **218**, 1135–40.

Ginsborg, B. L., and Hirst, G. D. S. (1972). The effect of adenosine on the release of the transmitter from the phrenic nerve of the rat. *J. Physiol. (Lond.)*, **224**, 629–45.

Gintzler, A. R., and Musacchio, J. M. (1974). Interaction between serotonin and morphine in the guinea-pig ileum. *J. Pharmac. Exp. Ther.*, **189**, 484–92.

Gintzler, A. R., and Musacchio, J. M. (1975). Interactions of morphine, adenosine, adenosine triphosphate and phosphodiesterase inhibitors on the field stimulated guinea-pig ileum. *J. Pharmac. exp. Ther.*, **194**, 575–82.

Giorguieff, M. F., Kemel, M. L., Glowinsky, J., and Besson, M. J. (1978). Stimulation of dopamine release by GABA in rat striatal slices. *Brain Res.*, **139**, 115–30.

Giorguieff, M. F., Lefloch, M. L., Westfall, T. C., Glowinsky, J., and Besson, M. J. (1976). Nicotinic effect of acetylcholine on the release of newly synthesized ^{3}H-dopamine in rat striatal slices and cat caudate nucleus. *Brain Res.*, **106**, 117–31.

Giorguieff, M. F., Kemel, M. L., Wandscheer, D., and Glowinsky, K. (1979). Attempts to localize the GABA receptors involved in the GABA induced release of newly synthesized ^{3}H-dopamine in rat striatal slices. *Brain Res.*, **175**, 383–6.

Gold, M. S., Redmond, D. E., and Kleber, H. D. (1978). Clonidine blocks acute opiate-withdrawal symptoms. *Lancet*, **2**, 599–602.

Goldberg, A. L., and Singer, J. J. (1969). Evidence for a role of cyclic AMP in neuromuscular transmission. *Proc. Natl Acad. Sci. USA*, **64**, 134–41.

Goldberg, L. I., Volkman, P. H., and Kohli, J. D. (1978). A comparison of the vascular dopamine receptor with other receptors. *Ann. Rev. Pharmacol. Tox.*, **18**, 57.

Goldman, J. E., and Schwartz, J. H. (1974). Cellular specificity of serotonin storage and axonal transport in identified neurones of *Aplysia californica*, *J. Physiol. (Lond.)*, **242**, 61–76.

Gooddale, D. B., Rusterholz, D. B., Long, J. P., Flynn, J. R., Walsh, B., Cannon, J. G., and Lee, T. (1980). Neurochemical and behavioral evidence for a selective presynaptic dopamine receptor agonist. *Science*, **210**, 1141–3.

Gordon-Weeks, P. R. (1981a). Properties of nerve endings with small granular vesicles in the distal colon and rectum of the guinea-pig. *Neuroscience*, **6**, 1793–1811.

Gordon-Weeks, P. R. (1981b). Are there noradrenergic synapses in Auerbach's plexus? *Scand. J. Gastroenterology*, **16**, 111–82.

Gordon-Weeks, P. R. (1983). Noradrenergic and non-noradrenergic nerves containing small granular vesicles in Auerbach's plexus of the guinea-pig. Evidence against the presence of noradrenergic synapses. (In press.)

Gordon-Weeks, P. R., and Hobbs, M. J. (1979). A non-adrenergic nerve ending containing small granular vesicles in the guinea pig gut. *Neuroscience Letts*, **12**, 81–6.

Göthert, M. (1980). Serotonin-receptor-mediated modulation of Ca^{2+} dependent 5-hydroxytryptamine release from neurones of the rat brain cortex. *Naunyn-Schmiedeberg's Arch. Pharmacol.*, **314**, 223–30.

Göthert, M., and Huth, H. (1980). Alpha-adrenoceptor-mediated modulation of 5-hydroxytryptamine release from rat brain cortex slices. *Naunyn-Schmiedeberg's Arch. Pharmacol.*, **313**, 21–6.

Göthert, M., and Wehking, E. (1980). Inhibition of Ca^{2+} induced noradrenaline release from central noradrenergic neurons by morphine. *Experientia*, **36**, 239.

Göthert, M., and Winheimer, G. (1979). Extracellular 5-hydroxytryptamine inhibits 5-hydroxytryptamine release from rat brain cortex slices. *Naunyn-Schmiedeberg's Arch. Pharmacol.*, **310**, 93–6.

Göthert, M., Huth, H., and Schlicker, E. (1981). Characterization of the receptor subtype involved in alpha-adrenoceptor-mediated modulation of serotonin release from rat brain cortex slices. *Naunyn-Schmiedeberg's Arch. Pharmacol.*, **317**, 199–203.

Gráf, L., Barát, E., and Patthy, A. (1976). Isolation of a COOH-β-lipotropin fragment (residues 61-91) with morphine-like analgesic activity from porcine pituitary glands. *Acta Biochim. Biophys.* (Acad. Sci. Hung.), **11**, 121–2.

Grafe, P., Mayer, C. J., and Wood, J. (1980). Synaptic modulation, of calcium-dependent potassium conductance in myenteric neurones in the guinea-pig. *J. Physiol. (Lond.)*, **305**, 235–48.

Grath, M. A., and Shepherd, J. T. (1976). Inhibition of adrenergic neurotransmission in canine vascular smooth muscle by histamine. Mediation by H2-receptor. *Circ. Res.*, **39**, 566–73.

Greden, J. F., Fontaine, P., Lubetsky, M., and Chamberlin, K. (1978). Anxiety and depression associated with caffeinism among psychiatric inpatients. *Amer. J. Phychiat.*, **135**, 963–6.

Greenacre, J. K., Teychenne, P. F., Petrie, A., Calne, D. B., Leigh, P. N., and Reid, J. L. (1976). The carciovascular effects of bromocriptine in Parkinsonism. *Br. J. Clin. Pharmac.*, **3**, 571–4.

Greenberg, D. A., Prichard, D. S., and Snyder, S. H. (1976). Alpha-noradrenergic receptor binding in mammalian brain: differential labelling of agonist antagonist states, *Life Sci.*, **19**, 69–76.

Greenfield, S. A., and Smith, A. D. (1976). Changes in acetylcholinesterase concentration in rabbit cerebrospinal fluid following central electrical stimulation. *J. Physiol. (Lond.)*, **258**, 108–9.

Greengard, P. (1978). *Cyclic Nucletides, Phosphorylated Proteins and Neuronal function*, Raven Press, New York.

Grofová, I., and Rinvik, E. (1970). An experimental electron microscopic study on the striato-nigral projection in the cat. *Exp. Brain, Res.*, **11**, 249–62.

Grossman, Y., Spira, M. E., and Parmas, I. (1973). Differential flow of information into branches of a single axon. *Brain Res.*, **64**, 379–86.

Groves, P. M. (1983). A theory of the functional organization of the neostriatum and the neostriatal control of voluntary movement. *Brain Res.*, **5**, 109–32.

Groves, P. M., Wilson, C. J., Young, S. J., and Rebec, G. V. (1975). Self-inhibition by dopaminergic neurons. *Science*, **109**, 522–9.

Groves, P. M., Wilson, C. J., and McGregor, R. J. (1972). In *International Symposium on Interactions among Putative Neurotransmitters in the Brain* (Eds Garattini, J. F. Pujol, and R. Samanin), Raven Press, New York.

Groves, P. M., Young, S. J., and Wilson, C. J. (1976). Self-inhibition by dopaminergic neurones: disruption by (+/−)-α methyl-p-tyrosine pre-treatment or anterior diencephalic lesion. *Neuropharmacology*, **15**, 755–62.

Guillemin, R. (1977). The expanding significance of hypothalamic peptides, or is endocrinology a branch of neuroendocrinology. *Rev. Progr. Horm. Res.*, **33**, 1–28.

Gustafsson, L. (1980). Studies on modulation of transmitter release and effector responsiveness in autonomic cholinergic neurotransmission. *Acta Physiol. Scand., Suppl.*, **489**, 1–28.

Gustafsson, L., Fredholm, B. B., and Hedqvist, P. (1980). Theophylline interferes with the modulatory role of endogenous adenosine on cholinergic neurotransmission in guinea-pig ileum. *Acta Physiol. Scand.*, **111**, 269–80.

Gustafsson, L., Hedqvist, P., Fredholm, B. B., and Lundgren, G. (1978). Inhibition of acetylcholine release in guinea-pig ileum by adenosine. *Acta Physiol. Scand.*, **104**, 469–78.

Gutman, Y., and Boonyaviroj, P. (1977). Inhibition of catecholamine release by alpha-adrenergic activation. Interaction with Na,K-ATPase. *J. Neural. Transm.*, **40**, 245–52.

Gyang, E. A., and Kosterlitz, H. W. (1966). Agonist and antagonist actions of morphine-like drugs on the guinea-pig isolated ileum. *Br. J. Pharmac. Chemother.*, **27**, 514–27.

Guyenet, P., Agid, Y., Javoy, F., Beaujouan, J. C., Rossier, J., and Glowinsky, J. (1975). Effect of dopaminergic receptor agonist and antagonists on the activity of the neostriatal cholinergic system. *Brain Res.*, **84**, 227–44.

Haas, H. L., Felix, D., and Davis, M. D. (1982). Angiotensin excites hippocampal pyramidal cells by two mechanisms. *Cell and Mol. Neurobiol.*, **2**, 21–32.

Haddy, F. J., and Scott, J. B. (1968). Metabolically linked vasoactive chemicals in local regulation of blood flow. *Physiol. Rev.*, **48**, 688–707.

Hadházy, P., and Szerb, J. C. (1977). The effect of cholinergic drugs on ^{3}H-acetylcholine release from slices of rat hippocampus, striatum and cortex. *Brain Res.*, **123**, 311–22.

Hadházy, P., and Vizi, E. S. (1974). The inhibitory action of noradrenaline on the chronotropic responses to vagal stimulation, in *Symposium on Drugs and Heart Metabolism* (Eds J. Knoll, L. Szekeres, and J. Papp), Akadémiai Kiadó, Budapest, pp. 181–9.

Hadházy, P., Illés, P., and Knoll, J. (1973). The effect of PGE$_1$ on responses to cardiac vagus nerve stimulation and acetylcholine release. *Eur. J. Pharmacol.*, **23**, 251–5.

Hadházy, P., Vizi, E. S., Magyar, K., and Knoll, J. (1976). Inhibition of adrenergic neurotransmission by prostaglandin E$_1$ (PGE$_1$) in the rabbit ear artery. *Neuropharmac.*, **15**, 245–50.

Haefely, W. E. (1969). Effects of catecholamines in the cat superior cervical ganglion and their postulated role as physiological modulators of ganglionic transmission. *Progr. Brain Res.*, **31**, 61–72.

Haeusler, G., Thoenen, H., Haefely, W., and Huerlimann, A. (1968). Electrical events in cardiac adrenergic nerves and noradrenaline release from the heart induced by acetylcholine and KC1. *Naunyn–Schmideberg's Arch. Pharmacol.*, **261**, 389–411.

Häggendal, J. (1969). On release of transmitter from adrenergic nerve terminals at nerve activity. *Acta Physiol. Scand., Suppl.*, **330**, 29.

Häggendal, J. (1970). Some further aspects on the release of the adrenergic transmitter, in *New Aspects of Storage and Release Mechanisms of Catecholamines* (Eds H. J. Schüman and G. Kroneberg), Springer-Verlag, Berlin, pp. 100–9.

Häggendal, J., Dahlström, A. B., and Saunders, N. R. (1973). Axonal transport and acetylcholine in rat preganglionic neurones. *Brain Res.*, **58**, 494–9.

Haigler, H. J., and Aghajanian, G. K. (1977). Serotonin receptors in the brain. *Fedn Proc.*, **36**, 2159–64.

Hajdu, F., Hassler, R., and Bak, I. J. (1973). Electron microscopic study of the substantia nigra: the strio-nigral projection in the rat. *Z. Zellforsch.*, **146**, 207–21.

Hammer, R., Berrie, C. P., Birdsall, N. J., Durgen, A. S. V., and Hulme, E. C. (1980). Pirenzepine distinguishes between different subclasses of muscarinic receptors. *Nature*, **283**, 90–2.

Hamon, M., Bourgoin, S., Jagger, J., and Glowinsky, J. (1974). Effects of LSD on synthesis and release of 5-HT in rat brain slices. *Brain Res.*, **69**, 265–80.

Hámori, J., and Szentágothai, J. (1965). The Purkinje cell baskets: ultrastructure of inhibitory synapse. *Acta biol. Acad. Sci. Hung.*, **15**, 465–79.

Hársing, L. G. Jr, and Vizi, E. S. (1979). Different sensitivity of pre- and postsynaptic dopamine and enkephalin receptors in rat striatum, in *Amingergic and Peptidergic Receptors*. (Eds E. S. Vizi and M. Wollemann), Pergamon Press, Oxford; Akadémiai Kiadó, Budapest, pp. 85–94.

Hársing, L. G. Jr, Vizi, E. S., and Knoll, J. (1978). Increase by enkephalin of acetylcholine release from striatal slices of the rat. *Pol. J. Pharmacol. Pharm.*, **30**, 387–95.

Hársing, L. G. Jr, Magyar, K., Tekes, K., Vizi, E. S., and Knoll, J. (1979a). Inhibition by deprenyl of dopamine uptake in rat striatum: a possible correlation between dopamine uptake and acetylcholine release inhibition. *Pol. J. Pharmacol. Pharm.*, **31**, 297–307.

Hársing, L. G. Jr, Illés, P., Fürst, Zs., Vizi, E. S., and Knoll, J. (1979b). The effect of prostaglandin E$_1$ on acetylcholine release from the cat brain. *Acta Physiol. Hung.*, **54**, 177–85.

Hársing, L. G. Jr, Yang, H. Y. T., and Costa, M. (1982). Evidence for a γ-aminobutyric acid

(GABA) mediation in the benzodiazepine inhibition of the release of Met[5]-enkephalin elicited by depolarization. *J. Pharmacol. Exp. Ther.*, **220**, 616–20.

Hatch, R. C., Booth, N. H., Derrel, J., Clark, M., Crawford, M., Kitzman, J. V., and Wallner, B. (1982). Antagonism of xylazine sedation in dogs by 4-aminopyridine and yohimbine. *Am. J. Vet. Research*, **43**, 1009–14.

Hattori, T., Fibiger, H. C., McGeer, P. L., and Maler, L. (1973). Analysis of the fine structure of the dopaminergic nigrostriatal projection by electron microscopic autoradiography. *Exp. Neurol.*, **41**, 599–611.

Hayashi, E., Mori, M., Yamada, S., and Kunitomo, M. (1978). Effects of purine compounds on cholinergic nerves. Specificity of adenosine and related compounds on acetylcholine release in electrically stimulated guinea-pig ileum. *Eur. J. Pharmacol.*, **48**, 297–307.

Hazra, J. (1975). Evidence against prostaglandin E having a physiological role in acetylcholine liberation from Auerbach's plexus of guinea-pig ileum. *Experientia*, **315**, 565–6.

Headley, P. M., Duggan, A. W., and Griersmith, B. T. (1978). Selective reduction by noradrenaline and 5-hydroxytryptamine of nociceptive response of cat dorsal horn neurones. *Brain Res.*, **145**, 185–9.

Hedler, I., Stamm, G., Weitzell, R., and Starke, K. (1981). Functional characterization of central α-adrenoceptors by yohimbine diastereoisomers. *Eur. J. Pharmacol.*, **70**, 43–52.

Hedqvist, P. (1969). Modulating effect of prostaglandin E_2 on noradrenaline release from the isolated cat spleen. *Acta Physiol. Scand.*, **75**, 511–12.

Hedqvist, P. (1970a). Studies on the effect of prostaglandins E_1 and E_2 on the sympathetic neuromuscular transmission in some animal tissues. *Acta Physiol. Scand. Suppl.*, **345**, 11–40.

Hedqvist, P. (1970b). Antagonism by calcium of the inhibitory action of prostaglandin E_2 on sympathetic neurotransmission in the cat spleen. *Acta Physiol. Scand.*, **80**, 269–75.

Hedqvist, P. (1972a). Prostaglandin-induced inhibition of neurotransmission in the isolated guinea-pig seminal vesicle. *Acta Physiol. Scand.*, **84**, 506–11.

Hedqvist, P. (1972b). Prostaglandin-induced inhibition of vascular tone and reactivity in the cat's hindleg *in vivo*. *Eur. J. Pharmacol.*, **17**, 157–62.

Hedqvist, P. (1973a). Prostaglandin mediated control of sympathetic neuroeffector transmission. *Adv. Biosci.*, **9**, 461–73.

Hedqvist, P. (1973b). Autonomic neurotransmission, in *The Prostaglandins*, (Ed. P. W. Ramwell), Vol. 1, Plenum Press, New York and London, pp. 101–31.

Hedqvist, P. (1973c). Prostaglandin as a tool for local control of transmitter release from sympathetic nerves. *Brain Res.*, **62**, 483–8.

Hedqvist, P. (1973d). Aspects on prostaglandin and α-receptor mediated control of transmitter release from adrenergic nerves, in *Catecholamine Research* (Eds E. Usdin, S. M. Snyder, and S. H. Frontiers), Pergamon Press, New York, pp. 583–7.

Hedqvist, P. (1974a). Restriction of transmitter release from adrenergic nerves mediated by prostaglandins and α-adrenoceptors. *Pol. J. Pharmac., Pharm.*, **26**, 119–25.

Hedqvist, P. (1977). Basic mechanism of prostaglandin action on autonomic neurotransmission. *Ann. Rev. Pharmacol.*, **17**, 259–79.

Hedqvist, P. (1981). Trans-synaptic modulation versus α-autoinhibition of noradrenaline secretion, in *Chemical Neurotransmission: 75 years* (Eds L. Stjärne, P. Hedqvist, H. Lagercrantz, and A. Wennmalm), Academic Press, London, pp. 223–48.

Hedqvist, P., and Brundin, J. (1969). Inhibition by prostaglandin E_1 or noradrenaline release and of effector response to nerve stimulation in the cat spleen. *Life Sci.*, **8**, 389–95.

Hedqvist, P., and Euler, U. S. von (1972). Prostaglandin-induced-neurotransmission failure in the nerve stimulated isolated vas deferens. *Neuropharmac.*, **11**, 177–87.

Hedqvist, P., and Fredholm, B. B. (1976). Effects of adenosine on adrenergic neurotransmission, prejunctional inhibition and postjunctional enhancement. *Naunyn–Schmiedeberg's Arch. Pharmacol.*, **293**, 217–23.

Hedqvist, P., and Moaward, A. (1975). Pre-synaptic α and β adrenoceptor mediated control of noradrenaline release in human oviduct. *Acta Physiol. Scand.*, **95**, 494–6.

Hedqvist, P., and Persson, N. A. (1975). Prostaglandin action on adrenergic and cholinergic responses in the rabbit and guinea-pig intestine, in *Chemical Tools, Catecholamine Research* (Eds O. Almgren, A. Carlsson, and J. Engel), Vol. 2, North-Holland, Amsterdam and Oxford, pp. 211–18.

Hedqvist, P., Fredholm, B., and Olundh, S. (1978). Antagonistic effects of theophylline and adenosine on adrenergic neuroeffector transmission in the rabbit kidney. *Circulation Res.*, **43**, 592–8.

Hefti, F., and Lichtensteiger, W. (1978). Subcellular distribution of dopamine in substantia nigra of the rat brain. Effects of butyrolactone and destruction of noradrenergic afferents suggest formation of particles from dendrites. *J. of Neurochem.*, **30**, 1217–30.

Heidenhain, R. (1872). Über die Wirkung einiger Gifte auf die Nerven der glandula submaxillaris. *Pflügers Arch.*, **5**, 309–18.

Heise, A., and Kroneberg, G. (1970). Periphere und zentrale Kreislaufwirkung des α-sympathicamimeticums 2-(-(2,6-xylidino)-5,6-dihydro-4H-1,3-thiazinhydrochlorid (Bayer, 1470). *Naunyn-Schmiedeberg's Arch. Pharmacol.*, **266**, 350.

Heise, A., Kroneberg, G., and Schlossmann, K. (1971). α-Sympathicomimetische Eigenschaften als Ursache der blutdrucksteigernden und blutdrucksenkenden Wirkung von Bayer 1470 (2-(2,6-xylidino)-5,6-dihydro-4H-1,3-thiazinhydrochlorid). *Naunyn-Schmiedeberg's Arch. Pharmacol.*, **268**, 348–60.

Hemsworth, B. A., and Neal, M. J. (1968). The effect of central stimulant drugs on acetylcholine release from rat cerebral cortex. *Br. J. Pharmacol.*, **34**, 543–50.

Henderson, G. (1976). Effect of normorphine and enkephalin on spontaneous potentials in the vas deferens. *Eur. J. Pharmacol.*, **39**, 409–12.

Henderson, G., and Hughes, J. (1976). The effects of morphine on the release of noradrenaline from the mouse vas deferens. *Br. J. Pharmacol.*, **57**, 551–7.

Henderson, G., Hughes, J., and Kosterlitz, H. W. (1978). *In vitro* release of Leu- and Met-enkephalin from the corpus striatum. *Nature*, **271**, 677–9.

Henderson, G., Hughes, J., and Thompson, J. W. (1972). The variation of noradrenaline output with frequency of nerve stimulation and the effect of morphine on the cat nictitating membrane and on the guinea-pig myenteric plexus. *Br. J. Pharmacol.*, **46**, 524–5P.

Henle, J. (1879). *Handbuch de Nervenlehre des Menschen*, F. Vieweg u. Sohn, Braunschweig.

Henn, F. A., Anderson, D. J., and Rustad, D. G. (1976). Glial contamination of synaptosomal fractions. *Brain Res.*, **101**, 341–4.

Henry, J. L. (1976). Effects of substance P on functionally identified units in cat spinal cord. *Brain Res.*, **114**, 439–51.

Henry, J. L. (1980). Substance P and pain: an updating *TINS*, **1**, 95–7.

Herman, H., Borvendég, J., Bajusz, S., Székely, J. I., and Rónai, A. Z. (1980). Effect of enkephalin analogs on prolaction and LH-release in rats. *Adv. Physiol. Sci.*, **14**, 333–53.

Hertting, A., Axelrod, J., Kopin, J. J., and Whitby, L. G. (1961a). Lack of uptake of catecholamines after chronic denervation of sympathetic nerves. *Nature*, **189**, 66.

Hertting, A., Axelrod, J., and Whitby, L. G. (1961b). Effect of drugs on the uptake and metabolism of ³H-norepinephrine. *J. Pharmac. Exp. Ther.*, **134**, 146–53.

Hertting, G., Zumstein, A., Jackish, R., Hottmann, I., and Starke, K. (1980). Modulation by endogenous dopamine of the release of acetylcholine in the caudate nucleus of the rabbit. *Naunyn-Schmiedeberg's Arch. Pharmacol.*, **315**, 111–7.

Hexum, T. D. (1978). Studies on the reaction catalyzed by transport (Na,K) adenosine triphosphatase-II. *In Vitro* and *in vivo* effects of phenoxybenzamine. *Biochem. Pharmacol.*, **27**, 2109–14.

Hicks, P. E. (1981). Antagonism of pre- and postsynaptic α-adrenoceptors by BE 2254 (HEAT) and prazosin. *J. Auton. Pharmacol.*, **1**, 391–7.

Hieble, J. P., and Pendleton, R. G. (1979). Effects of ring substitution on the pre- and postjunctional alpha-adrenergic activity of ariliminoimidazolidines. *Naunyn–Schmiedeberg's Arch. Pharmacol.*, **309**, 217–24.

Hilton, S. M., and Lewis, I. P. (1955). The cause of the vasodilatation accompanying activity in the submandibular salivary gland. *J. Physiol. (Lond.)*, **128**, 235–48.

Hirst, G. D. S., and McKirdy, H. C. (1974). Presynaptic inhibition at a mammalian peripheral synapse. *Nature*, **250**, 430–1.

Hisada, S., Fujimoto, S., Kamiya, T., Endo, Y., and Tsushima, M. (1977). Antidiuresis of centrally administered amines and peptides and release of antidiuretic hormone from isolated rat neurohypophysis. *Jap. J. Pharmacol.*, **27**, 153–61.

Hoffer, B. J., Siggins, G. R., and Bloom, F. E. (1971). Studies on norepinephrine-containing afferents to Purkinje cells of rat cerebellum. II. Sensitivity of Purkinje cells to norepinephrine and related substances administered by microiontophorcsis. *Brain Res.*, **25**, 523–34.

Hoffer, B. J., Siggins, G. R., Oliver, A. P., and Bloom, F. E. (1973). Activation of the pathway from locus coeruleus to rat cerebellar Purkinje neurons: pharmacological evidence of noradrenergic central inhibition. *J. Pharmacol. Exp. Ther.*, **184**, 553–69.

Hoffman, B. B., De Lean, A., Wood, C. L., Schocken, D., and Lefkowitz, R. J. (1979).

Alpha-adrenergic receptor subtypes: quantitative assessment by ligand binding. *Life Sci.*, **24**, 1739–46.

Hökfelt, T. (1968). *In vitro* studies on central and peripheral monoamine neurons at the ultrastructural level. *Z. Zellforsch.*, **91**, 1–74.

Hökfelt, T., Fuxe, K., Johansson, O., and Ljungdahl, A. (1974). Pharmacohistochemical evidence of the existence of dopamine nerve terminals in the limbic cortex. *Eur. J. Pharmacol.*, **25**, 98–112.

Hökfelt, T., Elde, R., Johansson, O., Terenius, L., and Stein, L. (1977a). The distribution of enkephalin-immunoreactive cell bodies in the rat central nervous system. *Neuroscience, Letters*, **5**, 25–31.

Hökfelt, T., Elfvin, L. G., Elde, R., Schultzberg, M., Goldstein, M., and Luft, R. (1977b). Occurrence of somatostatin-like immunoreactivity in some peripheral sympathetic noradrenergic neurons. *Proc. Natl Acad. Sci.*, **74**, 3587–91.

Hökfelt, T., Ljungdahl, A., Terenius, L., Elde, R., and Nilsson, G. (1977c). Immunohistochemical analysis of peptide pathways possibly related to pain and analgesia: enkephalin and substance P. *Proc. Natl Acad. Sci.*, **74**, 3081–5.

Hökfelt, T., Rehfeld, J. F., Skirboll, L., Ivemark, B., Goldstein, M., and Markey, K. (1980a). Evidence for coexistence of dopamine and CCK in mesolimbic neurons. *Nature*, **185**, 476–8.

Hökfelt, T., Skirboll, L., Rehfeld, J. F., Goldstein, M., Markey, K., and Dann, O. (1980b). A subpopulation of mesencephalic dopamine neurons projecting to limbic areas contains a cholecystokinin-like peptide: evidence from immunohistochemistry combined with retrograde tracing. *Neuroscience*, **5**, 2093–2124.

Hökfelt, T., Lundberg, J. M., Schultzberg, M., Johansson, O., Ljungdahl, A., and Rehfeld, J. F. (1980c). Coexistence of peptides and putative transmitters in neurons, in *Neural Peptides and Neuronal Communication (Adv. in Biochem. Psychopharmacol.*, **22**) (Eds E. Costa and M. Trabucchi), Raven Press, New York, pp. 1–23.

Hökfelt, T., Johansson, O., Ljungdahl, A., Lundberg, J. M., and Schultzberg, M. (1980d). Peptidergic neurones. *Nature*, **284**, 515–21.

Hökfelt, T., Lundberg, J. M., Schultzberg, M., Johansson, O., Skirboll, L., Anggard, A., Fredholm, B., Hamberger, B., Pernow, B., Rehfeld, J., and Goldstein, M. (1980c). Cellular localization of peptides in neural structures. *Proc. R. Soc. Lond.*, **210**, 63–83.

Holaday, J. W., and Faden, A. I. (1981a). Naloxone reverses the pathophysiology of shock through an antagonism of endorphin system. *Adv. Biochem. Psychopharm.*, **28**, 421–34.

Holaday, J. W., and Faden, A. I. (1981b). Naloxone treatment in shock (letter). *Lancet 2*, 201.

Holman, M. E., and Surprenant, A. (1980). Electrophysiological analysis of the effects of noradrenaline and α-receptor antagonists on neuromuscular transmission in mammalian muscular arteries. *Br. J. Pharmacol.*, **71**, 651–61.

Holman, M. E., Hirst, G. D. S., and Spence, I. (1972). Preliminary studies of the neurons of Auerbach's plexus using intracellular microelectrodes. *Aust. J. Exp. Biol. Med. Sci.*, **50**, 795–801.

Holton, F. A., and Holton, P. (1953). The possibility that ATP is a transmitter at sensory nerve endings. *J. Physiol. (Lond.)*, **119**, 50–1.

Holzbauer, M., Sharman, D. F., and Godden, U. (1978). Observation on the function of the dopaminergic nerves innervating the pituitary gland. *Neuroscience*, **3**, 1251–62.

Holzer, P., and Lembeck, F. (1980). Neurally mediated contraction of ileal longitudinal muscle by substance P. *Neurosci.*, *Lett.*, **17**, 101–5.

Hom, G. J., and Lokhandwala, M. F. (1981). Presynaptic inhibition of vascular sympathetic neurotransmission by adenosine. *Eur. J. Pharmacol.*, **69**, 101–6.

Hong, J. S., Yang, H. Y. T., Fratta, W., and Costa, E. (1977). Determination of methionine enkephalin in discrete regions of rat brain. *Brain Res.*, **134**, 383–6.

Honjin, R., Takahasni, A., Shimasaki, S., and Maruyama, H. (1963). Two types of synaptic nerve processes in the ganglia of Auerbach's plexus of mice, as revealed by electronmicroscopy. *J. Electron Microscopy*, **14**, 43–9.

Hope, W., Law, M., McGulloch, M. W., Rand, M. J., and Story, D. F. (1976). Effects of some catecholamines on noradrenergic transmission in the rabbit ear artery. *Clin. Exp. Pharmac. Phys.*, **3**, 15–28.

Hope, W., McCulloch, M. W., Story, D. F., and Rand, M. J. (1977). Effects of pimozide on noradrenergic transmission in rabbit isolated ear arteries. *Eur. J. Pharmacol.*, **46**, 101–11.

Hope, W., Majawski, H., McCulloch, M. W., Rand, M. J., and Story, D. F. (1979). Modulation of sympathetic transmission by neuronally-released dopamine. *Br. J. Pharmacol.*, **67**, 185–92.

Horn, J. P., and McAfee, D. A. (1980). Alpha-adrenergic inhibition of calcium dependent potentials in rat sympathetic neurones. *J. Physiol.*, **301**, 191–204.

Hornykiewicz, O. (1963). The topical localization and content of noradrenalin and dopamine (3-hydroxytyramine) in the substantia nigra of normal persons and patients with Parkinson's disease. *Wien. Klin. Wochenschr.*, **75**, 309–12.

Horton, E. W. (1973). Prostaglandins at adrenergic nerve-endings. *Brit. Med. Bull.*, **29**, 148–51.

Horton, E. W., and Main, I. H. M. (1963). A comparison of the biological activities of four prostaglandins. *Br. J. Pharmacol.*, **21**, 182–9.

Houston, M. C. (1981). Clonidine hydrochloride; review of pharmacologic and clinical aspects. *Prog. Cardiovasc., Dis.*, **23**, 337–50.

Hubbard, J. (1973). Microphysiology of vertebrate neuromuscular transmission. *Physiol. Rev.*, **53**, 674–723.

Hughes, J. (1975). Isolation of an endogenous compound from the brain with pharmacological properties similar to morphine. *Brain Res.*, **88**, 295–308.

Hughes, J., and Roth, R. J. (1971). Evidence that angiotensin enhances transmitter release during sympathetic nerve stimulation. *Br. J. Pharmacol.*, **41**, 239.

Hughes, J., and Vane, J. R. (1967). An analysis of the responses of the isolated protal vein of the rabbit to electrical stimulation and to drugs. *Br. J. Pharmacol.*, **30**, 46–66.

Hughes, J., Smith, T., Kosterlitz, H. W., Fothergill, L. A., Morgan, B., and Morris, H. R. (1975). Identification of two related pentapeptides from the brain with potent opiate agonist activity. *Nature*, **258**, 577–9.

Hughes, J., Kosterlitz, H. W., and Sosa, R. P. (1978). Enkephalin release from the myenteric plexus of the guinea-pig small intestine in the presence of cycloheximide. *Br. J. Pharmacol.*, **63**, 397.

Huidobro-Toro, J. P., Yoshimura, K., Lee, N. M., Loh, H. H., and Way, E. L. (1981). Dynorphin interaction at the κ-opiate site. *Eur. J. Pharmacol.*, **72**, 265–6.

Hulten, L., and Jodal, M. (1969). Extrinsic nervous control of colonic motility. *Acta Physiol. Scand., Suppl.*, **335**, 21–38.

Hunt, S. P., Kelly, J. S., and Emson, P. C. (1980). The electron microscopic localization of methionine–enkephalin within the superficial layers (I and II) of the spinal cord. *Neuroscience*, **5**, 1871–90.

Idowu, O. A., and Zar, M. A. (1977). The use of rat atria as a simple and sensitive *in vitro* preparation for detecting presynaptic actions of drugs on adrenergic transmission. *Br. J. Pharmacol.*, **61**, 157P.

Idowu, O. A., and Zar, M. A. (1978). Effects of clonidine and noradrenaline on the release of ^{3}H-noradrenaline from the rat anococcygeus. *Br. J. Pharmacol.*, **63**, 342P.

Illés, P., and North, R. A. (1982). Effects of divalent cations and normorphine on spontaneous excitatory junction potentials in the mouse vas deferens. *Br. J. Pharmacol.*, **75**, 599–604.

Illés, P., and Vyskocil, F. (1978). Calcium dependent inhibition by prostaglandin E_1 on spontaneous acetylcholine release from frog motor nerve. *Eur. J. Pharmacol.*, **48**, 455–7.

Illés, P., and Starke, K. (1983). An electrophysiological study of presynaptic α-adrenoceptors in the vas deferens of the mouse. *Br. J. Pharmacol.*, **78**, 365–73.

Illés, P., Hadházy, P., Torma, Z., Vizi, E. S., and Knoll, J. (1973). The effect of number of stimuli and rate of stimulation on the inhibition by PGE_1 of adrenergic transmission. *Eur. J. Pharmacol.*, **24**, 29–36.

Illés, P., Vizi, E. S., and Knoll, J. (1974). Adrenergic neuroeffector junctions sensitive and insensitive to the effect of PGE_1. *Polish J. Pharmacol. Pharmac.*, **26**, 127–36.

Illés, P., Zieglgänsberger, W., and Herz, A. (1979). Normorphine inhibits neurotransmission in the mouse vas deferens by a calcium-dependent mechanism. *Neurosci., Lett., Suppl.*, **3**, 238.

Innemee, H. C., de Jonge, A., van Meel, J. C., Timmermans, P. B., and van Zwieten, P. A. (1981). The effect of selective $α_1$ and $α_2$-adrenoceptor stimulation on intra-ocular pressure in the conscious rabbit. *Naunyn–Schmiedeberg's Arch. Pharmacol.*, **316**, 294–8.

Israel, M., Lesbats, B., Manaranche, R., Marsal, J., Mastour-Frachon, P., and Meunier, F. M. (1977). Related changes in amounts of ACh and ATP in resting and active Torpedo nerve electroplaque synapses. *J. Neurochem.*, **28**, 1259–67.

Israel, M., Lesbats, B., Meunier, F. M., and Stinnakre, J. (1976). Postsynaptic release of adenosine triphosphate induced by single impulse transmitter action. *Proc. R. Soc. B.*, **193**, 461–8.

Ivanov, A. Y., and Skok, V. I. (1980). Slow inhibitory postsynaptic potentials and hyperpolarization evoked by noradrenaline in the neurons of mammalian sympathetic ganglion. *J. Aut. Nerve, Syst.*, **1**, 255–63.

Iversen, L. L. (1978). Chemical messenger in the brain. *TINS*, **1**, 15–6.

Iversen, L. L. (1979). Neurotransmitter interactions in the substantia nigra a model for local circuit chemical interactions, in *Neuroscience Fourth Study Program* (Eds F. O. Schmitt and F. G. Worden), MIT Press, Cambridge, Mass., pp. 260–5.

Iversen, L. L., and Bloom, F. E. (1972). Studies on the uptake of ^{3}H-GABA and ^{3}H-glycine in slices and homogenates of rat brain and spinal cord by electron microscopic autoradiography. *Brain Res.*, **41**, 131–43.

Iversen, L. L., and Langer, S. Z. (1969). Effects of phenoxybenzamine on the uptake and metabolism of noradrenaline in the rat heart and vas deferens. *Brit. J. Pharmacol.*, **37**, 627–37.

Iversen, L. L., Iversen, S. D., Bloom, F. F., Varga, T., and Guillemin, R. (1978). Release of enkephalin from rat globus pallidum *in vitro*. *Nature*, **271**, 679–84.

Iversen, S. D., Kelley, A. E., and Stinus, L. (1979). Dose dependent behavioural stimulation after local infusion of substance P into the ventral segmental area in the rat. *Br. J. Pharmacol.*, **66**, 113–4.

Jacobowitz, D. (1965). Histochemical studies of the autonomic innervation of the gut. *J. Pharmac. Exp. Ther.*, **149**, 358–64.

Jacquet, Y. F., and Marks, N. (1976). The C-fragment of β-lipotropin: an endogenous neuroleptic or antipsychotogen. *Science*, **194**, 632–6.

Jaim-Etcheverry, G., and Zieher, L. M. (1975). Octopamine probably coexists with noradrenaline and serotonin in vesicles of pineal adrenergic nerves. *J. Neurochem.*, **25**, 915–7.

Jan, Y. N. and Jan. L. Y. (1983a). Coexistence and corelease of cholinergic and peptidergic transmitters in frog sympathetic ganglia. *Fedn. Proc.*, **42**, 2929–33.

Jan, Y. N., and Jan, L. Y. (1983b). A LHRH-like peptidergic neurotransmitter capable of action at a distance in autonomic ganglia. *TINS*, **3**, 320–6.

Jansson, G., and Martinson, J. (1966). Studies on the ganglionic site of action of sympathetic outflow to the stomach. *Acta Physiol. Scand.*, **68**, 184–92.

Jasper, H. H., Kahn, R. T., and Elliott, K. A. C. (1965). Amino acids released from the cerebral cortex in relations to its state of activation. *Science*, **147**, 1448–9.

Jeftinija, S., Miletic, V., and Randic, M. (1981). Cholecystokinin octapeptide excites dorsal horn neurons both *in vivo* and *in vitro*. *Brain Res.*, **213**, 231–6.

Jenden, D. J. (1974). Effect of chlorpromazine on acetylcholine turnover *in vitro*, in *Phenothiazines and Structurally Related Drugs* (Eds L. S. Forrest, C. J. Carr, and E. Usdin), Raven Press, New York.

Jenkins, D. A., Marshall, I., and Nasmyth, P. A. (1977). An inhibitory role for noradrenaline in the mouse vas deferens. *Br. J. Pharmacol.*, **61**, 649–55.

Jennewein, H. W. (1977). The effect of clonidine on gastric acid secretion in rats and dogs. *Naunyn-Schmiedeberg's Arch. Pharmacol*, **297**, 85–90.

Jessel, T. M., and Iversen, L. L. (1977a). Opiate analgetic inhibits substance P release from rat trigeminal nucleus. *Nature*, **268**, 549–51.

Jessel, T. M., and Iversen, L. L. (1977b). Inhibition of substance P release from the isolated rat substantia nigra by GABA. *Br. J. Pharmacol.*, **59**, 468–93.

Jessel, T. M., Iversen, L. L., and Cuello, A. C. (1978). Capsaicin-induced depletion of substance P from primary sensory neurones. *Brain Res.*, **152**, 183–8.

Jhamandas, K., and Dickinson, G. (1973). Modification of precipitated morphine and methadone abstinence in mice by acetylcholine antagonists. *Nature*, **245**, 219–21.

Jhamandas, K., and Sutak, M. (1976). Morphine–naloxone interaction in the central cholinergic system: the influence of subcortical lesioning and electrical stimulation. *Br. J. Pharmacol.*, **58**, 101–7.

Jhamandas, K., Pinsky, C., and Phillis, J. W. (1970). Effect of morphine and its antagonists on release of cerebral acetylcholine. *Nature*, **228**, 176–7.

Jhamandas, K., Phillis, J. W., and Pinsky, C. (1971). Effect of narcotic analgetics and antagonists on the *in vivo* release of acetylcholine from the cerebral cortex of the cat. *Br. J. Pharmacol.*, **43**, 53–66.

Johnson, D. A., and Pilar, G. (1980). The release of acetylcholine from post-ganglionic cell bodies in response to depolarization. *J. Physiol. (Lond.)*, **299**, 605–19.

Johnson, D. G., Thoa, N. B., Weinshilboum, R., Axelrod, J., and Kopin, I. J. (1971). Enhanced release of dopamine-β-hydroxylase sympathetic nerves by calcium and phenoxybenzamine and its reversal by prostaglandins. *Proc. Natl Acad. Sci., USA*, **68**, 2227–30.

Johnson, S. M., Katayama, Y., and North, R. A. (1980). Multiple actions of 5-hydroxytryptamine on myenteric neurones of the guinea-pig ileum. *J. Physiol. (Lond.)*, **304**, 459–70.

Johnsson, E. S. (1963). The origin of the acetylcholine released spontaneously from the guinea-pig isolated ileum. *Br. J. Pharmac. Chemother.*, **21**, 555–68.

Johnston, G. A. R., and Mitchell, J. F. (1971). The effect of bicuculline, metrazol, picrotoxin and strychnine on the release of ³H-GABA from rat brain slices. *J. Neurochem.*, **18**, 2441–6.

Jones, B. E., Guyenet, P., Chéramy, A., Gauchy, C., and Glowinsky, J. (1973). The *in vivo* release of acetylcholine from cat caudate nucleus after pharmacological and surgical manipulations of dopaminergic nigrostriatal neurons. *Brain Res.*, **64**, 355–69.

Junstad, M., and Wennmalm, A. (1973). On the release of prostaglandin E_2 from the rabbit heart following infusion of noradrenaline. *Acta Physiol. Scand.*, **87**, 573–4.

Jurna, I. (1966). Inhibition of the effect of repetitive stimulation on spinal meotoneurones of the cat by morphine and pethidine. *Int. J. Neuropharmac.*, **5**, 117–23.

Kadlec, O., Masek, K., and Seferna, I. (1974). Modulating role of prostaglandins in concentrations of the guinea-pig ileum. *Brit. J. Pharmacol.*, **51**, 565–70.

Kadlec, O., Masek, K., and Seferna, I. (1978). Modulation by prostaglandin of the release of acetylcholine and noradrenaline in guinea-pig isolated ileum. *J. Pharmac. Exp. Ther.*, **205**, 635–45.

Kalsner, S. (1979a). Single pulse stimulation of guinea-pig vas deferens and the presynaptic receptor hypothesis. *Br. J. Pharmacol.*, **66**, 343–9.

Kalsner, S. (1979b). Limitations of presynaptic adrenoceptor theory: the characteristics of the effects of noradrenaline and phenoxybenzamine on stimulation-induced efflux of ³H-noradrenaline in vas deferens. *J. Pharmacol. Exp. Ther.*, **212**, 232–9.

Kalsner, S. (1980a). The effects of (+) and (−)propranolol on ³H-transmitter efflux in guinea-pig atria and the presynaptic β-adrenoceptor hypothesis. *Br. J. Pharmacol.*, **70**, 491–8.

Kalsner, S. (1980b). Limitations of presynaptic adrenoceptor theory: the characteristics of the effect of noradrenaline and phenoxybenzamine on stimulation-induced efflux of ³H-noradrenaline in vas deferens. *J. Pharmacol. Exp. Ther.*, **212**, 232–9.

Kalsner, S. (1982a). The presynaptic receptor controversey. *TIPS*, **3**, 11–16.

Kalsner, S. (1982b). Evidence against the unitary hypothesis of agonist and antagonist action at presynaptic adrenoceptors. *Br. J. Pharmacol.*, **77**, 375–80.

Kalsner, S. (1982c). Positive feedback regulation of noradrenaline release from sympathetic nerves: a questionable hypothesis. *Can. J. Physiol. Pharmacol.*, **60**, 743–6.

Kalsner, S., and Chan, C. C. (1979). Adrenergic antagonists and the presynaptic receptor hypothesis in vascular tissue. *J. Pharmacol. Exp. Ther.*, **211**, 257–64.

Kalsner, S., Suleiman, M., and Dobson, R. E. (1980). Adrenergic presynaptic receptors: an overextended hypothesis? *J. Pharm. Pharmacol.*, **32**, 290–2.

Kamal, L., Arbilla, S., and Langer, S. Z. (1978). Effects of GABA-receptor agonists and antagonists on the potassium evoked release of ³H-GABA from the rat substantia nigra, in *Pre-synaptic Receptors* Symposium, Paris, *Abstracts*, p. 31.

Kamikawa, Y., Shimo, A., and Uchida, K. (1982). Inhibitory actions of catecholamines on electrically induced contractions of the submucous plexus–longitudinal muscularis mucosae preparation of the guinea-pig oesophagus. *Br. J. Pharmacol.*, **76**, 271–7.

Kanazawa, I., and Jessel, T. (1976). Post-mortem changes and regional distribution of substance P and mouse nervous system. *Brain Res.*, **117**, 362–70.

Kanazawa, I., Emson, P. C., and Cuello, A. C. (1977). Evidence for the existence of substance P containing fibres in striato-nigral and pallido-nigral pathways in the brain. *Brain Res.*, **119**, 447–56.

Kao, C. Y. (1966). Tetrodotoxin, saxitoxin and their significance in the study of excitation phenomena. *Pharmac. Rev.*, **18**, 997–1049.

Kása, P. (1968). Acetylcholinesterase transport in the central and peripheral nervous tissue: the role of tubules in the enzyme transport. *Nature*, **218**, 1265–7.

Kása, P., Mann, S., Karcus, S., Tóth, L., and Jordan, S. (1973). Transport of choline acetyltransferase and acetylcholinesterase in the rat sciatic nerve: a biochemical and electrohistochemical study. *J. Neurochem.*, **21**, 431–6.

Kato, A. C., Katz, H. S., and Collier, B. (1974). Absence of adenine nucleotide release from autonomic ganglion. *Nature*, **249**, 576–7.

Katsuragi, T., and Su, C. (1980). Purine release from vascular adrenergic nerves by high potassium and a calcium ionophrore, A-23187. *J. Pharmacol. Exp. Ther.*, **215**, 685–90.

Katsuragi, T., and Su, C. (1982). Augmentation by theophylline of ³H-purine release from vascular adrenergic nerves evidence for presynaptic autoinhibition. *J. Pharmacol. Exp. Ther.*, **220**, 152–6.

Katz, B. (1969). *The Release of Neural Transmitter Substances*, Sherrington Lecture No. 10, Liverpool University Press, Liverpool.

Katz, B., and Miledi, R. (1969). Spontaneous and evoked activity of motor nerve endings in calcium Ringer. *J. Physiol. (Lond.)*, **203**, 459–87.

Kayaalp, S. O., and McIsaac, R. J. (1970). Differential blockade and potentiation of transmission in a sympathetic ganglion. *J. Pharmac. Exp. Ther.*, **173**, 193–204.

Kazic, T. (1971). Effect of adrenergic factors on peristalsis and acetylcholine release. *Eur. J. Pharmacol.*, **16**, 367–73.

Kazic, T., and Josavljevic, M. (1976). Influence of adenosine cAMP and db-cAMP on responses of the isolated terminal guinea-pig ileum to electrical stimulation. *Arch. int. Pharmacodyn.*, **233**, 187–95.

Kebabian, J. W., and Calne, D. B. (1979). Multiple types of dopamine receptors. *Nature*, **177**, 93–6.

Kemp, J. M., and Powell, T. P. (1971). The synaptic organization of the caudate nucleus. *Phil. Trans. B.*, **262**, 403–12.

Kerwin, R., and Pycock, C. (1979). Effect of ω-amino acids on tritiated dopamine release from rat striatum: evidence for a possible glycinergic mechanism. *Biochem. Pharmacol.*, **28**, 2193–7.

Kewenter, J. (1965). The vagal control of the jejunal and ileal motility and blood flow. *Acta Physiol. Scand., Suppl.*, **257**, 1–68.

Khodorov, B. I., Timin, Y. N., Vilenken, S. Y., and Gulko, F. B. (1969). Theoretical analysis of the mechanisms of conduction of a nerve pulse over an inhomogeneous axon. I. Conduction through a portion with increased diameter (in Russian). *Biofizika*, **14**, 323–5.

Kilbinger, H. (1977). Modulation by oxotremorine and atropine of acetylcholine release evoked by electrical stimulation of the myenteric plexus of the guinea-pig ileum. *Naunyn-Schmiedeberg's Arch. Pharmacol.*, **300**, 145–51.

Kilbinger, H. (1982). The myenteric plexus–longitudinal muscle preparation, in *Progress in Cholinergic Biology*, Model Cholinergic Synapses (Eds I. Hanin and A. M. Goldberg), Raven Press, New York, pp. 137–67.

Kilbinger, H., and Wagner, P. (1975). Inhibition of oxotremorine of acetylcholine resting release from guinea-pig ileum longitudinal muscle strips. *Naunyn-Schmiedeberg's Arch. Pharmac.*, **287**, 47–60.

Kilbinger, H., and Wessler, I. (1979). Increase by α-adrenolytic drugs of acetylcholine release evoked by field stimulation of the guinea-pig ileum. *Naunyn-Schmiedeberg's Arch. Pharmacol.*, **309**, 255–7.

Kilbinger, H., and Wessler, I. (1980). Inhibition of acetylcholine of the stimulation-evoked release of ^{3}H-acetylcholine from the guinea-pig myenteric plexus. *Neuroscience*, **5**, 1331–40.

Kim, J. S., Bak, I. J., Hassler, R., and Okada, Y. (1971). Role of gamma-aminobutyric acid (GABA) in the extrapyramidal motor system. Some evidence for the existence of a type of GABA rich strio-nigral neurones. *Exp. Brain Res.*, **14**, 95–106.

Kirpekar, S. M., and Cervoni, P. (1963). Effect of cocaine, phenoxybenzamine and phentolamine on the catecholamine output from spleen and adrenal medulla. *J. Pharmac. Exp. Ther.*, **142**, 59–70.

Kirpekar, S. M., and Puig, M. (1971). Effect of flow stop on noradrenaline release from normal spleens and spleens treated with cocaine phentolamine or phenoxybenzamine. *Br. J. Pharmacol.*, **43**, 359–69.

Kirpekar, S. M., and Wakade, A. R. (1968). Release of noradrenaline from the cat spleen by potassium. *J. Physiol. (Lond.)*, **194**, 595–608.

Kirpekar, S. M., Wakade, A. R., and Prat, J. C. (1968). Properties of ligated sympathetic nerves. *Fedn Proc.*, **27**, 709–19.

Kirpekar, S. M., Prat, J. C., Puig, M., and Wakade, A. R. (1972). Modification of the evoked release of noradrenaline from the perfused cat spleen by various ions and agents. *J. Physiol. (Lond.)*, **221**, 601–15.

Kirpekar, S. M., Forchsott, R. P., Wakade, A. R., and Prat, J. C. (1973). Inhibition of sympathomimetic amines of the release of norepinephrine evoked by nerve stimulation in the cat spleen. *J. Pharmac. Exp. Ther.*, **187**, 529–38.

Kirpekar, S. M., Prat, J. C., and Wakade, A. R. (1975). Effect of calcium on the relationship between frequency of stimulation and release of noradrenaline from the perfused spleen of the cat. *Naunyn-Schmiedeberg's Arch. Pharmacol.*, **287**, 205–12.

Kleikues, P., Kobayashi, K., and Hossman, K. A. (1974). Purine nucleotide metabolism in the cat brain after one hour of complete ischemia. *J. Neurochem.*, **23**, 417–25.

Klein, R. L., Wilson, S. P,, Dzielak, D. J., Yang, W. H., and Viveros, O. H. (1982). Opioid peptides and noradrenaline co-exist in large dense-cored vesicles from sympathetic nerve. *Neurosci.*, **7**, 2255–61.

Kline, N. S., and Lehman, H. E. (1979). Therapy with β-endorphin in psychiatric patients, in *Endorphins in Mental Health Research* (Eds E. Usdin, W. E. Bunney, and N. S. Klina), Macmillan, London, pp. 500–17.

Knepel, W., and Reimann, W. (1982). Inhibition by morphine and β-endorphin of vasopressin release evoked by electrical stimulation of the rat medial basal hypothalamus *in vitro*. *Brain Res.*, **238**, 484–8.

Knoll, J. (1976). Neuronal peptide (enkephalin) receptors in the ear artery of the rabbit. *Eur. J. Pharmacol.*, **39**, 403–7.

Knoll, J. (1977). Two kinds of opiate receptor. *Pol. J. Pharmac., Pharm.*, **29**, 165–75.

Knoll, J., and Vizi, E. S. (1970). Presynaptic inhibition of acetylcholine release by endogenous and exogenous noradrenaline at high rate of stimulation. *Br. J. Pharmacol.*, **40**, 554–5.

Knoll, J., and Vizi, E. S. (1971). Effect of frequency of stimulation on the inhibition by noradrenaline of the acetylcholine output from parasympathetic nerve terminals. *Br. J. Pharmacol.*, **41**, 263–72.

Knoll, J., and Illés, P. (1978). The isolated nictitating membrane of the cat. A new model for the study of narcotic analgetics. *Pharmacology*, **17**, 215–25.

Knoll, J., Somogyi, G. T., Illés, P., and Vizi, E. S. (1972). Acetylcholine release from isolated vas deferens of the rat. *Naunyn-Schmiedeberg's Arch. Pharmacol.*, **274**, 198–202.

Kobinger, W. (1974). Medicinal chemistry related to the central regulation of blood pressure, II. Pharmacological part, in *Medicinal Chemistry*, IV (Ed. J. Maas). Proceedings of the 4th International Symposium on Medicinal Chemistry, Elsevier, Amsterdam, Oxford, and New York, pp. 107–20.

Koda, L. Y., Schulman, J. A., and Bloom, F. E. (1978). Ultrastructural identification of noradrenergic terminals in rat hippocampus after unilateral destruction of the locus with 6-hydroxydopamine. *Brain Res.*, **145**, 190–5.

Koelle, G. B. (1962). A new general concept of the neurohumoral function of acetylcholine and acetylcholinesterase. *J. Pharm. Pharmacol.*, **14**, 65–90.

Koiker, H., Einstadt, M., and Schwartz, J. H. (1972). Axonal transport of newly synthesized acetylcholine in an identified neuron of *Aplysia*. *Brain Res.*, **37**, 152–9.

Kondo, H., and Yui, E. (1981). An electron-microscopic study on substance P like immunoreactive nerve fibers in the celiac ganglion of guinea-pigs. *Brain Res.*, **222**, 134–7.

Konishi, S., and Otsuka, M. (1975). Excitatory action of hypothalamic substance P on spinal motoneurones of newborn rats. *Nature*, **252**, 734–5.

Konishi, S., Tsunoo, A., and Otsuka, M. (1979). Enkephalins presynaptically inhibit cholinergic transmission in sympathetic ganglia. *Nature*, **282**, 515–16.

Konishi, S., Tsunoo, A., Yanaihara, N., and Otsuka, M. (1980). Peptidergic excitatory and inhibitory synapses in mammalian sympathetic ganglia: roles of substance P and enkephalin. *Biomed. Res.*, **1**, 528–36.

Korf, J., Zieleman, M., and Westerink, B. H. C. (1976). Dopamine release in substantia nigra. *Nature (Lond.)*, **260**, 257–8.

Koritsánszky, M., Vizi, E. S., and Knoll, J. (1977). Acetylcholine release by naloxone from isolated guinea-pig ileum, in *Physiology and Pharmacology of Smooth Muscle* (Eds M. Papason and E. Atannossoz), Publishing House of the Bulgarian Academy of Sciences, Sofia, pp. 197–202.

Koslow, S. H., Cattabeni, F., and Costa, E. (1972). Norepinephrine and dopamine, assay by mass fragmentography in the picomole range. *Science*, **176**, 177–80.

Kostepoulos, G. K., Limacher, J. J., and Phillis, J. W. (1975). Action of various adenine derivatives on cerebellar Purkinje cells. *Brain Res.*, **88**, 162–5.

Kosterlitz, H. W., and Robinson, J. A. (1957). Inhibition of the peristaltic reflex of the isolated guinea-pig ileum. *J. Physiol. (Lond.)*, **136**, 249–62.

Kosterlitz, H. W., and Waterfield, A. A. (1975). *In vitro* models in the study of structure–activity relationships of narcotic analgetics. *Ann. Rev. Pharmac.*, **15**, 29–47.

Kosterlitz, H. W., Lyden, R. J., and Watt, A. J. (1970). The effects of adrenaline, noradrenaline and isoprenaline on inhibitory α and β-adrenoceptors in the longitudinal muscle of the guinea-pig ileum. *Br. J. Pharmacol.*, **39**, 389–413.

Kosterlitz, H. W., Lord, J. A. H., Paterson, S. J., and Waterfield, A. A. (1980). Effect of changes in the structure of enkephalin and of narcotic analgesic drugs on their interactions with μ and δ receptors. *Brit. J. Pharmacol.*, **68**, 333–42.

Kottegoda, S. R. (1969). An analysis of possible nervous mechanism involved in the peristaltic reflex. *J. Physiol. (Lond.)*, **200**, 687.

Kravitz, E. A., Kuffler, S. W., and Potter, D. D. (1963). Gamma-aminobutyric acid and other blocking compounds in crustacea, III. Their relative concentrations in separated motor and inhibitory axons. *J. Neurophysiol.*, **26**, 739–51.

Kreutzberg, G. W., and Schubert, P. (1975). The cellular dynamics of interneuronal transport, in *The Use of Axonal Transport for Studies of Neuronal Connectivity* (Eds W. M. Cowan and M. Cuenod), Elsevier, Amsterdam, pp. 183–212.

Kriz, N., Sykova, E., Vjec, E., and Vyklicky, L. (1974). Changes of extracellular potassium concentration induced neuronal activity in the spinal cord of the cat. *J. Physiol. (Lond.)*, **238**, 1–15.

Krnjevic, K. (1980). Principles of synaptic transmission, in *Antiepileptic Drugs: Mechanisms of Action* (Eds G. H. Glaser, J. K. Penry, and D. M. Woodbury), Raven Press, New York, pp. 127–54.

Krnjevic, K., and Miledi, R. (1958). Some effects produced by adrenaline upon neuromuscular propagation in the rat. *J. Physiol. (Lond.)*, **141**, 291–304.

Krnjevic, K., and Phillis, I. W. (1963). Actions of certain amines on cerebral cortical neurones. *Br. J. Pharmacol.*, **20**, 471–90.

Kuchii, M., Miyahara, J. T., and Shibata, S. (1973). ^{3}H-adenine nucleotide and ^{3}H-noradrenaline release evoked by electrical field stimulation; perivascular nerve stimulation and nicotine from the taenia of the guinea-pig caecum. *Br. J. Pharmacol.*, **49**, 258–67.

Kuffler, S. W. (1980). Slow synaptic responses in autonomic ganglia and the pursuit of a peptidergic transmitter. *J. Exp. Biol.*, **89**, 257–86.

Kuffler, S. W., and Nicholls, J. G. (1976). *From Neuron to Brain*, Sinauer Associates, Dunderland, Massachusetts.

Kuhar, M. J., Aghajanian, G. K., and Roth, R. H. (1972a). Tryptophan hydroxylase activity and synaptosomal uptake of serotonin in discrete brain regions after midbrain raphé lesions: correlations with serotonin levels and histochemical fluorescence. *Brain Res.*, **44**, 165–76.

Kuhar, M. J., Roth, R. H., and Aghajanian, G. K. (1972b). Synthesis of catecholamines in the locus coeruleus from ^{3}H-tyrosine *in vivo*. *Biochem. Pharmacol.*, **21**, 2280–82.

Kuhar, M. J., Simon, J. P., and Taylor, N. (1975). Serotonergic synaptosomes from rat hippocampus: lack of acetylcholinesterase. *Brain Res.*, **99**, 415–18.

Kunos, G., Farsang, C., and Gonzales, R. (1981). β-endorphin possible involvement in the antihypertensive effect of central α-receptor activation. *Science*, **211**, 82–4.

Kuroda, Y., and McIlwain, H. (1974). Uptake and release of ^{14}C-adenine derivatives at beds of mammalian cortical synaptosomes in a superfusion system. *J. Neurochem.*, **22**, 691–9.

Kurtzman, N. A., and Boonjarern, S. (1975). Physiology of antidiuretic hormone and the interrelationship between the hormone and the kidney. *Nephron*, **15**, 167–85.

Kyncl, J. J., Minard, F. N., and Jones, P. H. (1979). Physiological and pharmacological role of regulations of noradrenaline release by presynaptic dopamine receptors in the peripheral nervous system, in *Peripheral Dopaminergic Receptors* (Eds J. L. Imbs, and J. Schwartz), Pergamon Press, Oxford, pp. 233–40.

Lagercrantz, H. (1976). On the composition and function of large dense core vesicles in sympathetic nerves. *Neuroscience*, **1**, 81–92.

Lagercrantz, H., and Stjärne, L. (1974). Evidence that most noradrenaline is stored without ATP in sympathetic large dense core nerve vesicles. *Nature (Lond.)*, **249**, 843–4.

Langen, C. D. J., Hogentrom, F., and Mulder, A. H. (1979). Presynaptic noradrenergic α-receptors and modulation of ^{3}H-noradrenaline release from rat brain synaptosomes. *Eur. J. Pharmacol.*, **60**, 79–89.

Langer, S. Z. (1973). The regulation of transmitter release elicited by nerve stimulation through a presynaptic feedback mechanism, in *Frontiers in Catecholamine Research* (Eds E. Usdin and S. Snyder), Pergamon Press, New York, pp. 543–9.

Langer, S. Z. (1974). Presynaptic regulation of catecholamine release. *Biochem. Pharmacol.*, **23**, 1793–1800.

Langer, S. Z. (1975). Prejunctional regulatory mechanisms for noradrenaline release elicited by nerve stimulation, in *Chemical Tools in Catecholamine Research* (Ed. T. Malmfors), North-Holland, Amsterdam.

Langer, S. Z. (1976). The role of α and β presynaptic receptors in the regulation of noradrenaline release elicited by nerve stimulation. *Clin. Sci. Mol. Med.*, **51**, 4238–65.

Langer, S. Z. (1977). Presynaptic receptors and their role in the regulation of transmitter release, Sixth Gaddum Memorial Lecture. *Br. J. Pharmacol.*, **60**, 481–97.

Langer, S. Z. (1981a). Presynaptic regulation of the release of catecholamines. *Pharmacol. Rev.*, **32**, 337–62.

Langer, S. Z. (1981b). Presence and physiological role of presynaptic inhibitory α_2-adrenoceptors in guinea-pig atria. *Nature*, **294**, 671–2.

Langer, S. Z., and Dubocovich, M. L. (1981). Cocaine and amphetamine antagonize the decrease of noradrenergic neurotransmission elicited by oxymetazoline but potentiate the inhibition by α-methylnorepinephrine in the perfused cat spleen. *J. Pharmacol. Exp. Ther.*, **216**, 162–71.

Langer, S. Z., and Pinto, J. E. B. (1976). Possible involvement of a transmitter different from norepinephrine in the residual responses to nerve stimulation of the cat nictitating membrane after pretreatment with reserpine. *J. Pharmacol. Exp. Ther.*, **196**, 697–713.

Langer, S. Z., Adler, E., Enero, M. A., and Stefano, J. F. E. (1971). The role of the receptor in regulating noradrenaline overflow by nerve stimulation. *Proc. Int. Union Physiol. Sci.*, **9**, 335.

Langer, S. Z., Dubocovich, M. L., and Celuch, S. M. (1975). Pre-junctional regulatory mechanisms for noradrenaline release elicited by nerve stimulation, in *Chemical Tools in Catecholamine Research* (Eds J. Almgren, O. Carlsson, and A. J. Engel), Vol. 2, North-Holland, Amsterdam and Oxford, pp. 183–91.

Langer, S. Z., Adler-Graschinsky, E., and Giorgi, O. (1977). Physiological significance of the alpha-adrenoceptor mediated negative feedback mechanism that regulates noradrenaline release during nerve stimulation. *Nature*, **265**, 648–50.

Langley, J. N. (1878). On the physiology of the salivary secretion. *J. Physiol. (Lond.)*, **1**, 339–64.

Langley, J. N. (1905). On the reaction of cells and of nerve endings to certain poisons, chiefly as regards the reaction of striated muscle to nicotine and to curari. *J. Physiol. (Lond.)*, **33**, 374.

Langley, J. N. (1921). *The Autonomic Nervous System*, Heffer, Cambridge.

Langley, J. N., and Anderson, H. K. (1895). The innervation of the pelvic and adjoining viscera. Part V. Position of the nerve cells on the course of the efferent nerve fibres. *J. Physiol. (Lond.)*, **19**, 131–9.

Lapierre, Y., Beaudet, A., Demianczuk, N., and Descarries, L. (1973). Noradrenergic axon terminals in the cerebral cortex of rat. II. Quantitative data revealed by light and electron microscope radioautography of the frontal cortex. *Brain Res.*, **63**, 175–82.

Lars, J., Nilsson, G., and Carlsson, A. (1982). Dopamine receptor agonist with apparent selectivity for autoreceptors, a new principle for antipsychotic action? *TIPS*, **36**, 322–5.

Larsson, L. I., Faherkrug, J., Schaffalitzky De Muckadell, O., Suncler, F., Hakansson, R., and Rehfeld, J. F. (1976). Localization of vasoactive intestinal VIP polypeptide to central and peripheral neurones. *Proc. Natl Acad. Sci. USA*, **73**, 3197–200.

Laubie, M., Schmitt, H., and Falo, E. (1977). Dopamine receptors in the fenoral vascular bed of the dog as mediators of a vasodilator symphathoinhibitory effect. *Eur. J. Pharmacol.*, **42**, 307–10.

Leedham, J. A., and Pennefather, J. N. (1982). Dopamine acts at the same receptors as noradrenaline in the rat isolated vas deferens. *Br. J. Pharmacol.*, **77**, 293–9.

Leeman, S. E., and Carraway, R. E. (1976). Discovery of a sialogic peptide in bovine hypothalamic extracts, isolation, characterization as substance P structure and synthesis, in *Substance P* (Eds U. S. von Euler and B. Pernow), Raven Press, New York, pp. 5–14.

Leeman, S. E., and Hemmershlag, R. (1967). Stimulation of salivary secretion by a factor extracted from hypothalamic tissue. *Endocrinology*, **81**, 803–10.

Lefevre-Borg, F., Roach, A. G., and Cavero, I. (1979). Comparison of cardiovascular actions of dihydrazine, phentolamine, and prazosin in spontaneously hypertensive rats. *J. Cardivasc. Pharmacol.*, **1**, 19–29.

Leger, L., and Descarries, L. (1978). Serotonin nerve terminals in the locus coeruleus of adult rat, a radioautographic study. *Brain Res.*, **145**, 1–13.

Lembeck, F. N. S. (1953). Zur Frage zentralen Übertragung afferenter impulse. *Archiv. Pharmac.*, **19**, 197–213.

Lev-Tov, A. (1978). Tetanic and post-tetanic potentiation at the frog neuromuscular synapse. PhD, thesis, The Hebrew University of Jerusalem, Jerusalem.

Levy, R. A. (1980). Presynaptic control of input to the central nervous system. *Can. J. Physiol. Pharmac.*, **58**, 751–66.

Li, C. H., and Chung, D. (1976). Isolation and structure of a nutriakontapeptide with opiate activity from camel pituitary glands. *Proc. Natl. Acad. Sci.*, **73**, 1145.

Libet, B. (1970). Generation of slow inhibitory and excitory postsynaptic potential. *Fed. Proc.*, **29**, 1945–56.

Lindl, T., and Cramer, H. (1975). Evidence against dopamine as the mediator of the rise of cyclic AMP in the superior cervical ganglion of the rat. *Biochem., Biophys. Res. Commun.*, **65**, 731–9.

Lindmar, R., Löffelholz, K., and Muscholl, E. (1968). A muscarinic mechanism inhibiting the release of noradrenaline from peripheral adrenergic nerve fibres by nicotinic agents. *Br. J. Pharmacol., Chemother.*, **32**, 280–94.

Lindvall, O., Björklund, A., Moore, R. Y., and Stenevi, U. (1974). Mesencephalic dopamine neurons projecting to neocortex. *Brain Res.*, **81**, 325–31.

Lissak, K. (1939). Liberation of acetylcholine and adrenaline by stimulating isolated nerves. *Am. J. Physiol.*, **127**, 263–71.

Lister, J. (1858). Preliminary account of an inquiry into the functions of the visceral nerves, with special reference to the 'so called' inhibitory system. *Proc. Roy. Soc.*, **9**, 367–80.

Llewellyn-Smith, I. J., Wilson, A. J., Furness, J. B., Costa, M., and Rush, R. A. (1981). Ultrastructural identification of noradrenergic axons and their distribution within the enteric plexuses of the guinea-pig small intestine. *J. Neurocytol.*, **10**, 331–52.

Llinas, R., and Hess, R. (1976). Tetrodotoxin-resistant dendritic spikes in avian Purkinje cells. *Proc. Natl Acad. Sci., USA*, **73** (7), 2520–3.

Loewi, O. (1921). Über humorale Übertragbarkeit der Herznervenwirkung. *Pfügers Arch. Gesamte Physiol.*, **189**, 239–42.

Löffelholz, K., and Muscholl, E. (1969). A muscarinic inhibition of the noradrenaline release evoked by postganglionic sympathetic nerve stimulation. *Arch. Exp. Path. Pharmac.*, **265**, 1–15.

Löffelholz, K., and Muscholl, E. (1970). Inhibition by parasympathetic nerve stimulation of the release of the adrenergic transmitter. *Naunyn-Schmiedeberg's Arch. Pharmacol.*, **267**, 181–4.

Loh, H. H., Brase, D. A., Sampath-Khanna, S., Mar, J. B., Way, E. L., and Li, C. H. (1976). β-endorphin *in vitro* inhibition of striatal dopamine release. *Nature*, **264**, 567–8.

Lokhandwala, M. F., Tadepalli, A. S., and Jandhyala, B. S. (1979). Cardiovascular actions of bromocriptine: evidence for a neurogenic mechanism. *J. Pharmacol. Exp. Ther.*, **211**, 620.

Lokhandwala, M. F., and Buckley, J. P. (1976). Effect of presynaptic α-adrenoceptor blockade on responses to cardiac nerve stimulation in anaesthetized dogs. *Eur. J. Pharmacol.*, **40**, 183–6.

Londos, C., Cooper, D. M. F., and Wolff, J. (1980). Subclasses of external adenosine receptors. *Proc. Natl Acad. Sci., USA*, **77**, 2551–4.

Long, J. P., Heintz, S., Cannon, J. G., and Kim, J. (1975). Inhibition of the sympathetic nervous system by 5,6 dihydroxy-2-dimethylaminotetralin (M-7) apomorphine and dopamine. *J. Pharmac. Exp. Ther.*, **192**, 336.

Lord, J. A. H., Waterfield, A. A., Hughes, J., and Kosterlitz, H. W. (1977). Endogenous opioid peptides: multiple agonists and receptors. *Nature. (Lond.)*, **267**, 495–9.

Lord, J. A. H., Waterfield, A. A., Hughes, J., and Kosterlitz, H. W. (1976). Multiple opiate receptors, in *Opiates and Endogenous Opioid Peptides* (Ed. H. W. Kosterlitz), Elsevier, Amsterdam, pp. 275–80.

Lubinska, L., and Niemierko, S. (1971). Velocity and intensity of bidirectional migration of acetylcholinesterase in transected nerves. *Brain Res*, **27**, 329–42.

Luchelli-Fortis, M. A., Fredholm, B. B., and Langer, S. Z. (1979). Release of radioactive purine from the ^{3}H-adenine prelabelled cat nictitating membrane by nerve stimulation noradrenaline tyramine, acetylcholine and ATP. *Eur. J. Pharmacol.*, **58**, 389–97.

Lundberg, A. (1952). Adrenaline and transmission in the sympathetic ganglion of the cat. *Acta Physiol. Scand.*, **26**, 252–63.

Lundberg, J. M. (1981). Evidence for coexistence of vasoactive intestinal polypeptide and ACh in neurons of cat exocrine glands. *Acta Physiol. Scand.*, 496–502.

Lundberg, J. M., Anggard, A., Fahrenkrug, J., Johansson, O., and Hökfelt, T. (1981). Vasoactive intestinal polypeptide VIP in cholinergic neurons of exocrine glands, in *Vasoactive Intestinal Polypeptide* (Ed. S. Said), Raven Press, New York. (In press.)

Lundberg, J. M., Rokaeus, A., Hökfelt, T., Rosell, S., Brosn, M., and Goldstein, M. (1982). Neurotensin-like immunoreactivity in the preganglionic sympathetic nerves and in the adrenal medulla of the cat. *Acta Physiol. Scand.*, **114**, 153–5.

Lundberg, J. M., Hökfelt, T., Anggard, A., Uvnas-Wallensten, K., Brimijoin, S., Brodin, E., and Fahrenkrug, J. (1980). Peripheral peptide neurons distribution, axonal transport and some aspects on possible function, in *Neuronal Peptides and Neuronal Communication* (Eds E. Costa, and M. Trabucchi), Raven Press, New York, pp. 25–36.

Lutz, E. G. (1978). Restless legs, anxiety and caffeinism. *J. Clin. Psychiat.*, **39**, 691–8.

Machova, J., and Boska, D. (1963). The effect of 5-hydroxytryptamine, dimethylphenylpiperazinium and acetylcholine on transmission and surface potential in the cat sympathetic ganglion. *Eur. J. Pharmacol.*, **7**, 152–8.

MacIntosh, F. C. (1941). The distribution of acetylcholine in the peripheral and the central nervous system. *J. Physiol. (Lond.)*, **99**, 436–42.

MacIntosh, F. C. (1963). Synthesis and storage of acetylcholine in nervous tissue. *Canad. J. Biochem. Physiol.*, **41**, 2555–71.

MacIntosh, F. C., and Oborin, P. E. (1953). Release of acetylcholine from intact cerebral cortex. *Abstr. XIX. Physiol. Congr.*, pp. 580–1.

Magyar, K., Vizi, E. S., Ecseri, Z., and Knoll, J. (1967). Comparative pharmacological analysis of the optical isomers of phenylisopropylphenyl-propinylamine (E-250). *Acta Physiol. Hung.*, **32**, 377–87.

Malcolm, J. L., Saraiva, P., and Spear, P. J. (1967). Cholinergic and adrenergic inhibition of the rat cerebral cortex. *Int. J. Neuropharmac.*, **6**, 509–27.

Manber, L., and Gershon, M. D. (1979). A reciprocal adrenergic–cholinergic axo-axonic synapse in the mammalian gut. *Am. J. Physiol.*, **236**, 738–45.

Marazzi, A. S. (1939). Electrical studies on the pharmacology of autonomic synapses. II. The action of a sympathomimetic drug (epinephrine) on sympathetic ganglia. *J. Pharm. Exp. Ther.*, **63**, 394–404.

Marchbanks, R. M. (1979). Role of storage vesicles in synaptic transmission, in *Secretory Mechanisms* (Eds C. R. Hopkins and C. J. Duncan), Cambridge University Press, Cambridge, pp. 251–76.

Markiewicz, M., Marshall, I., and Nasmyth, R. A. (1979). Lack of feedback via pre-synaptic α-adrenoceptors by noradrenaline released by a single pulse. *Br. J. Pharmacol.*, **70**, 343P.

Marshall, J. (1983). Stimulation-evoked release of ^{3}H-noradrenaline by 1, 10 or 100 pulses and its modification through presynaptic α_2-adrenoceptors. *Br. J. Pharmacol.*, **78**, 221–31.

Martin, I. L., and Mitchell, P. R. (1980). Effects of some amino acid on K$^+$-induced release of ^{3}H-DA from rat striatal tissue. *Br. J. Pharmacol.*, **68**, 162–3.

Martin, W. R., Eades, C. G., Thompson, J. A., Huppler, R. E., and Gilberg, P. E. (1976). The effects of morphine- and nalorphine-like drugs in the nondependent and morphine-dependent chronic spinal dog. *J. Pharmac. Exp. Ther.*, **197**, 517–32.

Martinez, A. E., and Adler-Graschinsky, E. (1980). Release of norepinephrine induced by preganglionic stimulation of the isolated superior cervical ganglion of the rat. *J. Pharm. Exp. Ther.*, **212**, 527–32.

Marx, J. L. (1979). Brain peptides. Is substance P a transmitter of pain signals? *Science*, **31**, 886–9.

Massingham, R., Dubocovich, M. L., and Langer, S. Z. (1980). The role of presynaptic receptors in the cardiovascular actions of N,N-di-u-propyldopamine in the cat and dog. *Naunyn-Schmiedeberg's Arch. Pharmacol.*, **314**, 17–28.

Matthews, J. D., Labrecque, G., and Domino, E. F. (1973). Effects of morphine, nalorphine and naloxone on neocortical release of acetylcholine in the rat. *Psychopharmacologia*, **29**, 113–20.

Matz, R., Rich, W., Thompson, H., and Gershon, S. (1974). Clozapine, a potential antipsychotic agent without extra-pyramidal manifestations. *Curr. Ther. Res.*, **16**, 687–95.

McAfee, D. A., Henon, B. K., Horn, J. P., and Yarowsky, P. (1981). Calcium currents modulated by adrenergic receptors in sympathetic neurons. *Fedn Proc.*, **40**, 2246–9.

McCulloch, M. W., Rand, M. J., and Story, D. F. (1972). Inhibition of ^{3}H-noradrenaline release from sympathetic nerves of guinea-pig atria by a presynaptic α-adrenoceptor mechanism. *Br. J. Pharmacol.*, **46**, 523–4.

McCulloch, M., Rand, M., and Story, D. (1973). Evidence for a dopaminergic mechanism for modulation of adrenergic transmission in the rabbit heart artery. *Br. J. Pharmacol.*, **49**, 141–56.

McCulloch, M., Rand, M., and Story, D. F. (1974). Resting and stimulation-induced efflux of tritium from guinea-pig atria incubated with ³H-noradrenaline. *Clin. exp. Pharmac., Physiol.*, **1**, 275–89.

McDougal, M. D., and West, G. B. (1952). The action of isoprenaline on intestinal muscle. *Arch. int. Pharmacodyn.*, **90**, 86–92.

McDougal, M. D., and West, G. B. (1954). The inhibition of the peristaltic reflex by sympatho-mimetic animals. *Br. J. Pharmacol.*, **9**, 131–7.

McGrath, M. A., and Shepherd, J. T. (1967a). Inhibition of canine adrenergic neurotransmission in vascular smooth muscle by 5-hydroxytryptamine. *Circulation*, **53** and **54**, 11–20.

McGrath, M. A., and Shepherd, J. T. (1976b). Inhibition of adrenergic neurotransmission in canine vascular smooth muscle by histamine. Mediation by H2-receptor. *Circulation Res.*, **39**, 566–73.

McIlwain, H. (1972). Regulatory significance of the release and action of adenine derivatives in cerebral system. *Biochem. Soc. Symp.*, **36**, 69–85.

McLenna, A., and York, D. H. (1966). Cholinergic mechanism in the caudate nucleus. *J. Physiol. (Lond.)*, **187**, 163–75.

Medgett, I. C. (1983). Modulation of transmission in rat sympathetic ganglia by activation of presynaptic α and β-adrenoceptors. *Br. J. Pharmacol.*, **78**, 17–27.

Medgett, I. C., McCulloch, M. W., and Rand, M. J. (1978). Partial agonist action of clonidine on pre-junctional and post-junctional α-adrenoceptors. *Naunyn-Schmiedeberg's Arch. Pharmacol.*, **304**, 215, 221.

Meites, J., Bruni, J. F., Van Vugt, D. A., and Smit, A. F. (1979). Relation of endogenous opioid peptides and morphine to neuroendocrine functions. *Life Sci.*, **24**, 1325–36.

Melamed, E., Lakov, M., and Atlas, D. (1976). Direct localization of β-adrenoceptor sites in rat cerebellum by a new fluorescent analogue of propranolol. *Nature*, **261**, 420–2.

Mercer, L., Del Fiacco, M., and Cuello, A. C. (1979). The smooth endoplasmic reticulum as a possible storage site for dendritic dopamine in substantia nigra neurones. *Experientia*, **35**, 101–3.

Merrillees, N. C., Burnstock, G., and Holman, M. E. (1963). Correlation of fine structure and physiology of the innervation of smooth muscle in the guinea-pig vas deferens. *J. Cell Biol.*, **19**, 529–50.

Meunier, F. M., Israel, M., and Lesbats, B. (1975). Release of ATP from stimulated nerve electroplaque junctions. *Nature*, **257**, 407–8.

Meves, H., and Pichon, Y. (1977). The effect of internal and external 4-aminopyridine on the potassium currents in intracellularly perfused squid giant axons. *J. Physiol.*, **268**, 511–32.

Mey, I., Burnstock, G., and Vanhoutte, P. M. (1979). Modulation of the evoked release of noradrenaline in canine saphenous vein via presynaptic receptors for adenosine but not ATP. *Eur. J. Pharmacol.*, **55**, 401–5.

Meyer, F. M., and Cooper, J. R. (1981). Correlations between Na⁺,K⁺ ATPase activity and acetylcholine release in rat cortical synaptosomes. *J. Neurochem.*, **36**, 467–75.

Miach, P. J., Causse, J. P., and Meyer, P. (1978). Direct biochemical demonstration of two types of α-adrenoceptor in rat brain. *Nature*, **274**, 492–4.

Michelot, R., Leviel, V., Giorguieff-Chesselet, M. F., Chéramy, A., and Glowinski, J. (1979). Effects of the unilateral nigral modulation of substance P transmission on the activity of the two nigra-striatal dopaminergic pathways. *Life Sci.*, **24**, 715–23.

Miletic, V., Kovács, M., and Randic, M. (1977). Effect of substance P and somatostatin on activity of cut dorsal horn neurons activated by noxious stimuli. *Fedn Proc.*, **36**, 1014.

Miledi, R. (1973). Transmitter release induced by injection of calcium ions into nerve terminals. *Proc. R. Soc. B.*, **183**, 421–3.

Miledi, R., Molinaar, P. C., and Polak, R. L. (1978). α-bungarotoxin enhances transmitter released at the neuromuscular junction. *Nature*, **272**, 641–3.

Miller, J. C., and Friedhoft, A. J. (1979). Dopamine receptor coupled modulation of the K⁺-depolarized overflow of ³H-acetylcholine from rat striatal slices: alteration after chronic haloperidol and alpha-methyl-p-tyrosine pretreatment. *Life. Sci.*, **25**, 1249–56.

Miller, R. J., and Hiley, C. R. (1974). Anti-muscarinic properties of neuroleptics and drug induced Parkinsonism. *Nature*, **248**, 596–7.

Miller, J. J., Richardson, T. L., Fibger, H. C., and McLennan, H. (1975). Anatomical and electrophysiological identification of a projection from the mesencephalic raphé to the caudate-putamen in the rat. *Brain Res.*, **97**, 133–8.

Mitchell, J. F. (1963). The spontaneous and evoked release of acetylcholine from the cerebral cortex. *J. Physiol. (Lond.)*, **165**, 98–116.

Mitchell, J. F., and Srinivasan, V. (1969). Release of ³H-γ-aminobutyric acid from the brain during synaptic inhibition. *Nature, (Lond.)*, **224**, 663–6.

Mitchell, D. R., and Fleetwood-Walker, S. (1981). Substance P, but not TRH, modulates the 5-HT autoreceptor in ventral lumbar spinal cord. *Eur. J. Pharmacol.*, **76**, 119–20.

Mitchell, P. R., and Martin, I. L. (1978). Is GABA release modulated by pre-synaptic receptors? *Nature*, **274**, 904–5.

Miyamoto, M. D., and Breckenridge, B. McL. (1974). A cyclic adenosine monophosphate link in the catecholamine enhancement of transmitter release at the neuromuscular junction. *J. gen. Physiol.*, **63**, 609–24.

Molenaar, P. C., and Polak, R. L. (1970). Stimulation by atropine of acetylcholine release and synthesis in cortical slices from rat brain *Br. J. Pharmacol.*, **40**, 406–17.

Molenaar, P. C., and Polak, R. L. (1980). Inhibition of acetylcholine release by activation of acetylcholine receptors. *Progr. in Pharmacology*, **3**, 39–44.

Montel, H., Starke, K., and Weber, F. (1974a). Influence of morphine and naloxone on the release of noradrenaline from rat brain cortex slices. *Naunyn-Schmiedeberg's Arch. Pharmacol.*, **283**, 357.

Montel, H., Starke, K., and Weber, F. (1974b). Influence of fentanyl levorphanol and pethidine on the release of noradrenaline from rat brain cortex slices. *Naunyn-Schmiedeberg's Arch. Pharmacol.*, **283**, 371–80.

Montel, H., Starke, K., and Taube, H. D. (1975). Influence of morphine and naloxone on the release of noradrenaline from rat cerebellar cortex slices. *Naunyn-Schmiedeberg's Arch. Pharmacol.*, **288**, 427–33.

Moore, R. Y., and Bloom, F. E. (1979). Central catecholamine neuron systems, anatomy and physiology of the morepinephrine and epinephrine systems. *Ann. Rev. Neurosci.*, **2**, 163–8.

Morita, K., and North, R. A. (1981). Clonidine activates membrane potassium conductance in myenteric plexus. *Br. J. Pharmacol.*, **74**, 419–28.

Morita, K., North, R. A., and Tokimasa, T. (1982). Muscarinic pre-synaptic inhibition of synaptic transmission in myenteric plexus of guinea-pig ileum. *J. Physiol. (Lond.)*, **333**, 141–9.

Moritoki, H., Morita, M., and Kanbe, T. (1976). Effect of methylxanthines and imidazole on the contractions of guinea-pig ileum induced by transmural stimulation. *Eur. J. Pharmacol.*, **35**, 185–9.

Morley, J. S. (1983). Chemistry of opioid peptides. *Brit. Med. Bull.* (in press).

Moroni, F., Cheney, D., and Costa, E. (1977a). Inhibition of acetylcholine turnover in rat hippocampus by intraseptal injections of β-endorphin and morphine. *Naunyn-Schmiedeberg's Arch. Pharmacol.*, **299**, 149–53.

Moroni, F., Cheney, D., and Costa, E. (1977b). Endorphin inhibits ACh turnover in nuclei of rat brain. *Nature*, **267**, 267–8.

Mudge, A., Leeman, S., and Fischbach, G. (1979). Enkephalin inhibits release of substance P from sensory neurones in culture and decreases action potential duration. *Proc. Natl Acad. Sci., USA*, **76**, 526–30.

Mulder, A. H., DeLangen, C. D., de Regt, V., and Hogenboom, F. (1978). Alpha-receptor-mediated modulation of ³H-noradrenaline release from rat brain cortex synaptosomes. *Naunyn-Schmiedeberg's Arch. Pharmacol.*, **303**, 193–6.

Muscholl, E. (1970). Cholinomimetic drugs and release of the adrenergic transmitter, in *New Aspects of Storage and Release Mechanisms of Catecholamines* (Eds H. J. Schümann and G. Kroneberg), Springer-Verlag, Berlin, pp. 168–86.

Muscholl, E. (1972). Adrenergic false transmitter. In *Handbook of Experimental Pharmacology* (Eds H. Blaschko and E. Muscholl) Springer-Verlag, Berlin and New York, pp. 618–60.

Muscholl, E. (1973a). Muscarinic inhibition of the norepinephrine release from peripheral sympathetic fibres, in *Pharmacology and the Future of Man*, Proc. 5th Int. Pharmacol., Vol. 4, Karger, Basel, pp. 440–57.

Muscholl, E. (1973b). Regulation of catecholamine release. The muscarinic inhibitory mechanism, in *Frontiers in Catecholamine Research* (Eds E. Usdin and S. H. Snyder), Pergamon Press, New York, pp. 537–43.

Muscholl, E. (1980). Presynaptic muscarinic control of noradrenaline release, in *Modulation of Neurochemical Transmission* (Ed. E. S. Vizi), Pergamon Press, Oxford; Akadémiai Kiadó, Budapest, pp. 1–12.

Musick, J., and Hubbard, J. I. (1972). Release of protein from mouse motor nerve terminals. *Nature*, **327**, 279–81.

Nagy, J. I., Lee, T., Seeman, P., and Fibiger, H. C. (1978). Direct evidence for presynaptic and postsynaptic dopamine receptors in brain. *Nature*, **274**, 278–81.

Nagy, J. I., Vincent, S. R., Staines, W. A., Fibiger, H. C., Reisine, T. D., and Yamamura, T. (1980). Neurotoxic action of capsaicin on spinal substance P neurons. *Brain Res.*, **186**, 435–44.

Nakamura, S., Tepper, J. M., Young, S. J., and Groves, P. M., (1981). Neurophysiological consequences of presynaptic receptor activation: changes in noradrenergic terminals excitability. *Brain Res.*, **226**, 155–70.

Nakazato, Y., Ohga, A., and Onoda, Y. (1978). The effect of ouabain on noradrenaline output from peripheral adrenergic neurones of isolated guinea-pig vas deferens. *J. Physiol. (Lond.)*, **278**, 45–54.

Narahashi, R., Moore, J. W., and Scott, W. R. (1964) Tetrodotoxin blockade of sodium conductance increase in lobster giant axons. *J. gen. Physiol.*, **49**, 963–74.

Nicholls, J., and Wallace, B. G. (1978). Modulation of transmission at an inhibitory synapse in the central nervous system of the leech. *J. Physiol. (Lond.)*, **281**, 157–70.

Nicoll, R. A., and Alger, B. E. (1979). Presynaptic inhibition transmitter and ionic mechanisms. *Int. Rev. Neurobiol.*, **21**, 217–58.

Nicoll, R. A., and Jahr, C. E. (1982). Self-excitation of olfactory bulb neurones. *Nature*, **296**, 441–4.

Nicoll, R. A., Alger, B. E., and Jahr, C. E. (1980). Enkephalin blocks inhibitory pathways in the vertebrate CNS. *Nature*, **287**, 22–5.

Nieoullon, A., Chéramy, A., and Glowinsky, J. (1977). Release of dopamine *in vivo* from cat substantia nigra. *Nature*, **266**, 375–7.

Nilsson, G., Hökfelt, T., and Pernow, B. (1974). Distribution of substance P-like immunoreactivity in the rat central nervous system as revealed by immunohistochcmistry. *Med. Biol.*, **52**, 424–7.

Nilsson, J. L. G., and Carlsson, A. (1982). Dopamine receptor agonist with apparent selectivity for autoreceptors: a new principle for antipsychotic action? *TIPS*, **3**, 322–35.

Nishi, S., and North, R. A. (1973a). Presynaptic action of noradrenaline in the myenteric plexus. *J. Physiol. (Lond.)*, **231**, 29–30.

Nishi, S., and North, R. A. (1973b). Intracellular recording from the myenteric plexus of the guinea-pig ileum. *J. Physiol. (Lond.)*, **231**, 471–91.

Nishi, S., Minota, S., and Karczmar, A. G. (1974). Primary afferent neurones: the ionic mechanism of GABA mediated depolarization. *Neuropharmacology*, **13**, 215–19.

Nistri, A. (1976). The effect of electrical stimulation and drugs on the release of acetylcholine from the frog spinal cord. *Naunyn-Schmiedeberg's Arch. Pharamacol.*, **293**, 269–76.

Noon, J. P., and Roth, R. H. (1975). Some physiological and pharmacological characteristics of the stimulus induced release of norepinephrine from the rabbit superior cervical ganglion. *Naunyn-Schmiedeberg's Arch. Pharmacol.*, **291**, 163–74.

Norberg, K. A. (1964). Adrenergic innervation of the intestinal wall studied by fluorescence microscope. *Int. J. Neuropharmacol.*, **3**, 379–82.

Norberg, K. A., and Sjögvist, F. (1966). New possibilities for adrenergic modulation of ganglionic transmission. *Pharmac. Rev.*, **18**, 743–51.

Nordström, O. and Bártfai, T. (1981). 8-Br-cyclic-GMP mimics activation of muscarinic autoreceptor and inhibits acetylcholine release from rat hippocampal slices. *Brain Res.*, **213**, 467–71.

North, R. A. (1979). Opiates, opioid peptides and single neurones. *Life Sci.*, **24**, 1527–46.

North, R. A., and Henderson, G. (1975). Action of morphine on guinea-pig myenteric plexus and mouse vas deferens studied by intracellular recording. *Life Sci.*, **17**, 63–9.

North, R. A., and Nishi, S. (1974). Properties of the ganglion cells of the myenteric plexus of the guinea-pig ileum determined by intracellular recording, in *Proc. 4th Int. Symp. Gastrointestinal Motility*, New York, pp. 667–76.

North, R. A., and Tonini, M. (1976). Hyperpolarization by morphine of myenteric neurones, in *Opiates and Endogenous Opioid Peptides* (Ed. H. W. Kosterlitz), Elsevier/North-Holland, Biochemical Press/Amsterdam, pp. 205–12.

North, R. A., and Tonini, M. (1977). The mechanism of action of narcotic analgesics in the guinea-pig ileum. *Br. J. Pharmacol.*, **61**, 541–9.

North, R. A., and Williams, J. T. (1977). Extracellular recording from the guinea-pig myenteric plexus and the action of morphine. *Eur. J. Pharmacol.*, **45**, 23–33.

North, R. A., Henderson, G., Katayama, Y., and Johnson, S. M. (1980). Electrophysiological evidence of presynaptic inhibition of ACh release by 5-hydroxytryptamine in the enteric

238

nervous system. *Neuroscience*, **5**, 581–6.

Nowycki, M. C., and Roth, R. H. (1978). Dopaminergic neurons. Role of presynaptic receptors in the regulation of transmitter biosynthesis. *Prog. Neuro-Psychopharmacol.*, **2**, 139–58.

O'Dea, R. F., and Zatz, M. (1976). Catecholamines stimulated cyclic GMP accumulation in the rat pineal. Apparent presynaptic site of action. *Proc. Natl Acad. Sci.*, **73**, 3398–3402.

Okwuasaba, F. K., and Cook, M. A. (1980). The effect of theophylline and other methylxanthines on presynaptic inhibition of the longitudinal smooth muscle of the guinea-pig ileum induced by purine nucleotides. *J. Pharmac., exp. Ther.*, **215**, 704–9.

Okwuasaba, F. K., Hamilton, J. T., and Cook, M. A. (1977). Antagonism by methylxanthines of purine nucleotide- and dipyridamole-induced inhibition of peristaltic activity of the guinea-pig ileum. *Eur. J. Pharmacol.*, **43**, 181–94.

Olney, J. W., and de Gutareff, T. (1978). The fate of synaptic receptors in the kainate lesioned striatum. *Brain Res.*, **140**, 340–3.

Olpe, H. R., and Koella, W. P. (1977). The response of striatal cells upon stimulation of the dorsal and median raphé nuclei. *Brain Res.*, **122**, 357–60.

Olson, L., and Fuxe, K. (1971). On the projections from the locus coeruleus norepinephrine neurons. *Brain. Res.*, **28**, 165–8.

Orkand, R. K., Nicholls, J. G., and Kuffler, S. W. (1966). The effect of nerve impulses on the membrane potential of glial cells in the central nervous system of amphibia. *J. Neurophysiol.*, **29**, 788–805.

Orrego, F. (1979). Criteria for the identification of central neurotransmitters, and their application to studies with some nerve tissue preparations *in vitro*. *Neuroscience*, **4**, 1037–57.

Osborn, R. H., Bradford, H. F., and Jones, D. G. (1973). Patterns of amino acid release from nerve-endings isolated from spinal cord and medulla. *J. Neurochem.*, **21**, 407–19.

Osborne, N. N. (1977) Do snail neurones contain more than one neurotransmitter? *Nature*, **270**, 622–3.

Osborne, N. N. (1981). Communication between neurones: current concepts. *Neurochem. International*, **3**, 3–16.

Osborne, H., and Herz, A. (1980). K+ evoked release of Met-enkephalin from rat striatum *in vitro*: effect of putative neurotransmitters and morphine. *Naunyn-Schmiedeberg's Arch. Pharmacol.*, **310**, 203–9.

Otsuka, M., and Konishi, S. (1976). Release of substance P immunoreactivity from isolated spinal cord of newborn rat. *Nature*, **264**, 81–4.

Otsuka, M., Iversen, L. L., Hall, R. W., and Kravitz, E. A. (1966). Release of gamma-aminobutyric acid from inhibitory nerves of lobster. *Proc. Natl Acad. Sci.*, **56**, 1110–15.

Oyama, T., Jin, T., Yamaya, R., (1980). Profound analgesic effects of β-endorphin in man, *Lancet*, 122–3.

Paden, C., Wilson, C. J., and Groves, P. M. (1976). Amphetamine induced release of dopamine from the substantia nigra *in vitro*, *Life Sci.*, **19**, 1499–1506.

Palaty, V. (1981). Release of noradrenaline from the rat tail artery induced by inhibition of the sodium pump in calcium-free solution. *Can. J. Physiol. Pharmacol.*, **59**, 347–50.

Palkovits, M., Bronwnstein, M., and Saavedra, J. M. (1974). Serotonin content of the brain stem nuclei in the rat. *Brain Res.*, **80**, 237–49.

Papanicolaou, J., Summers, R. J., Vajda, F. J. E., and Louis, W. J. (1982). The relationship between α_2-adrenoceptor selectivity and anticonvulsant effect in a series of clonidine-like drugs. *Brain Res.*, **241**, 393–7.

Pasik, T., and Pasik, P. (1982). Serotoninergic afferents in the monkey neostriatum. *Acta Biol. Hung.*, **33**, 277–88.

Passo, S. S., Thornborough, J. R., and Ferris, C. F. (1981). A functional analysis of dopaminergic innervation of the neurohypophysis. *Am. J. Physiol.* **241**, E186–90.

Paton, W. D. M. (1957). The action of morphine and related substances on contraction and on acetylcholine output of coaxically stimulated guinea-pig ileum. *Br. J. Pharmac. Chemother.*, **12**, 119–27.

Paton, W. D. M. (1958). Central and synaptic transmission in the nervous system. Pharmacological aspects. *J. Rev. Physiol*, **20**, 431–70.

Paton, W. D. M. (1960). Discussion comments, in *CIBA Symposium: Adrenergic Mechanisms* (Ed. J. R. Vane), Churchill, London, pp. 124–7.

Paton, W. D. M., and Thompson, J. W. (1958). The mechanism of action of adrenaline on the superior cervical ganglion of the cat, XIX. Int. Cong. of Physiol., Montreal, *Abstracts of Communications*, pp. 664–5.

239

Paton, W. D. M., and Vizi, E. S. (1969). The inhibitory action of noradrenaline and adrenaline on acetylcholine output by guinea-pig longitudinal muscle strip. *Br. J. Pharmacol.* **35**, 10–28.

Paton, W. D. M., and Zar, M. A. (1968). The origin of acetylcholine released from guinea-pig intestine and longitudinal muscle strip. *J. Physiol. (Lond.)*, **194**, 13–33.

Paton, D. M., Bar, H. P., Clanachan, A. S., and Lauson, P. A. (1978). Structure–activity relations for inhibition of neurotransmission in rat vas deferens by adenosine. *Neuroscience*, **3**, 65–70.

Paton, W. D. M., Vizi, E. S., and Zar, M. A. (1971). The mechanism of acetylcholine release from parasympathetic nerves. *J. Physiol. (Lond.)*, **215**, 819–48.

Patterson, P. H. (1979). Environmental determination of autonomic neurotransmitter functions. *Ann. Rev. Neurosci.*, **1**, 1–18.

Pearse, A. G. E. (1976). Peptides in brain and intestine. *Nature*, **262**, 92–4.

Pearse, A. G. E. (1977). The APUD concept and its implications: related endocrine peptides in brain, intestine, pituitary, placenta and anuran cutaneous glands. *Med. Biol.*, **55**, 115–25.

Pelayo, F., Dubocovich, M. L., and Langer, S. Z. (1977). Regulation of noradrenaline release from the rat pineal through presynaptic adrenoceptors. Possible involvement of cyclic nucleotides. *Nature (Lond.)*, **274**, 76–8.

Pelayo, F., Dubocovich, M. L., and Langer, S. Z. (1980). Inhibition of neuronal uptake reduces the presynaptic effects of clonidine but not of α-methylnoradrenaline on the stimulation evoked release of ^{3}H-noradrenaline from rat occipital cortex slices. *Eur. J. Pharmacol*, **64**, 143–55.

Pelhate, M., and Picho, Y. (1974). Selective inhibition of potassium current in the giant axon of the cockroach. *J. Physiol. (Lond.)*, **242**, 90.

Pepeu, G., and Bartholini, A. (1968). Effect of psychoactive drugs on the output of acetylcholine from the cerebral cortex of the cat. *Eur. J. Pharmacol.*, **4**, 254–63.

Pepeu, G., Garau, L., Mulas, M. L., and Pepeu, I. (1975). Stimulation by morphine of acetylcholine output from the cerebral cortex of septal rats. *Brain Res.*, **100**, 677–80.

Perkins, M. N., and Stone, T. W. (1979). Adenosine may mediate neuronal depressant effects of morphine. *Br. J. Pharmacol.*, **67**, 476–7.

Perkins, M. N., and Stone, T. W. (1980). Blockade of striatal neurone responses to morphine by aminophylline: evidence for adenosine mediation of opiate action. *Br. J. Pharmacol.*, **69**, 131–7.

Persson, C. G. A. (1982). Xanthines for asthma-present status. *TIPS*, **3**, 312–13.

Persson, C. G. A., and Kjellin, G. (1981). Enprofylline, a principally new antiasthmatic xanthine. *Acta Pharmacol, Toxicol*, **49**, 313–16.

Petri, G., Szenohradszky, J., and Porszász-Gibiszer, K. (1971). Sympatholytic treatment of 'paralytic' ileus, *Surgery*, **70**, 359–67.

Pert, C. B., and Snyder, S. H. (1973). Opiate receptor: Demonstration in nervous tissue. *Science*, **1979**, 1011–14.

Pflünger, E. (1857). *Über des Hemmung Nervensystem für die peritaltischen Bewegungen der Gedarme*, Hirschwald, Berlin.

Phillis, J. W. (1968). Acetylcholine release from the cerebral cortex: its role in cortical arousal. *Brain Res.*, **7**, 378–89.

Phillis, J. W. (1970). *The Pharmacology of Synapses*, Pergamon Press, Oxford.

Phillis, J. W., and Kostopoulos, G. K. (1975). Adenosine as a putative transmitter in the cerebral cortex. Studies with potentiators and antagonists. *Life Sci.*, **17**, 1083–94.

Phillis, J. W., and Kostopoulos, G. K. (1977). Activation of a noradrenergic pathway from the brain stem to rat cerebral cortex. *Gen. Pharmac.*, **8**, 207–11.

Phillis, J. W., and Wu, P. H. (1981). The role of adenosine and its nucleotides in central synaptic transmission, *Progr. in Neurobiol.*, **16**, 187–239.

Phillis, J. W., Edström, J. P., Kostopoulos, G. K., and Kirkpatrick, J. R. (1979). Effect of adenosine and adenine neucleotides on synaptic transmission in the cerebral cortex. *Can. J. Physiol. Pharmacol.*, **57**, 1289–1312.

Phillis, J. W., Hang, Z. G., Chelack, B. J., and Wu, P. H. (1980a). Morphine enhances adenosine release from the *in vivo* rat cerebral cortex. *Eur. J. Pharmacol.*, **65**, 97–100.

Phillis, J. W., Hang, Z. G., Chelack, G. J., and Wu, P. H. (1980b). The effect of morphine on purine and acetylcholine release from rat cerebral cortex: evidence for a purinergic component in morphine's action. *Pharmac., Biochem., and Behav.*, **13**, 421–7.

Phillis, J. W., Kostopoulos, G. K., and Limacher, J. J. (1974). Depression of cortocospinal cells by various purines and pyramidines. *Can. J. Physiol. Pharmac.*, **52**, 1226–9.

Phillis, J. W., Kostopoulos, G. K., and Limacher, J. J. (1975). A potent depressant action of adenine derivatives on cerebral cortical neurones. *Eur. J. Pharmacol.*, **30**, 125–9.

Phillis, J. W., Tebecis, A. K., and York, D. H. (1968a). Depression of spinal motoneurones by noradrenaline, 5-hydroxytryptamine and histamine. *Eur. J. Pharmacol.*, **4**, 471–5.

Phillis, J. W., Tebecis, A. K., and York, D. H. (1968b). Histamine and some antihistamines, their actions on cerebral cortical neurones. *Br. J. Pharmacol.*, **33**, 426–40.

Pickel, V. M., Joh, T. H. Reis, D. J., Leeman, S. E., and Miller, R. J. (1979). Electron microscopic localization of substance P and enkephalin in axon terminals related to dendrites of catecholaminergic neurons. *Brain Res.*, **160**, 387–400.

Pickel, V. M., Joh, T. H., and Reis, D. J. (1977). A serotonergic innervation of noradrenergic neurons in nucleus locus coeruleus, demonstration by immunocytochemical localization of the transmitter specific enzymes tyrosine and tryptophan hydroxylase, *Brain Res.*, 131, 197–214.

Pickel, V. M., Segal, M., and Bloom, F. E. (1974). A radio-autographic study of the efferent pathways of the nucleus locus coeruleus. *J. Comp. Neurol.*, **155**, 15–42.

Pickles, H. B., and Simmonds, M. A. (1976). Possible presynaptic inhibition in rat olfactory cortex. *J. Physiol. (Lond.)*, **260**, 475–86.

Plotsky, P. M., Wightman, R. M., Chey, W., and Adams, R. N. (1977). Liquid chromatographic analysis of endogenous catecholamine release from brain slices. *Science*, **904**–6.

Pohl, J., and Brock, N. (1974). Vergleichende Untersuchungen zur Hemmung des Adenosinabbaus *in vitro* durch Dilazep. *Arzneim, Frosch.* 24, 1901–5.

Polak, R. L. (1965). Effect of hyoscine on the output of acetylcholine into perfused cerebral ventricles of cats. *J. Physiol., (Lond.)*, **181**, 317–23.

Polak, R. L. (1969). The influence of drugs on the uptake of acetylcholine by slices of rat cerebral cortex. *Br. J. Pharmacol.*, **36**, 144–52.

Polak, R. L. (1971). The stimulating action of atropine on the release of acetylcholine by rat cerebral cortex *in vitro*. *Br. J. Pharmacol.*, **41**, 600–6.

Polak, R. L., and Meeuws, M. M. (1966). The influence of atropine on the release and uptake of acetylcholine by the isolated cerebral cortex of the rat. *Biochem. Pharmac.*, **15**, 989–92.

Pollard, H., Lorens-Cortes, C., and Schwartz, J. C. (1977). Enkephalin receptors on dopaminergic neurones in rat striatum. *Nature*, **268**, 745–7.

Potter, L. T. (1970). Synthesis, storage and release of ^{14}C-acetylcholine in isolated rat diaphragm muscles. *J. Physiol. (Lond.)*, **216**, 145–66.

Potter, P., and White, T. D. (1980). Release of adenosine 5-triphosphate from synaptosomes from different regions of rat brain. *Neuroscience*, **5**, 1351–6.

Powis, D. A. (1981). Does Na,K-ATPase play a role in the regulation of neurotransmitter release by prejunctional α-adrenoceptors? *Biochem. Pharm.*, **30**, 2389-97.

Praag, H. M., and Verhoeven, W. M. A. (1980). Neuropeptides, A new dimension in biological pyschiatry, in *Enzymes and Neurotransmitters in Mental Disease* (Eds E. Usdin, T. L. Sourkes, and M. B. H. Youdim), John Wiley, Chichester, pp. 3–34.

Priestley, J. V., Somogyi, P., and Cuello, A. C. (1982). Immuno-cytochemical localization of substance P in the spinal trigeminal nucleus of the rat; a light and electron microscopic study. *J. Comp. Neurol.*, **211**, 31–49.

Pull, I., and McIlwain, H. (1971). Metabolism of ^{14}C- and derivatives by cerebral tissues superfused and electrically stimulated. *Biochem.J.*, **126**, 965–73.

Racagni, G., Cheney, D. L., Trabucchi, M., and Costa, E. (1975). *In vivo* actions of clozapine and haloperidol on the turnover rate of acetylcholine in rat striatum. *J. Pharmac. exp. Ther*, **196**, 323–32.

Racké, K., Ritzel, H., Trapp, B., and Muscholl, E. (1982). Dopaminergic modulation of evoked vasopressin release from the isolated neurohypophysis of the rat. *Naunyn-Schmiedeberg's Arch. Pharmacol.*, **319**, 56–65.

Rahamimoff, R., Lev Tov, A., Meiri, H., Rahamimoff, H., and Nussinovitsh, I. (1980). Regulation of acetylcholine liberation from presynaptic nerve terminals. *Monogr. Neural. Sci.*, **7**, 3–18.

Raiteri, M., Cervoni, A. M., Del Carmine, R., and Levi, G. (1978). Do presynaptic autoreceptors control dopamine release? *Nature*, **274**, 706–8.

Rand, M. J., McCulloch, M. W., and Story, D. F. (1975). Prejunctional modulation of noradrenergic transmission by noradrenaline, dopamine and acetylcholine, in *Central Action of Drugs in Blood Pressure Regulation* (Eds S. S. Davies and J. L. Reid), Pitman, London, p. 94.

Rand, M. J. McCulloch, M. W., and Story, D. F. (1980). Catecholamine receptors on nerve terminals, in *Adrenergic Activators and Inhibitors* (Ed. L. Szekeres), Springer-Verlag, Berlin, pp. 223–66.

Rand, M. J., Story, D. F., Allen, G. S., Glover, A. B., and McCulloch, M. W. (1973). Pulse-to-pulse modulation of noradrenaline release through a prejunctional receptor auto-inhibitory mechanism, in *Frontiers in Catecholamine Research* (Eds E. Usdin and S. H. Snyder), Pergamon Press, New York, pp. 579–81.

Rang, H. P., and Ritchie, J. M. (1968). On the electrogenic sodium pump in mammalian nonmyelinated nerve fibres and its activation by various external cations. *J. Physiol. (Lond.)*, **196**, 183–221.

Ranish, N., and Ochs, S. (1972). Fast axoplasmic transport of acetylcholinesterase in mammalian nerve fibres. *J. Neurochem.*, **19**, 2641–9.

Reader, T. A., De Champlain, Z., and Jasper, H. (1976). Catecholamines released from cerebral cortex in the cat: decreases during sensory stimulation. *Brain Res.*, **111**, 95–108.

Reading, H. W., and Isbir, T. (1980). The role of cation-activated ATPases in transmitter release from the rat iris. *Q. J. Exp. Physiol.*, **65**, 105–16.

Reichardt, L. F., and Patterson, P. H. (1977). Neurotransmitter synthesis and uptake by isolated neurons in microcultures. *Nature*, **270**, 147–51.

Reichenbacher, D., Reimann, and K. Starke. (1982). α-adrenoceptor mediated inhibition of noradrenaline release in rabbit brain cortex slices. *Naunyn-Schmiedeberg's Arch. Pharmacol.*, **319**, 71–7.

Reubi, J. C., and Sandri, C. (1979). Ultrastructural observations on intercellular contacts of nigral dendrites. *Neurosci. Lett.* **13** (2), 183–8.

Reimann, W., Zumstein, A., Jackish, R., Starke, K., and Hertting, G. (1979). Effect of extracellular dopamine on the release of dopamine in the rabbit caudate nucleus. Evidence for a dopaminergic feedback inhibition. *Naunyn-Schmiedeberg's Arch. Pharmacol*, **306**, 53–60.

Reimann, W., Zumstein, A., and Starke, K. (1982). γ-aminobutyric acid can both inhibit and facilitate dopamine release in the caudate nucleus of the rabbit. *J. Neurochem.*, **39**, 961–9.

Réthelyi, M. (1977). Preterminal and terminal axon arborizations in the substantia gelatinosa of cat's spinal cord. *J. Comp. Neur.*, **172**, 511–28.

Réthelyi, M., and Szentágothai, J. (1965). On a peculiar type of synaptic arrangement in the substantia gelatinosa of Rolandi. *8th Int. Congr. of Anatomists, Stuttgart, Georg Thieme*, p. 99.

Réthelyi, M., Light, A. R., and Perl, E. R. (1982). Synaptic complexes formed by functionally defined primary afferent units with fine myelinated fibers. *J. Comp. Neurology*, **207**, 381–93.

Reubi, J. C., Iversen, L. L., and Jessel, T. M. (1977). Dopamine selectively increases ^{3}H-GABA release from slices of rat substantia nigra *in vitro*. *Nature*, **268**, 652–4.

Ribeiro, J. A. (1978). ATP: related nucleotides and adenosine on neurotransmission. *Life Sci.*, **22**, 1373.

Ribeiro, J. A. (1981). The modulation of transmission by purinergic substances at the neuromuscular junction. *Ann. N.Y. Acad. Sci.*, **377**, 874–6.

Ribeiro, J. A., and Dominquez, M. L. (1978). Mechanisms of depression of neuromuscular transmission by ATP and adenosine. *J. Physiol. Paris*, **74**, 491–6.

Ribeiro, J. A., and Walker J. (1973). Action of adenosine triphosphate on endplate potentiale recorded from muscle fibres of the rat diaphragm and frog sartorius. *Br. J. Pharmacol.*, **49**, 724–5.

Ribeiro, J. A., and Walker, J. (1975). The effects of ATP and ADP on transmission at the rat and frog neuromuscular junctions. *Br. J. Pharmacol.*, **54**, 213–18.

Ribeiro, J. A., Almeida, S. A., Namorado, J. M. (1979). Adenosine and adenosine triphosphate decrease ^{45}Ca uptake by synaptosomes stimulated by potassium. *Biochem. Pharmac.*, **28**, 1297–1300.

Richards, J. G., Lorez, H. P., and Tranzer, J. P. (1973). Indolealkylamine nerve terminals in cerebral ventricles, identification by electron microscopy and fluorescence histochemistry. *Brain Res.*, **57**, 277–88.

Richardson, K. C. (1964). The fine structure of the albino rabbit iris with special reference to the identification of adrenergic and cholinergic nerves and nerve endings in its intrinsic muscles. *Am. J. Anat.*, **114**, 173–205.

Richter, J. A. Wesche, D. L., and Frederickson, R. C. A. (1979). K-stimulated release of Leu- and Met-enkephalin from rat striatal slices: lack of effect of morphine and naloxone. *Eur. J. Pharmacol.*, **56**, 105–13.

Roach, A. G., Lefevre-Borg, F., and Cavero, I. (1978). Effects of prazosin and phentolamine on cardiac presynaptic α-adrenoceptors in the cat, dog and rat. *Clin. exp. Hypertension*, **1**, 87–101.

Roamer, D., Buescher, H. H., Hill, R. C., Pless, J., Bauer, W., Cardinaux, F., Closse, A., Hauser, D., and Huguenin, R. (1977). A synthetic enkephalin analogue with prolonged parenteral and oral analgesic activity. *Nature*, **268**, 547–9.

Roberts, P. J., and Anderson, S. D. (1979). Stimulatory effect of 1-glutamate and related amino acids on ^{3}H-dopamine release from rat striatum: an *in vitro* model glutamate actions. *J. Neurochem.*, **32**, 1539–45.

Roberts, P. J., and Hillier, K. (1976). Facilitation of noradrenaline release from rat brain synaptosomes by prostaglandin E_2. *Brain Res.*, **112**, 425–8.

Robinson, R. G., and Gershon, M. D. (1971a). Synthesis and uptake of 5-hydroxytryptamine by the myenteric plexus of the guinea-pig ileum: a histochemical study. *J. Pharmac. Exp. Ther.*, **178**, 311–24.

Robinson, R., and Gershon, M. D. (1971b). Synthesis and uptake of 5-hydroxytryptamine by the myenteric plexus of the small intestine of the guinea pig. *J. Pharmacol. exp. Ther.*, **179**, 29–41.

Robson, R. D., and Antonaccio, M. J. (1974). Effect of clonidine on response to cardiac nerve stimulation as a function of impulse frequency and stimulus duration in vagotomized dogs. *Eur. J. Pharmacol.*, **29**, 182–6.

Römer, D., Büscher, H., Hill, R. C., Maurer, R., Petcher, T. J., Welle, H. B., Bakel, H. C., and Akkerman, A. M. (1980). Bremacozine, a potent, long-acting opiate kappa-agonist. *Life Sci.*, **27**, 971–8.

Rónai, A. Z., Berzétei, I. P., Székely, J. I., and Bajusz, S. (1978). The effect of synthetic and natural opioid peptides in isolated organs, in *Characteristics and Function of Opioids* (Eds J. M. Van Ree and L. Terenius), Elsevier, Amsterdam, pp. 493–4.

Rónai, A. Z., Gráf, L., Székely, J. I., Dunai-Kovács, Zs., and Bajusz, S. (1977). Differential behaviour of LPH (61–69) peptide in different model systems. Comparison of the opioid activities of LPH (61–69) peptide and its fragments. *Febs Lett.*, **74**, 182–4.

Rónai, A. Z., Berzétei, I. P., and Bajusz, S. (1977). Differentiation between opioid peptides by naltrexone. *Eur. J. Pharmacol.*, **45**, 393–4.

Rónai, A. Z., Hársing, L. G., Berzétei, I. P., Bajusz, S., and Vizi, E. S. (1982). Met^5-enkephalin-Arg-Phe acts on vascular opiate receptors. *Eur. J. Pharmacol.*, **79**, 337–8.

Rosell, S. (1980). Substance P and neurotensin in the control of gastrointestinal function, in *Peptides: Integrators of Cell and Tissue Function* (Ed. F. E. Bloom), Raven Press, New York, pp. 147–62.

Rosell, S. U., Björkroth, D., Cahng, I. Yamaguchi, A. P., Wan, G., Rackur, G., Fisher, and Folkeres, G. (1977). Effects of substance P and analogs on isolated guinea pig ileum, in *Substance P* (Eds U. S. von Euler and B. Pernow) Raven Press, New York, pp. 83–90.

Rosenblueth, A. (1950). *The Transmission of Nerve Impulses at Neuroeffector Junctions and Peripheral Synapses*, The Technology Press of Massachusett Institute of Technology and John Wiley and Sons, New York.

Ross, I. L., and Gershon, M. D. (1970). Adrenergic innervation of the myenteric plexus of the guinea-pig: ultrastructural, biochemical, and histofluorometric studies using 6-hydroxydopamine. *Anat. Res.*, **47**, 175.

Roth, R. H., Walters, J. R., Murrin, L. C., and Morgenroth, V. H. (1975). Dopamine neurons: role of impulse flow and presynaptic receptors in the regulation of tyrosine hydroxylase, in *Pre-and Postsynaptic Receptors* (Eds E. Usdin and W. E. Bunney), Marcel Dekker, New York, pp. 5–46.

Rotsztejn, W. H. (1980). Neuromodulation in neuroendocrinology. *TINS*, **2**, 67–9.

Rowe, J. N., Van Dyke, K., and Stitzel, R. E. (1978). Purine salvage pathways for the biosynthesis *in vitro* of adenine nucleotides in the guinea-pig vas deferens. *Biochem. Pharmacol.*, **27**, 45–51.

Rutherford, A., and Burnstock, G. (1978). Neuronal non-neuronal components in the overflow of labelled adenyl compounds from guinea-pig taenia coli. *Eur. J. Pharmacol.*, **48**, 195–202.

Saavedra, I. M., Palkovits, M., Kizer, I. S., Brownstein, M., and Zivin, J. A. (1975). Distribution of biogenic amines and related enzymes in the rat pituitary gland. *J. Neurochem.*, **25**, 257–60.

Salt, P. J. (1972). Inhibition of noradrenaline uptake in the isolated rat heart by steroids, clonidine and methoxylated phenylethylamines. *Eur. J. Pharmacol.*, **20**, 329–40.

Sastry, B. S., and Phillis, J. W. (1976). Depression of rat cerebral cortical neurones by H_1 and H_2 histamine receptor agonists. *Eur. J. Pharmacol.*, **38**, 269–73.

Sastry, B. S., and Phillis, J. W. (1977). Antagonism by biogenic amine-induced depression of cerebral cortical neurones by Na+, K+-ATPase inhibitors. *Can. J. Physiol. Pharmac.*, **55**, 170–9.

Satchell, D. G., and Burnstock, G. (1971). Quantitative studies of the release of purine compounds following stimulation of non-adrenergic inhibitory nerves in the stomach. *Biochem. Pharmacol.*, **20**, 1694–7.

Sato, T., Takayanagi, I., and Takagi, K. (1973). Pharmacological properties of electrical activities obtained from neurons of Auerbach's plexus. *Jap. J. Pharmac.*, **23**, 665–73.

Sato, T., Takayanagi, I., and Takagi, K. (1974). Effects of acetylcholine releasing drugs on electrical activities obtained from Auerbach plexus in the guinea-pig ileum. *Jap. J. Pharmac.*, **24**, 447–51.

Saum, W. R., and DeGroat, W. C. (1973). The actions of 5-hydroxytryptamine on the urinary bladder and on vesical autonomic ganglia in the cat. *J. Pharmacol. exp. Ther.*, **185**, 70–83.

Sawynok, J., and Jhamandas, K. H. (1976). Inhibition of acetylcholine release from cholinergic nerves by adenosine, adenine nucleotides and morphine: antagonism by theophylline. *J. Pharmacol. exp. Ther.*, **197**, 379–90.

Sawynok, J., and Jhamandas, K. H. (1977). Muscarinic feedback inhibition of acetylcholine release from the myenteric plexus in the guinea-pig ileum and its status after chronic exposure to morphine. *Can. J. Physiol. Pharmacol.*, **55**, 909–16.

Sawynok, J., and Jhamandas, K. H. (1979). Interactions of methylxanthines, nonxanthine phosphodiesterase inhibitors, and calcium with morphine in the guinea-pig myenteric plexus. *Can. J. Physiol. Pharmacol.*, **57**, 853–9.

Sawynok, J., Labella, F. S., and Pinsky, C. (1980). Effects of morphine and naloxone on the K+-stimulated release of methionine–enkephalin from slices of rat corpus striatum. *Brain Res.*, **189**, 483–93.

Scatton, B., Dedek, J., and Ziwkovic, B. (1983). Lack of involvement of α_2 -adrenoceptors in the regulation of striatal dopaminergic transmission. *Eur. J. Pharmacol.*, **86**, 427–33.

Scatton, B., Pelayo, F., Dubocovich, M. L., Langer, S. Z. and Bartholini, P. (1979). Effect of clonidine on utilizaton and potassium evoked release of adrenaline in rat brain areas. *Brain Res.*, **176**, 197–201.

Schaefer, A., Unyi, G., and Pfeifer, A. K. (1972). The effects of a soluble factor and of catecholamine on the activity of adenosine triphosphatase in subcellular fractions of rat brain, *Biochem. Pharmac.*, 21, 2289–94.

Schaefer, A., Komlós, M., and Seregi, A. (1979). Studies on the effect of catecholamines and chelating agents on the synaptic membrane Na+, K+, ATPase activity in the presence and absence of hydroxylamine. *Biochem. Pharmacol.*, **28**, 2307–12.

Schildkraut, J. J. (1965). The catecholamine hypothesis of affective disorders: a review of supporting evidence. *Am. J. Psychiat.*, **122**, 509–22.

Schmidt, R. F. (1971). Presynaptic inhibition in the vertebrate central nervous system. *Ergebn. Physiol.*, **63**, 20–101.

Schoffelmeer, A. N. M., and Mulder, A. H. (1982). Presynaptic alpha-adrenoceptors and opiate receptors inhibition of 3H-noradrenaline release from rat cerebral cortex slices by different mechanisms. *Eur. J. Pharmacol.*, **79**, 329–32.

Scholtysik, G. (1978). Dopamine receptor mediated inhibition by bromocriptine of accelerator nerve stimulation effects in the pithead cat. *Br. J. Pharmacol.*, **62**, 379–85.

Schubert, P., and Kreutzberg, G. W. (1976). Communication between the neuron and the vessels, in *The Cerebral Vessel Wall* (Eds. J. Cervos-Navarro, E. Betz, F. Matakas, and R. Wüllenweber), Raven Press, New York, pp. 207–13.

Schulman, J. D., Greene, M. L., Fujimoto, W. Y., and Seegmiller, J. E. (1971). Adenine therapy for Lesch–Nyhan syndrome. *Pediat. Res.*, **5**, 77–82.

Schultzberg, M., Hökfelt, T., Lundberg, J. M. Terenius, L., Elfvin, L. B., and Elde, R. (1978a). Enkephalin-like immunoreactivity in nerve terminals in sympathetic ganglia and adrenal medulla and in adrenal medullary gland cells, *Acta Physiol. Scand.*, **103**, 475–83.

Schultzberg, M., Lundberg, J. M., Hökfelt, T., Terenius, L. Brandt, J., Elde, R. P., and Goldstein, M. (1978b). Enkephalin-like immunoreactivity in gland cells and nerve terminals of the adrenal medulla. *Neuroscience*, **3**, 1169–72.

Schultzberg, M., Hökfelt, T., Nilsson, G., Terenius, L., Rehfeld, J. F., Brown, M., Elde, R., Goldstein, M., and Said, S. (1980). Distribution of peptide and catecholamine-containing neurons in the gastrointestinal tract of rat and guinea-pig: immunohistochemical studies with antisera to substance P vasoactive intestinal polypeptide enkephalin, somatostatin gastrin (cholecystokinin, neurotensin and dopamine β-hydroxylase. *Neuroscience*, **5**, 689–93.

Schulz, R., Wüster, M., Simatov, R., Snyder, S., and Herz, A. (1977). Electrically stimulated release of opiate-like material from the myenteric plexus of the guinea-pig ileum. *Eur. J. Pharmacol.*, **41**, 347–8.

Schwartz, A., Lindenmayer, G. E., and Allen, I. C. (1975). The sodium–potassium adenosine triphosphatase: pharmacological, physiological and biochemical aspects, *Pharmac. Rev.*, **27**, 3–134.

Scott, F. H. (1905). On the metabolism and action of nerve cells. *Brain*, **28**, 506–26.

Scriabine, A., and Stavorski, J. M. (1973). Effect of clonidine on cardiac acceleration in vagotomized dogs. *Eur. J. Pharmacol.*, **24**, 101–24.

Sedvall, G. (1975). Receptor feedback and dopamine turnover in CNS, in *Handbook of Psychopharmacology* (Eds L. L. Iversen, S. D., Iverson, and S. H. Snyder). Vol. 6, Plenum Press, New York, pp. 127–77.

Seegmiller, J. E. (1979). Abnormalities of purine metabolism in human immunodeficiency diseases, in *Physiological and Regulatory Functions of Adenosine and Adenine Nucleotides* (Eds H. P. Baer and G. I. Drummond), Raven Press, New York, pp. 395–408.

Seeman, P. (1980). Brain dopamine receptors. *Pharm. Rev.*, **32**, 229–313.

Segal, M. (1975). Physiological and pharmacological evidence for a serotonergic projection to the hippocampus. *Brain Res.*, **94**, 115–31.

Segal, M. (1979). Serotonergic innervation of the locus coeruleus from the dorsal raphé and its action on responses to noxious stimuli. *J. Physiol. (Lond.)*, **286**, 401–16.

Segal, M., and Bloom, F. E. (1974a). The action of norepinephrine in the rat hippocampus. I. Iontophoretic studies. *Brain Res.*, **72**, 79–97.

Segal, M., and Bloom, F. E. (1974b). The action of norepinephrine in the rat hippocampus. II. Activation of the input pathology. *Brain Res.*, **72**, 99–114.

Seno, N., Nakazato, Y., and Ohga, A. (1978). Presynaptic inhibitory effect of catecholamines on cholinergic transmission in the smooth muscle of the chick stomach, *Eur. J. Pharmacol.*, **51**, 229–37.

Serfözö, P., and Vizi, E. S. (1983). Effect of noradrenaline and vanadate on Na, K-activated ATPase of human brain cortical homogenate. *Neurochemistry Internat.* **5**, 237–44.

Sethy, V. H., and Van Woert, M. H. (1974). Regulation of striatal acetylcholine concentration by dopamine receptors. *Nature*, **251**, 529–30.

Shapiro, E., Castellucci, V. F., and Kandel, E. R. (1980). Presynaptic inhibition in *Aplysia* involves a decrease in the Ca^{2+} current of the presynaptic neuron. *Proc. Natl Acad. Sci., USA*, **77**, 1185–9.

Shepherd, G. M. (1974). *The Synaptic Organization of the Brain*, Oxford University Press, London.

Shepherd, J. T., and Vanhoutte, P. M. (1981). Local modulation of adrenergic neurotransmission. *Circulation*, **64**, 655–66.

Shepherd, J. T., and Vanhoutte, P. M. (1979). *The Human Cardiovascular System: Facts and Concepts*, Raven Press, New York.

Sherrington, C. (1906). *The Integrative Action of the Nervous System*, Scribner Press, New York.

Shimahara, T., and Tauc, L. (1975). Multiple interneuronal afferents to the giant cells in *Aplysia*. *J. Physiol.*, **247**, 299–319.

Shoji, T., Daiku, Y., and Igarashi, T. (1980). α-adrenoceptor blocking properties of a new antihypertensive agent, 2- (4- (n-butyryl)-homopiperazine-1-yl)-4-amino-6, 7-dimethoxyquinazoline (E-643). *Jap. J. Pharmac.*, **30**, 763.

Siggins, G. R., and Zieglgänsberger, W. (1981). Morphine and opioid peptides reduce inhibitory potentials in hippocampal pyramidal cells *in vitro* without alteration of membrane potential. *Proc. Natl Acad. Sci., USA*, **78**, 5235–9.

Silbergeld, E. K., and Walters, J. R. (1979). Synaptosomal uptake and release of dopamine in substantia nigra: effects of γ-aminobutyric acid and substance P. *Neurosci., Lett.*, **12**, 119–26.

Silinsky, E. M. (1974). A simple, rapid method for detecting the efflux of small quantities of adenosine triphosphate from biological tissues. *Comp. Biochem.*, **48A**, 561–71.

Silinsky, E. M. (1975). On the association between transmitter secretion and the release of adenine nucleotides from mammalian motor nerve terminals. *J. Physiol. (Lond.)*, **247**, 145–62.

Silinsky, E. M., and Hubbard, J. I. (1973). Release of ATP from rat motor nerve terminals. *Nature, Lond.*, **243**, 404–5.

Simantov, R., Kuhar, M. J., Uhl, G. R., and Snyder, S. H. (1977). Opioid peptide enkephalin:

immunohistochemical mapping in rat central nervous system. *Proc. Natl Acad. Sci., USA*, **74**, 2167–71.

Singh, I. (1964). Seasonal variations in the nature of neurotransmitters in a frog-vagal-stomach muscle preparation. *Arch. Int. Physiol. Biochem.*, **72**, 843–51.

Skangiel-Kramska, J., and Niemierko, S. (1975). Soluble and particle-bound acetylcholinesterase and its isoenzymes in peripheral nerves. *J. Neurochem.*, **24**, 1135–41.

Skirboll, L. R., Grace, A. A., and Bunney, B. S. (1979). Dopamine auto- and postsynaptic receptors: electrophysiological evidence for differential sensitivity to dopamine agonists. *Science*, **206**, 80–2.

Skok, V. I., and Selyanko, A. A. (1979). Acetylcholine and serotonin receptors in mammalian sympathetic ganglion neurons, in *Integrative Functions of the Autonomic Nervous System* (Eds. C. McC. Brooks, K. Koizumi, and A. Sato), University of Tokyo Press, Tokyo, pp. 248–53.

Smith, A. D. (1971). Summing up: some implications of the neuron as a secreting cell. *Phil. Trans. R. Soc., Lond., B.*, **261**, 423–37.

Smith, A. D. (1972). Subcellular localization of noradrenaline in sympathetic neurons. *Pharmacol. Rev.*, **24**, 435–56.

Smith, A. D. (1973). Mechanisms involved in the release of noradrenaline from sympathetic nerves. *Br. Med. Bull.*, **29**, 123–9.

Smith, T. W., Hughes, J., Kosterlitz, H. W., and Sosa, R. P. (1976). Enkephalins: isolation distribution and function, in *Opiates and Endogenous Opidoid Peptides* (Ed. H. H. Kosterlitz), Elsevier/North-Holland/Biomedical Press, Amsterdam, pp. 57–62.

Smith, R. C., Tamminga, C., and Davis, J. M. (1977). Effect of apomorphine on schizophrenic symptoms, *J. Neural. Transm.*, **40**, 171–6.

Somogyi, P. (1977). A specific axo-axonal interneuron in the visual cortex of the rat. *Brain Res.*, **136**, 345–50.

Somogyi, P., and Hámori, J. (1976). A quantitative electron microscopic study of the Purkinje cell axon initial segment. *Neuroscience*, **1**, 361–5.

Somogyi, P., Freund, T. F., and Cowey, A. (1982a). The axo-axonic interneuron in the cerebral cortex of the rat, cat and monkey. *Neuroscience*, **7**, 2577–607.

Somogyi, P., Priestley, J. V., Cuello, A. C., Smith, A. D., and Takagi, H. (1982b). Synaptic connections of enkephalin-immunoreactive nerve terminals in the neostriatum: a correlated light and electron microscopic study. *J. of Neurocytology*, **11**, 779–807.

Somogyi, P., Nunzi, M. G., Gorio, A., and Smith, A. D. (1982c). A new type of specific interneuron in the monkey hippocampus forming synapses exclusively with the axon initial segments of pyramidal cells. *Brain Res.* (submitted).

Somogyi, G. T., Vizi, E. S., and Knoll, J. (1970). The effects of drugs on the contractions of the vas deferens in response to single stimuli, in *Abstract of Regional Congress of the IUPS*, Brasov, N314.

Somogyi, G. T., Vizi, E. S., and Knoll, J. (1977). Effect of hemicholinium-3 on the release and net synthesis of acetylcholine in Auerbach's plexus of guinea-pig ileum. *Neuroscience*, **2**, 791–6.

Spehlmann, R., Daniels, M. B., and Chang, C. M. (1971). Acetylcholine and the epileptiform activity of chronically isolated cortex. I. Marcoelectride studies, *Arch. Neurol.*, **24**, 401–8.

Spano, P. F., DiChiarro, G., Tonon, G. L., and Trabucchi, M. (1975). A dopamine stimulated adnylate cyclase in rat substantia nigra. *J. Neurochem.*, **27**, 1565–8.

Spyraki, C., and Fibiger, H. C. (1982). Clonidine-induced sedation in rats: evidence for mediation by postsynaptic α_2-adrenoceptors. *J. Neural. Transm.*, **54**, 153–63.

Stadler, H., Lloyd, K. G., Gadea-Ciria, M., and Bartholini, G. (1973). Enhanced striatal acetylcholine release by chlorpromazine and its reversal by apomorphine. *Brain Res.*, **55**, 476–80.

Standaert, F. G., Dretchen, G., Skirboll, L. R., and Morgenroth, V. H. III. (1976). A role of cyclic nucleotides in neuromuscular transmission. *J. Pharmac. exp. Ther*, **199**, 553–64.

Starke, K. (1971). Influence of α-receptor stimulants on noradrenaline release. *Naturwissenschaften*, **58**, 420–5.

Starke, K. (1972a). Alpha sympathomimetic inhibition of adrenergic and cholinergic transmission in the rabbit heart. *Naunyn-Schmiedeberg's Arch. Pharmacol.*, **274**, 18–45.

Starke, K. (1972b). Influence of extracellular noradrenaline on the stimulation evoked secretion of noradrenaline from sympathetic nerves: evidence for an alpha-receptor mediated feedback inhibition of noradrenaline release. *Naunyn-Schmiedeberg's Arch. Pharmak.*, **275**, 11–23.

246

Starke, K. (1973). Influences of phenylephrine and orciprenaline on the release of noradrenaline. *Experientia*, **29**, 579–80.

Starke, K. (1977). Regulation of noradrenaline release by presynaptic receptor systems. *Rev. Physiol. Biochem., Pharmacol.*, **77**, 1–124.

Starke, K. (1981). Presynaptic receptors. *Ann. Rev. Pharmacol., Toxicol.*, **21**, 7–30.

Starke, K., and Altman, K. P. (1973). Inhibition of adrenergic neurotransmission by clonidine. An action on prejunctional α-receptors. *Neuropharmacol.*, **12**, 339–47.

Starke, K., and Langer, S. Z. (1979). A note on terminology for presynaptic receptors, in *Presynaptic Receptors*, Advances in the Biosciences, Vol. 188 (Eds S. Z. Langer, K. Starke, and M. L. Dubocovich,) Pergamon Press, Oxford, pp. 1–3.

Starke, K., and Montel, H. (1973a). Involvement of α-receptors in clonidine-induced inhibition of transmitter release from central monoamine neurones, *Neuropharmacology.* **12**, 1073–80.

Starke, K., and Montel, H. (1973b). Alpha-receptor mediated modulation of transmitter release from central noradrenergic neurones. *Naunyn-Schmiedeberg's Arch. Pharmacol.*, **279**, 53–60.

Starke, K., and Montel, H. (1974). Influence of drugs with affinity for α-adrenoceptors on noradrenaline release by potassium, tyramine and dimethylphenylpiperazinium. *Eur. J. Pharmacol.*, **27**, 273–80.

Starke, K. and Weitzell, R. (1980). γ-aminobutyric acid and postganglionic sympathetic transmission in the pulmonary artery of the rabbit. *J. Auton. Pharmacol.*, **1**, 445–520.

Starke, K., Endo, T., and Taube, H. D. (1975a). Pre- and post-synaptic components in effect of drugs with α-adrenoceptor affinity. *Nature*, **254**, 440–1.

Starke, K, Endo, T., and Taube, H. D. (1975b). Relative pre- and postsynaptic potencies of α-adrenoceptor agonists in the rabbit pulmonary artery. *Naunyn-Schmiedeberg's Arch. Pharmacol*, **291**, 55–78.

Starke, K., Borowski, E., and Endo, T. (1975c). Preferential blockade of presynaptic α-adrenoceptors by yohimbine. *Eur. J. Pharmacol.*, **34**, 385–8.

Starke, K., Endo, T., Taube, H. D., and Borowski, E. (1975d). Presynaptic receptor systems on noradrenergic nerves, in *Chemical Tools in Catecholamine Research* (Eds O. Almgenr, A. Carlsson, and J. Engel), Vol. 2, North-Holland, Amsterdam and Oxford, pp. 193–200.

Starke, K., Montel, H., Gayk, W., and Merker, R. (1974). Comparison of the effect of clonidine on pre- and postsynaptic adrenoceptors in the rabbit pulmonary artery. *Naunyn-Schmiedeberg's Arch. Pharmacol.*, **285**, 133–50.

Starke, K., Montel, H., and Schümann, H. J. (1971). Influence of cocaine and phenoxybenzamine on noradrenaline uptake and release. *Naunyn-Schmiedeberg's Arch. Pharmacol.*, **270**, 210–14.

Starke, K., Montel., H., and Wagner, J. (1971). Effect of phentolamine on noradrenaline uptake and release. *Naunyn-Schmiedeberg's Arch. Pharmacol.*, **271**, 181–92.

Starke, K., Taube, H. D., and Borowski, E. (1977). Presynaptic receptor system in catecholaminergic transmission. *Biochem. Pharmac.*, **26**, 259–68.

Starke, K., Wagner, J., and Schümann, H. J. (1972). Adrenergic neuron blockade by clonidine: comparison with guanethidine and local anaesthetics. *Arch. int. Pharmacodyn.*, **195**, 291–308.

Starke, K., Tanaka, I., and Weitzell, R. (1980). Pre- and post-synaptic receptors, physiological and pharmacological implications, in *Modulation of Neurochemical Transmission* (Ed. E. S. Vizi), Pergamon Press, Oxford; Akadémiai Kladó, Budapest, pp. 37–46.

Starr, M. S. (1978). GABA potentiates potassium-stimulated ^{3}H-dopamine release from slices of rat substantia nigra and corpus striatum. *Eur. J. Pharmacol.*, **53**, 325–8.

Starr, M. S. (1979). GABA mediated potentiation of amine release from nigrostriatal dopamine neurones *in vitro*, *Eur. J. Pharmacol.*, **53**, 215–26.

Steinsland, O. S., Furchigott, R. F., and Kirpekar, S. M. (1973). Inhibition of adrenergic neurotransmission by para-sympathomimetics in the rabbit ear artery. *J. Pharmac. Exp. Ther.*, **184**, 346–56.

Steinsland, O. S., Kirpekar, S. M., and Furchigott, R. F. (1971). Inhibition of adrenergic neurotransmission by cholinergic agents in the isolated perfused rabbit ear artery. *Pharmacologist*, **13**, 250–6.

Stewart, J. M., Getto, C. L., Neldner, K., Reeve, E. B., Krivoy, W. A., and Zimmerman, E. (1976). Substance P and analgesia. *Nature*, **262**, 784–5.

Stjärne, L. (1972). Prostaglandin E restricting noradrenaline secretion—neural in origin? *Acta Physiol. Scand.*, **86**, 574–6.

Stjärne, L. (1973a). α-adrenoceptor mediated feed-back control of sympathetic neurotransmitter secretion in guinea-pig vas deferens. *Nature New Biol.*, **241**, 190–1.

Stjärne, L. (1973b). Inhibitory effect of prostaglandin E_2 on noradrenaline secretion from sympathetic nerves as a function of external calcium. *Prostaglandins*, **3**, 105–9.

Stjärne, L. (1973c). Kinetics of secretion of sympathetic neurotransmitter as a function of external calcium: mechanism of inhibitory effect of prostaglandin E. *Acta Physiol. Scand.*, **87**, 428–30.

Stjärne, L. (1973d). Comparison of secretion of sympathetic neurotransmitter induced by nerve stimulation with that evoked by high potassium as triggers of dual α-adrenoceptor mediated negative feed-back control of noradrenaline secretion. *Prostaglandins*, **3**, 421–6.

Stjärne, L. (1973e). Frequency dependence of dual negative feedback control of secretion of sympathetic neurotransmitter in guinea-pig vas deferens. *Br. J. Pharmacol.*, **40**, 358–60.

Stjärne, L. (1974). Stereoselectivity of presynaptic α-adrenoceptors involved in feedback control of sympathetic neurotransmitter secretion. *Acta physiol. Scand.*, **90**, 286–8.

Stjärne, L. (1975a). Basic mechanism and local feedback control of secretion of adrenergic and cholinergic neurotransmitters, in *Handbook of Psychopharmacology* (Eds L. L. Iversen, S. D. Iversen, and S. H. Snyder), Vol. 6. Plenum Press, New York and London, pp. 179–233.

Stjärne, L. (1975b). Adrenoceptor mediated positive and negative feedback control of noradrenaline secretion from human vasoconstrictor nerves. *Acta Physiol. Scand.*, **95**, 18–19.

Stjärne, L., and Brundin, J. (1976). β-adrenoceptors facilitating noradrenaline secretion from human vasoconstrictor nerves. *Acta Physiol. Scand.*, **97**, 88–93.

Stjärne, L. (1976). The twitch contraction in guinea-pig isolated vas deferens: result of cooperative interaction of potassium and noradrenaline released from adrenergic nerves in close junctions? *Acta Physiol. Scand.*, **98**, 512–14.

Stjärne, L. (1977). Do potassium ions released from nerves modulate the sensitivity to transmitter in close neuro-effector junctions of the vas deferens? *Neuroscience*, **2**, 373–81.

Stjärne, L. (1978). Facilitation and receptor-mediated regulation of noradrenaline secretion by control of recruitment of varicosities as well as by control of electro-secretory coupling. *Neuroscience*, **3**, 1147–55.

Stjärne, L. (1980). Frequency dependence of presynaptic inhibition of transmitter secretion, in *Modulation of Neurochemical Transmission* (Ed. E. S. Vizi), Pergamon Press, Oxford, pp. 27–36.

Stjärne, L. (1981). On sites and mechanisms of presynaptic control of noradrenaline secretion, in *Chemical Neurotransmission: 75 years* (Eds L. Stjarne, P. Hedqvist, H. Lagercrantz, and A. Wennmalm), Academic Press, London, pp. 257–73.

Stjärne, L., and Brundin, J. (1975). Affinity of noradrenaline and dopamine for neural receptors mediating negative feedback control of noradrenaline secretion in human vasoconstrictor nerves. *Acta Physiol. Scand.*, **95**, 89–94.

Stjärne, L., and Brundin, J. (1976). Prostaglandin E_2 inhibits nerve stimulation-induced smooth muscle contraction in isolated human omental vein depressing secretion of noradrenaline. *Acta Physiol. Scand.*, **97**, 526–8.

Stjärne, L., and Gripe, L. (1973). Prostaglandin-dependent and independent feedback control of noradrenaline secretion in vasoconstrictor nerve of normotensive human subject. A preliminary report. *Naunyn-Schmiedeberg's Arch. exp. Path. Pharmac.*, **280**, 441–6.

Stjärne, L. Albert, P., and Bartfai, T. (1979). Models of regulation of norepinephrine secretion by prejunctional receptors and by facilitation. Role of calcium and cyclic nucleotides, in *Catecholamines, Basic and Clinical Frontiers* (Eds. E. Usdin, I. J. Kopin, and J. Barchas), p. Pergamon Press, New York and Oxford, pp. 292–7.

Stjärne, L., Bartfai, T., and Alberts, P. (1981). Is the alpha-adrenoceptor mediated inhibition of noradrenaline secretion a negative feedback control in the strict sense? in *Physiology of Excitable Membranes* (Ed. J. Salánki), Pergamon Press, Oxford; Akadémiai Kiadó, Budapest, p. 247.

Stoff, J. C., den Breejen, E. J. C., and Mulder, A. H. (1979). GABA modulates the release of dopamine and acetylcholine from rat caudate nucleus slices. *Eur. J. Pharmacol.*, **57**, 35–42.

Stone, T. W. (1973). Pharmacology of pyramidal tract cells in the cerebral cortex. *Naunyn-Schmiedeberg's Arch. Pharmacol.*, **278**, 333–46.

Stone, T. W. (1981). 4-aminopyridine and quinidine effects on adenosine inhibition of vas deferens contractions. *Br. J. Pharmacol*, **74**, 296P.

Stone, T. W. (1982). Cell-membrane receptors for purines. *Bioscience Reports*, **2**, 77–90.

Stone, T. W., and Taylor, D. A. (1977). The effect of cyclic nucleotides on excitability of neurones in rat cerebral cortex. *J. Physiol.*, **266**, 523–43.

Stone, T. W., and Taylor, D. A. (1978). An electrophysiological demonstration of a synergetic interaction between norepinephrine and adenosine in the cerebral cortex. *Brain Res.*, **147**, 396–400.

Stoof, J. C., Horn, A. S., Mulder, A. H. (1980). Simultaneous demonstration of the activation of presynaptic dopamine autoreceptors and postsynaptic dopamine receptors *in vitro* by N, N-dipropyl-5,6-ADTN. *Brain Res*, **196**, 276–81.

Stoof, J. C., Thieme, R. E. Vrijmoed-deVries, M. C., and Mulder, A. H. [1979]. *In vitro* acetylcholine release from rat caudate nucleus as a new model for testing drugs with dopamine-receptor activity. *Naunyn-Schmiedeberg's Arch. Pharmacol.*, **309**, 119–24.

Story, D. F., Briley, M. S., and Langer, S. Z. (1979). The effects of chemical sympathectomy with 6-hydroxydopamine on α-adrenoceptor and muscarinic cholinoceptor binding in rat heart ventricle. *Eur. J. Pharmacol.*, **57**, 423–6.

Strömbom, U. (1976). Catecholamine receptor agonists. Effects on motor activity and rate of tyrosine hydroxylation on mouse brain. *Naunyn-Schmiedeberg's Arch. Pharmacol.*, **292**, 167–76.

Studler, J. M., Simon, H., Cesselin, F., Legrand, J. C., Glowinsky, J., and Tassin, J. P. (1981). Biochemical investigation on the localization of the cholecystokinin octapeptide in dopaminergic neurons originating from the ventral tegmental area of the rat. *Neuropeptides*, **2**, 131–9.

Su, C. (1974). Vasodilator action of adenine compounds on the rabbit portal vein. *Pharmacologist*, **16**, 289–93.

Su, C. (1975). Neurogenic release of purine compound in blood vessels. *J. Pharmacol, Exp. Ther.*, **195**, 159–66.

Su, C., Bevan, J., and Burnstock, G. (1971). ^{3}H-adenosine triphosphate: release during stimulation of enteric nerves. *Science*, **173**, 337–9.

Subramanian, N., Mitznegg, P., Sprügel, W., Domschke, W., Wünsch, E., and Demling, L. (1977). Influence of enkaphalin on K^+-evoked efflux of putative neurotransmitter in rat brain. *Naunyn-Schmiedeberg's Arch. Pharmacol.*, **299**, 163–5.

Subramanian, N., and Mulder, A. H. (1977). Modulation by histamine of the efflux of radiolabeled catecholamines from rat brain slices. *Eur. J. Pharmacol.*, **43**, 143–52.

Sulakne, P. V., and Phillis, J. W. (1975). The release of ^{3}H-adenosine and its derivatives from cat sensorimotor cortex. *Life Sciences*, **17**, 551–66.

Svensson, T. H., Bunney, B. S., and Aghajanian, G. K. (1975). Inhibition of both noradrenergic and serotonergic neurons in brain by the α-adrenergic agonist clonidine. *Brain Res.*, **92**, 291–306.

Svoboda, P., and Mosinger, B. (1981). Catecholamines and the brain microsomal Na, K-adenosinetriphosphatase-II. The mechanism of action. *Biochem. Pharmacol.*, **30**, 433–9.

Swanson, L. W. (1976). The locus coeruleus: a cytoarchitectonic, Golgi and immunohistochemical study in the albino rat. *Brain Res.*, **110**, 39–56.

Szabadi, E., Bradshaw, C. M., and Bevan, P. (1977). Excitatory and depressant neuronal responses to noradrenaline, 5-hydroxytryptamine and mescaline: the role of the baseline firing rate. *Brain Res.*, **126**, 580–3.

Szabolcsi, I., Vizi, E. S., and Knoll, J. (1974). Inhibition by different drugs of neurochemical transmission in the guinea-pig oesophagus; morphological and functional study, in *Symposium on Current Problems in the Pharmacology of Analgesics* (Ed. E. S. Vizi), Publishing House of the Hungarian Academy of Science, Budapest, pp. 57–64.

Szenohradszky, J., Kerecsen, L., and Vizi, E. S. (1979). Reduction by β-receptor blockade of the inhibitory effect of sympathetic stimulation on cholinergic transmission in the Fnkleman preparation, in *Modulation of Neurochemical Transmission* (Ed. E. S. Vizi), Pergamon Press, Oxford; Akadémiai Kiadó, pp. 169–88.

Szentágothai, J. (1968). Synaptic structure and the concept of presynaptic inhibition, in *Structure and Function of Inhibitory Neuronal Mechanisms* (Eds U. S. von Euler, D. Skoglund, and V. Söderberg), Pergamon Press, Oxford, pp. 15–31.

Szentágothai, J., and Arbib, M. A. (1974). Conceptual models of neural organization. *Neurosci., Res. Prog. Bull.*, **12** (3) 305–10.

Szerb, J. C. (1961). The effect of morphine on the adrenergic nerves of the isolated guinea-pig jejunum. *Br. J. Pharmacol.*, **16**, 23–31.

Szerb, J. C. (1964). The effect of tertiary and quaternary atropine on cortical acetylcholine output and the electro-encephalogram in cats, *Can. J. Physiol. Pharmac.*, **42**, 303–14.

Szerb, J. C., (1975). Endogenous acetylcholine release and labelled acetylcholine formation from ³H-choline in the myenteric plexus of the guinea-pig ileum. *Can. J. Physiol. Pharmac.*, **53**, 566–74.

Szerb, J. C. (1976). Storage and release of labelled acetylcholine in the myenteric plexus of the guinea-pig ileum. *Can. J. Physiol. Pharmac.*, **54**, 12–22.

Szerb, J. C., and Somogyi, G. T. (1973). Depression of acetylcholine release from cerebral cortical slices by cholinesterase inhibition and by oxotermorine. *Nature*, **241**, 121–2.

Szerb, J. C., Hadházy, P., and Dudar, J. D. (1977). Release of ³H-acetylcholine from rat hippocampal slices: effect of septal lesion and of graded concentrations of muscarinic agonists and antagonists. *Brain Res.*, **128**. 285–91.

Tagerud, S. E. O., and Cuello, A. C. (1979). Dopamine release from the rat substantia nigra *in vitro*. Effect of raphé lesions and veratridine stimulation. *Neuroscience*, **4**, 2021–9.

Takayanagi, I., Sato, T., and Takagi, K. (1977). Effects of sympathetic nerve stimulation on electrical activity of Auerbach's plexus and intestinal smooth muscle tone. *J. Pharm. Pharmac.*, **29**, 376–7.

Takeuchi, A., and Takeuchi, N. (1966). A study of the inhibitory action of gamma-aminobutyric acid on neuromuscular transmission in the crayfish. *J. Physiol. (Lond.)*, **183**, 418–32.

Tamminga, C. A., Schaffer, M. H., Smith, R. C., and Davis, J. M. (1978). Schizophrenic symptoms improve with apomorphine. *Science*, **200**, 567–8.

Tang. J. H., Yang, Y. T., and Costa, E. (1982). Distribution of met-enkephalin Arg⁶-Phe⁷ in various tissues of rats and guinea-pig. *Neuropharmacology*, **21**, 595–600.

Taube, H. D., Borowski, E., Endo, T., and Starke, K. (1976). Enkephalin: a potential modulator of noradrenaline in rat brain. *Eur. J. Pharmacol..*, **38**, 377–80.

Taube, H. D., Starke, K., and Borowski, E. (1977). Presynaptic receptor systems on the noradrenergic neurones of rat brain. *Naunyn-Schmiedeberg's Arch. Pharmacol.*, **299**, 123–41.

Taxi, J. (1965). Contribution a l'étude des connexions des neurones moteurs du système nerveux autonome. *Annls Sci. nat. (Zool.) Sér. 12*, **7**, 413–674.

Taxi, J., and Droz, B. (1969). Radioautographic study of the accumulation of some biogenic amines in the autonomic nervous system, in *Cellular Dynamics of the Neuron* (Ed. S. H. Braondes), Academic Press, New York, p. 175–90.

Taylor, J. A., and Nabi Mir, G. (1982). Alpha adrenergic receptors and gastric function, *Drug. Development Research*, **2**, 105–22.

Tayo, F. M. (1979). Prejunctional inhibitory alpha-adrenoceptors and dopamiceptors of the rat vas deferens and the guinea-pig ileum *in vitro*. *Eur. J. Pharmacol.*, **58**, 189–95.

Teitelman, G., Joh, T. H., and Reis, D. J. (1981). Linkage of the brain–skin–gut axis: islet cells originate from dopaminergic precursors. *Peptides*, **2**, 157–68.

Tennyson, V. M., Heikkila, R., Mytilineon, C., Coté, L., and Cohen, G. (1974). 6-hydroxy-dopamine 'tagged' neuronal boutons in rabbit neostriatum: interrelationship between vesicles and axonal membrane. *Brain Res.*, **82**, 341–8.

Terenius, L. (1977). Opioid peptides and opiates differ in receptor selectivity, *Psychoneuroendocrinology*, **2**, 53–8.

Terenius, L. (1978). Endogenous peptides and analgesia. *Ann. Rev. Pharmacol. Toxicol.*, **18**, 189–204.

Terenius, L., Wahlström, A., Lindström, L., and Widerlov, E. (1976). Increased levels of endorphins in chronic psychosis. *Neurosci. Lett.*, **3**, 157–62.

Ternaux, J. P., Héri, F., Borugoin, S., Adrien, J., Glowinsky, J., and Hamon, M. (1977). The topographical distribution of serotoninergic terminals in the neostriatum of the rat and the caudate nucleus of the cat. *Brain Res.*, **121**, 311–26.

Thierry, A. M., Stinus, L., Blanc, G., and Glowinsky, J. (1973). Some evidence for the existence of dopaminergic neurons in rat cortex. *Brain Res.*, **50**, 230–4.

Thoa, N. B., Eccleston, D., and Axelrod, J. (1969). The accumulation of ¹⁴C-serotonin in the guinea pig vas deferens. *J. Pharmac. exp. Ther.*, **157**, 523–40.

Thorner, M. O. (1975). Dopamine—an important neurotransmitter in the autonomic nervous system. *Lancet*, **I**, 662–4.

Timmermans, P. B. M. W. M., and van Zwieten, P. A. (1982). α_2-adrenoceptors: classification, localization, mechanisms and targets for drugs. *J. Med. Chem.*, **25**, 1389–1401.

Timmermans, P. B. M. W. M., Schoop, A. M. C., and Kwa, H. Y. (1981). Characterization of α-adrenoceptors participating in the central hypotensive and sedative effects of clonidine using yohimbine, rauwolscine and corynanthine. *Eur. J. Pharmacol.*, **70**, 7–15.

Tomita, L., and Watanabe, H. (1973). A comparison of the effects of adenosine triphosphate with noradrenaline and with the inhibitory potential of the guinea-pig taenia coli. *J. Physiol.*

(Lond.), **231**, 167–77.

Török, T. L., Rubányi, G., Vizi, E. S., and Magyar, K. (1982). Stimulation by vanadate of ³H-noradrenaline release from rabbit pulmonary artery and its inhibition by noradrenaline. *Eur. J. Pharmacol.*, **84**, 93–7.

Trabucchi, M., Cheney, D. L., Racagni, G., and Costa, E. (1974). Involvement of brain cholinergic mechanism in the action of chlorpromazine. *Nature*, **249**, 664–6.

Tranzer, I. P., and Thoenen, H. (1967). Significance of 'empty vesicles' in post-ganglionic sympathetic nerve terminals. *Experientia* **23**, 123–4.

Tremblay, J. P., and Plowde, G. (1975). Presynaptic modulating effects of GABA on depression, facilitation and post-tetanic potentiation of a cholinergic synapse in *Aplysia californica*. *Brain Res.*, **88**, 455–74.

Tremblay, J. P., Schlapfer, W. T., Woodson, P. B. J., and Barondes, S. H. (1974). Morphine and related compounds: evidence that they decrease available neurotransmitter in *Aplysia californica*, *Brain Res.*, **81**, 107–18.

Tremblay, J. P., Woodson, P. B. J., Schlapfer, W. T., and Barondes, S. H. (1976). Dopamine, serotonin and related compounds: presynaptic effects on synaptic depression, frequency facilitation and post-tetanic potentiation at a synapse in *Aplysia californica*. *Brain Res.*, **109**, 61–81.

Tucek, S. (1975). Transport of choline acetyltransferase and acetylcholinesterase in the central stem and isolated segments of a peripheral nerve. *Brain Res.*, **86**, 259–70.

Tucek, S., and Cheng, S. C. (1974). Provenance of the acetyl group of acetylcholine and compartmentation of acetyl-CoA and Krebs cycle intermediates in the brain *in vivo*. *J. Neurochem.*, **27**, 893–914.

Twarog, B., and Page, J. H. (1953). Serotonin content of some mammalian tissue and urine and method for its determination. *Am. J. Physiol.*, **175**, 157–61.

Ungerstedt, U. (1971). Sterotaxic mapping of the monoamine pathways in the rat brain. *Acta Physiol. Scand. Suppl.* **367**, 1–48.

U'Prichard, D. C., and Snyder, S. H. (1979). Distinct α-noradrenergic receptors differentiated by binding and physiological relationships. *Life Sci.*, **24**, 79–88.

U'Prichard, D. C., Greenberg, D. A., and Snyder, S. H. (1977). Binding characteristic of a radiolabeled agonist and antagonist at central nervous system alpha noradrenergic receptors. *Mol. Pharmacol.*, **13**, 454–73.

U'Prichard, D. C., Charness, M. E., Robertson, and Snyder, S. H. (1978). Prazosin: differential affinities for two populations α-noradrenergic receptor binding sites. *Eur. J. Pharmacol.*, **50**, 87–9.

Van Calker, D., Muller, M., and Hamprecht, B. (1979). Adenosine regulates via two different types of receptors: the accumulation of cyclic AMP in cultured brain cells. *J. Neurochem.*, **33**, 999–1005.

Van Hee, H., and Vanhoutte, P. M. (1978). Cholinergic inhibition of adrenergic neurotransmission in the canine gastric artery. *Gastroenterology*, **74**, 1266–70.

Vanhoutte, P. M. (1977). Cholinergic inhibition of adrenergic transmission. *Fed. Proc.*, **36**, 2444–9.

Vanhoutte, P. M., and Shepherd, J. T. (1973). Venous relaxation caused by acetylcholine acting on the sympathetic nerves. *Circ. Res.*, **32**, 259–67.

Vanhoutte, P. M., and Janssens, W. J. (1980). Thermosensitivity of cutaneous vessels and Raynaud's disease. *Am. Heart J.*, **100**, 263–5.

Van Zwieten, P. A. (1980). Pharmacology of centrally acting hypotensive drugs. *Br. J. Clin. Pharmacol.*, **10**, 13–20.

Vercruysse, P., Bossuyt, P., Hanergreefs, G., Verveuren, T. J., and Vanhoutte, P. M. (1979). Gallamine and pancuronium inhibit pre- and postjunctional muscarinic receptors in canine saphenous veins. *J. Pharmacol., Exp. Ther.* **209**, 225–30.

Verhaeghe, R., Vanhoutte, P. M., and Shepherd, J. T. (1977). Inhibition of sympathetic neurotransmission in canine blood vessels by adenosine and adenine nucleotides. *Circulat. Res.*, **40**, 208–15.

Verhoeven, W. M., Van Praag, H. M., van Ree, J. M., and De Wied, D. (1979). Improvement of schizophrenic patients treated with Des-Tyr¹-gamma-endorphin DT-E. *Arch. Gen. Psychiatry*, **36**, 294–8.

Viveros, O. H., Diliberto, E. J., Haum E., and Chang, K. J. (1979). Opiate-like material in the adrenal medulla: evidence for storage and secretion with catecholamines. *Mol. Pharm.*, **16**, 1101–8.

Vizi, E. S. (1968). The inhibitory action of noradrenaline and adrenaline on release of acetylcholine from guinea-pig ileum longitudinal strips. *Arch. exp. Path. Pharmac.*, **259**, 199–200.

Vizi, E. S. (1972). Stimulation by inhibition of (Na$^+$, K$^+$-Mg^{2+}) activated ATPase of acetylcholine release in cortical slices from rat brain. *J. Physiol. (Lond.)*, **226**, 95–117.

Vizi, E. S. (1973a). Acetylcholine release from guinea-pig ileum by parasympathetic ganglion stimulants and gastrin-like polypeptides. *Br. J. Pharmacol.*, **47**, 765–77.

Vizi, E. S. (1973b). Does stimulation of Na$^+$, K$^+$-Mg^{2+} activated ATPase inhibit acetylcholine release from nerve terminals? *Br. J. Pharmacol.*, **48**, 346–7.

Vizi, E. S. (1974a). Interaction between adrenergic and cholinergic systems: presynaptic inhibitory effect of noradrenaline on acetylcholine release. *J. Neural Transmission, Suppl.*, **11**, 61–78.

Vizi, E. S. (1974b). Possible connection between the release of acetylcholine and the activity of Na$^+$, K$^+$-activated ATPase, in *Neurobiological Basis of Memory Formation* (Ed. H. Matthies), VEB Verlag Volk und Gesundheit, Berlin, pp. 96–116.

Vizi, E. S. (1974c). Inhibitory effect of morphine and noradrenaline on neurochemical transmission, in *Symposium on Current Problems in the Pharmacology of Analgetics* (Ed. E. S. Vizi), Akadémiai Kiadó, Budapest, pp. 19–38.

Vizi, E. S. (1975a). Release mechanisms of acetylcholine and the role of Na$^+$, K$^+$-activated ATPase, in *Cholinergic Mechanism* (Ed. P. G. Waser), Raven Press, New York, pp. 199–211.

Vizi, E. S. (1975b). The role of Na$^+$, K$^+$-activated ATPase in transmitter release: acetylcholine release from basal ganglia and its inhibition by dopamine and noradrenaline, in *Subcortical Mechanisms and Sensorimotor Activities* (Ed. T. L. Frigyesi), Hans Huber, Bern, Stuttgart, and Vienna, pp. 63–87.

Vizi, E. S. (1976). The role of α-adrenoceptors situated in Auerbach's plexus in the inhibition of gastrointestinal motility, in *Physiology of Smooth Muscle* (Eds E. Bülbring and M. F. Shuba), Raven Press, New York, pp. 357–67.

Vizi, E. S. (1977a). Dale's principle today: cholinergic tissues, in *Neuron Concepts Today* (Eds J. Szentágothai, J. Hámori, and E. S. Vizi), Akadémiai Kiadó, Budapest, pp. 63–81.

Vizi, E. S. (1977b). Termination of transmitter release stimulation of sodium-potassium activated ATPase. *J. Physiol. (Lond.)*, **267**, 261–80.

Vizi, E. S. (1978). Na$^+$, K$^+$-activated adenosinetriphosphatase as a trigger in transmitter release. *Neuroscience*, **3**, 367–84.

Vizi, E. S. (1979). Presynaptic modulation of neurochemical transmission, *Progr. in Neurobiol.*, **12**, 181–290.

Vizi, E. S. (1980a). Non-synaptic modulation of transmitter release: pharmacological implications, *Trends in Pharm. Sci.*, **2**, 172–5.

Vizi, E. S. (1980b). Modulation of cortical release of acetylcholine by noradrenaline released from nerves arising from the rat locus coeruleus. *Neuroscience*, **5**, 2139–44.

Vizi, E. S. (1980c). Non-synaptic interaction of neurotransmitters: presynaptic inhibition and disinhibtion, in *Modulation of Neurochemical Transmission* (Ed. E. S. Vizi), Pergamon Press, Oxford and Akadémiai Kiadó, Budapest, pp. 13–25.

Vizi, E. S. (1981a). Non-synaptic modulation of chemical neurotransmission, in *Chemical Neurotransmission: 75 years* (Eds L. Stjärne, P. Hedqvist, H. Lagercrantz, and A. Wennmalm), Academic Press, London, pp. 235–48.

Vizi, E. S. (1981b). Non-synaptic interneuronal communication, in *Physiology of Excitable Membranes* (Ed. J. Salánki), Pergamon Press, Oxford; Akadémiai Kiadó, Budapest, pp. 223–34.

Vizi, E. S. (1981c). Na$^+$, K$^+$-activated ATPase and non-quantal cytoplasmic release of acetylcholine, in *Cholinergic Mechanisms* (Eds G. Pepeu and H. Ladinsky), Plenum Publishing Corporation, New York, pp. 187–95.

Vizi, E. S. (1982a). Non-synaptic intercellular communication: presynaptic inhibition. *Acta Biol. Acad. Sci., Hung.*, **33**, 331–51.

Vizi, E. S. (1982b). Release modulating adrenoceptors, in *Adrenoceptors* (Ed. G. Kunos), Wiley, London, pp. 65–107.

Vizi, E. S. (1983). Non-synaptic interneuronal communication: physiological and pharmacological implications, in *Dale's Principle and Communication between Neurones* (Ed. N. N. Osborne), Pergamon Press, Oxford, pp. 83–111.

Vizi, E. S., and Knoll, J. (1971). The effects of sympathetic nerve stimulation and guanethidine on parasympathetic neuroeffector transmission: the inhibition of acetylcholine release, *J. Pharm. Pharmacol.*, **23**, 918–25.

Vizi, E. S., and Knoll. J. (1972). The mechanism of acetylcholine release by nicotine, in *Nicotine*

(Ed. P. Stern), Academic Press, Bosnia and Hercegovina, Sarajevo, pp. 131–50.

Vizi, E. S., and Knoll, J. (1976). The inhibitory effect of adenenosine and related nucleotides on the release of acetylcholine. *Neuroscience*, **1**, 391–8.

Vizi, E. S., and Palkovits, M. (1977). Acetylcholine content in different regions of the rat brain. *Brain Res. Bull.*, **3**, 93–6.

Vizi, E. S., and Vyskocil, F. (1979). Changes in total and quantal release of acetylcholine in the mouse diaphragm during activation and inhibition of membrane ATPase. *J. Physiol., (Lond.)*, **286**, 1–14.

Vizi, E. S., and Ludvig, N. (1983). (-)-Amidephrine, a selective agonist for α_1-adrenoceptors, *Eur. J. Pharmacol.*, **91**, 377–81.

Vizi, E. S., and Volbekas, V. (1980a). Inhibition by dopamine of oxytocin release from isolated posterior lobe of the hypophysis of the rat, disinhibitory effect of β-endorphin/enkephalin. *Neuroendocr.*, **31**, 46–52.

Vizi, E. S., and Volbekas, V. (1980b). Disinhibitory role of endorphins/enkephalins in the release of oxytocin and vasopressin controlled by dopamine: effects on axon terminals, in *Neuropeptides and Neural Transmission* (Eds C. Ajmone Marsan and W. Z., Traczyk), Raven Press, New York, pp. 257–70.

Vizi, E. S., and Hársing, L. (1980c). Presynaptic inhibitory effect of noradrenaline and dopamine in the central nervous system and the role of membrane ATPase, in *Presynaptic Receptors* (Ed. S. Z. Langer), Pergamon Press, Oxford, pp. 151–7.

Vizi, E. S., and Hámori, J. (1981). Acetylcholine release from vertebrate central nervous system: comparative study, in *Neurotransmitters in Invertebrates* (Ed. K. S. Rózsa), Pergamon Press, Oxford, pp. 145–57.

Vizi, E. S., and Pásztor, E. (1981). Release of acetylcholine from isolated human cortical slices: inhibitory effect of norepinephrine and phenytoin. *Experimental Neurology*, **73**, 144–53.

Vizi, E. S., and Stjärne, L. (1981). Concluding remarks on the symposium Neurochemical Transmission–Modulation, in *Physiology of Excitable Membranes* (Ed. J. Salánki), Pergamon Press, Oxford, pp. 287–94.

Vizi, E. S., Bertaccini, G., Impicciatorre, M., and Knoll, J. (1972). Acetylcholine-releasing effect of gastrin and related polypeptides. *Eur. J. Pharmacol.*, **17**, 175–8.

Vizi, E. S., Bertaccini, M. D., Impicciatori, J., and Knoll, J. (1973a). Evidence that acetylcholine released by gastrin and related polypeptides contributes to their effect on gastrointestinal motility. *Gastroenterology*, **64**, 268–77.

Vizi, E. S., Somogyi, G. T., Hadházy, P., and Knoll, J. (1973b). Effect of duration and frequency of stimulation on the presynaptic inhibition by α-adrenoceptor stimulation of the adrenergic transmission. *Naunyn-Schmiedeberg's Arch. Pharmacol.* **280**, 79–91.

Vizi, E. S., Rónai, A. Z., and Knoll, J. (1974). The inhibitory effect of dopamine on acetylcholine release, *Naunyn-Schmiedeberg's Arch. Pharmacol.*, **284**, R86.

Vizi, E. S., Török, T. L., and Knoll, J. (1976). Calcium dependent depolarization by PGE$_1$ on intestinal smooth muscle, in *Advances in Prostaglandin and Tromboxane Research* (Eds B. Samuelsson and R. Paoletti), Raven Press, New York, pp. 884–5.

Vizi, E. S., Hársing, L. G., Jr, and Knoll, J. (1977a). Presynaptic inhibition leading to disinhibition of acetylcholine release from interneurons of the caudate nucleus: effects of dopamine β-endorphin and D-Ala^2-Pro^5-enkephalinamide, *Neuroscience*, **2**, 953–61.

Vizi, E. S., Hársing, L. G., Jr., and Knoll, J. (1977b). Inhibitory effect of dopamine on acetylcholine release from caudate nucleus. *Pol. J. Pharmac.*, **29**, 201–11.

Vizi, E. S., Rónai, A. Z., Hársing, L. G. Jr, and Knoll, J. (1977c). Presynaptic modulation by norepinephrine and dopamine of acetylcholine release in the peripheral and central nervous system, in *Cholinergic Mechanisms and Psychopharmacology* (Ed. D. J. Jenden), Plenum Press, New York, pp. 587–603.

Vizi, E. S., Hársing, L. G. Jr, and Zsilla, G. (1981a). Evidence of the modulatory role of serotonin in acetylcholine release from striatal interneurons. *Brain Res.*, **212**, 89–99.

Vizi, E. S., Somogyi, G. T., and Magyar, K. (1981b). Evidence that morphine and opioid peptides do not share a common pathway with adenosine in inhibiting acetylcholine release from isolated intestine. *J. Auton. Pharmac.*, **1**, 413–19.

Vizi, E. S. Török, T., Seregi, A., Serfözö, P. and Adam-Vizi V. (1982). Na-K-activated ATPase and the release of acetylcholine and noradrenaline. *J. Physiol. (Paris)*, **78**, 399–406.

Vizi, E. S., Gyires, K., Somogyi, G. T., and Ungváry, G. (1983a). Evidence that transmitter can be released from regions of the nerve cell other than presynaptic axon terminal: axonal release of acetylcholine without modulation. *Neuroscience*, **10**, 967–72.

Vizi, E. S., Somogyi, G. T., and Magyar, K. (1983b). Presynaptic control by adenosine of acetylcholine release: inhibitory effect of norepinephrine and opioid peptides as an independent action, in *Physiology and Pharmacology of Adenosine* (Eds J. W. Daly, Y. Kuroda, J. W. Phillis, H. Shimizu, and M. Ui), Raven Press, New York, pp. 209–17.

Vizi, E. S., Ludvig, N., Rónai, A. Z. and Folly, G. (1983c). Dissociation of presynaptic α2-adrenoceptors following prazosin administration: presynaptic effect of prazosin. *Eur. J. Pharmacol.* **95**, 287–90.

Vizi, E. S., Török, T., and Magyar K. (1984). Effect of potassium on the release of ³H-noradrenaline from rabbit and human pulmonary artery. *J. Neurochem.*, **42**, 670–6.

Von Graffenried, B., mel Pozo, E., Roubicek, J., Krebs, E., Pöldinger, W., Burmeister, P. and Kerp, L. (1978). Effects of the synthetic enkephalin analogue FK-33-824 in man. *Nature*, **272**, 729–30.

Votavova, M., Boullin, D. J., and Costa, E. (1971). Specificity of action of 6-hydroxydopamine in peripheral cat tissues: depletion of noradrenaline without depletion of 5-hydroxy-tryptamine. *Life Sciences*, **10**, 87–91.

Vyklicky, L., Sykova, E., and Kriz, N. (1975). Slow potentials induced by changes of extracellular potassium in the spinal cord of cat. *Brain Res.*, **87**, 77–80.

Vyklicky, L., Sykova, E., Kriz, N., and Ujec, E. (1972). Post-stimulation changes of extracellular potassium concentration in the spinal cord of rat. *Brain Res.*, **45**, 608–11.

Wagner, H., Haberle, H., Maier, V., and Lang, R. E. (1978). Transmitter mediated arginine vasopressin release from superfused hypothalamus and pituitary gland. *J. Endocrinol. Invest.*, **1**, 215–20.

Wahl, M., and Kushinsky, W. (1976). The dilatatory action of adenosine on pial arteries of cats and its inhibition by theophylline. *Pflügers Arch.*, **362**, 55–9.

Wakade, A. R. (1980). Modulation of ³H-norepinephrine release in the rat hypothalamus and cortex by adenosine, in *Presynaptic Receptors* (Ed. S. Z. Langer), Pergamon Press, Oxford and New York, pp. 377–83.

Wakade, A. R. (1981). Facilitation of secretion on catecholamines from rat and guinea-pig adrenal glands in potassium-free medium or after ouabain. *J. Physiol. (Lond.)*, **313**, 481–98.

Wakade, A. R., and Wakade, T. D. (1977a). Another endogenous modulator or sympathetic neurotransmission at the synaptic level, adenosine. *Pharmacologist*, **19**, 129.

Wakade, A. R., and Wakade, T. (1977b). Factors controlling the output of norepinephrine from the rat vas deferens. *Fedn Proc.*, **36**, 327.

Wakade, A. R., and Wakade, T. D. (1978). Inhibition of noradrenaline release by adenosine. *J. Physiol. (Lond.)*, **282**, 35–49.

Waterfield, A. A., and Kosterlitz, H. W. (1975). Stereospecific increase by narcotic antagonists of evoked acetylcholine output in guinea-pig ileum. *Life Sci.*, **16**, 1777–92.

Waterfield, A. A., Smockum, O. W. J., Hughes, J., Kosterlitz, H. W., and Henderson, G. (1977). *In vitro* pharmacology of opioid peptides, enkephalins and endorphins. *Eur. J. Pharmacol.*, **43**, 107–16.

Wallis, D. I., and Woodward, B. (1975). Membrane potential changes induced by 5-hydroxytryptamine in the rabbit superior cervical ganglion. *Br. J. Pharmacol.*, **55**, 199–212.

Watkins, L. R., and Mayer, D. J. (1982). Organization of endogenous opiate and nonopiate pain control systems. *Science*, **216**, 1185–92.

Watson, S. J., Berger, P. A., Akil, H., Mills, M. J., and Barchas, J. D. (1978). Effect of naloxone on schizophrenia: reduction in hallucinations in a subpopulation of subjects. *Science*, **201**, 73–5.

Watt, A. J. (1971). The inhibitory effect of perivascular stimulation on the response of the guinea-pig isolated ileum. *J. Physiol. (Lond)*, **219**, 38–9.

Weight, F. F., and Erulkar, S. D. (1976). Modulation of synaptic transmitter release by repetitive postsynaptic action potentials. *Science*, **193**, 1023–5.

Weinrich, P., and Seeman, P. (1981). Binding of adrenergic ligands ³H-clonidine and ³H-WB-4101 to multiple sites in human brain. *Biochem. Pharm.*, **30**, 3115–20.

Weinreich, P., and Seeman, P. (1982). Effect of kainic acid on striatal dopamine receptors. *Brain Res.*, **198**, 491–6.

Weitzell, R., Tanaka, T., and Starke, K. (1979). Pre- and postsynaptic effects of yohimbine stereoisomers on noradrenergic transmission in the pulmonary artery of the rabbit. *Naunyn-Schmiedeberg's Arch. Pharmacol.*, **308**, 127&36.

Welsh, J. H., and Williams, L. D. (1970). Monoamine-containing neurons in planarea. *J. comp. Neurol.*, **138**, 103–16.

254

Wemer, J., and Mulder, A. (1981). Postnatal development of presynaptic alpha-adrenoceptors in rat cerebral cortex: studies with brain slices and synaptosomes. *Brain Res.*, **208**, 200–309.

Wemer, J., Frankhuyzen, A. L., and Mulder, A. H. (1982). Pharmacological characterization of presynaptic alpha-adrenoceptors in the nucleus tractus solitarii and the cerebral cortex of the rat. *Neuropharm*, **21**, 499–506.

Wemer, J., van der Lught, J. C., DeLangen, C. D., and Mulder, A. H. (1979). On the capacity of presynaptic alpha receptors to modulate norepinephrine release from slices of rat neocortex and the affinity of some agonists and antagonists for these receptors. *J. Pharmacol.*, **211**, 445–51.

Wennmalm, A. (1975). Prostaglandin release and mechanical performance in the isolated rabbit heart during induced changes in the internal environment. *Acta Physiol.*, **93**, 15–24.

Wennmalm, A., and Hedqvist, P. (1971). Inhibition by prostaglandin E_1 of parasympathetic neurotransmission in the rabbit heart. *Life Sci.*, **10**, 465–70.

Werner, U., Starke, K., and Schümann, H. J. (1972). Wirkung von Clonidin (ST 155) und Bay a 6781 auf das isolierte Kaninchenberz. *Naunyn-Schmiedeberg's Arch. Pharmacol.*, **266**, 474.

Westfall, T. C. (1974a). Effect of nicotine and other drugs on the release of ^{3}H-norepinephrine and ^{3}H-dopamine from rat brain slices. *Neuropharmac.*, **13**, 693–700.

Westfall, T. C. (1974b). Effect of muscarinic agonists on the release of ^{3}H-norepinephrine and ^{3}H-dopamine by potassium and electrical stimulation from rat brain slices. *Life Sci.*, **14**, 1641–52.

Westfall, T. C. (1974c). The effect of cholinergic agents on the release of ^{3}H-dopamine from the rat striatal slices by nicotine, potassium and electrical stimulation. *Fedn Proc.*, **33**, 524.

Westfall, T. C. (1977). Local regulation of adrenergic neurotransmission. *Physiol. Rev.*, **57**, 659–728.

Westfall, T. C., and Brasted, M. (1974). Specificity of blockade of the nicotine-induced release of ^{3}H-norepinephrine from adrenergic neurons of the guinea-pig heart by various pharmacological agents. *J. Pharmac. exp. Ther*, **189**, 659–64.

Westfall, T. C., and Kitay, D. (1977). The effect of prostaglandins on the release of ^{3}H-dopamine from superfused slices of rat striatum following electrical stimulation, *Proc. Soc. Exp. Biol. Med.*, **155**, 305–7.

Westfall, T. C., and Tittermary, V. (1982). Inhibition of the electrically induced release of ^{3}H-dopamine by serotonin from superfused rat striatal slices. *Neuroscience Lett.*, 28, 205–9.

Westfall, T. C., Kitay, D., and Whal, G. (1976). The effect of cyclic nucleotides on the release of ^{3}H-dopamine from rat striatal slices. *J. Pharmac. exp. Ther.*, **199**, 149–57.

Westfall, T. C., Peach, M. J., and Tittermary, V. (1979a). Enhancement of electrically induced release of norepinephrine from the rat portal vein: mediation by β_2-adrenoceptors. *Eur. J. Pharmacol.*, **58**, 67–74.

Westfall, T. C., Perkins, N. A., and Paul, C. (1979b). Role of presynaptic receptors in the synthesis and release of dopamine in the mammalian central nervous system, in *Presynaptic Receptors* (Eds S. Z. Langer, K. Starke, and M. L. Dubocovich), Pergamon Press, Oxford, pp. 243–8.

Westfall, D. P., Stitzel, R. E., and Rowe, J. N. (1978). The postjunctional effects and neural release of purine compounds in the guinea-pig vas deferens. *Eur. J. Pharmacol.*, **50**, 27–38.

White, T. D. (1977). Direct detection of depolarisation-induced release of ATP from a synaptosomal preparation. *Nature*, **267**, 67–8.

White, T. D. (1978). Release of ATP from a synaptosomal preparation by elevated extracellular K^+ and by veratridine. *J. Neurochem.*, **30**, 329–36.

White, T. D. (1982). Release of ATP from isolated myenteric varicosities by nicotinic agonists. *Eur. J. Pharmacol.*, **79**, 333–4.

White, T. D., and Leslie, R. A. (1982). Depolarization-induced release of adenosine 5-triphosphate from isolated varicosities derived from the myenteric plexus of the guinea-pig small intestine. *J. Neurosci.*, **2**, 206–15.

White, T. D., Potter, P., Moody, C., and Burnstock, G. (1981). Tetrodotoxin-resistant release of ATP from guinea-pig taenia coli and vas deferens during electrical field stimulation in the presence of luciferin–luciferase. *Can. J. Physiol. Pharm.*, **59**, 1094–1100.

White, T., Potter, P., and Wonnacott, S. (1980). Depolarisation induced release of ATP from cortical synaptosomes is not associated with acetylcholine release. *J. Neurochem.*, **34**, 1109–12.

Wikberg, J. E. S. (1977). Localization of adrenergic receptors in guinea-pig ileum and rabbit jejunum to cholinergic neurons and to smooth muscle cells. *Acta Phsyiol. Scand.*, **99**, 190–207.

Wikberg, J. E. S. (1978a). Pharmacological classification of adrenergic α-receptors in the guinea-pig ileum. *Nature*, **273**, 164–6.

Wikberg, J. E. S. (1978b). Differentiation between pre- and postjunctional α-receptors in guinea-pig ileum and rabbit aorta. *Acta Physiol. Scand.*, **103**, 225–39.

Wikberg, J. E. S. (1979). The pharmacological classification of adrenergic α_1 and α_2 receptors and their mechanisms of action. *Acta Physiol. Scand. Suppl.*, **468**, 1–99.

Wilson, D. F. (1973). Effects of caffeine on neuromuscular transmission in the rat. *Am. J. Physiol.*, **225**, 862.

Wilson, D. F. (1974). The effects of dibutyryl cyclic adenosine 3, 5 monophosphate, theophylline and aminophylline on neuromuscular transmission in the rat, *J. Pharmac. exp. Ther.*, **188**, 447.

Wilson, D. F. (1982). Influence of presynaptic receptors on neuromuscular transmission in rat. *Am. J. Physiol.*, **242**, 366–72.

Wilson, C. J., Groves, P. M., and Fifkova, K. (1977). Monoaminergic synapses including dendro–dendritic synapses in the rat substantia nigra. *Exp. Brain Res.*, **30**, 161–74.

Wilson, S. P., Klein, R. L., Chang, K. J., Gasparis, M. S., Viveros, O. H., and Yang, W. H. (1980). Are opioid peptides co-transmitters in noradrenergic vesicles of sympathetic nerves? *Nature*, **288**, 707–9.

Wood, J. D. (1975). Neurophysiology of Auerbach plexus and control of intestinal motility. *Physiol. Rev.*, **55**, 307–24.

Wood, J. D. (1979). Neurophysiology of the enteric nervous system, in *Integrative Functions of the Autonomic Nervous System* (Eds C. McBrooks, K. Koizumi, and A. Sato), Elsevier/North-Holland, Amsterdam, pp. 177–93.

Wood, J. D. (1981). Intrinsic neural control of intestinal motility. *Ann. Rev. Physiol.*, **43**, 39–51.

Wood, J. D., and Mayer, C. J. (1979). Serotonergic activation of tonic-type enteric neurons in guinea-pig small bowel. *J. Neurophysiol.*, **42**, 582–92.

Wood, C. L., Arnett, C. D., Clarke, W. R., Tsai, B. S., and Lefkowitz, R. J. (1979). Subclassification of alpha-adrenergic receptors by direct binding studies. *Biochem. Pharmacol.*, **28**, 1277–82.

Woodson, P. B. J., Schlapfer, W. T., Tremblay, J. P., and Barondes, S. H. (1975). Cholinergic agents effect two receptors that modulate transmitter release at a central synapse in *Aplysia californica*. *Brain Res.*, **88**, 455–74.

Wüster, M., Schulz, R., and Herz, A. (1978). Specificity of opioids towards the μ, δ and ε opiate receptors. *Neuroscience Letters*, **15**, 193–8.

Wüster, M., Rubini, P., and Schule, R. (1981). The preference of putative pro-enkephalins for different types of opiate receptor. *Life Sci.*, **29**, 1219–27.

Yaksh, T. L., and Yamamura, H. I. (1975). Blockade by morphine of acetylcholine release from the caudate nucleus in midpontine pretrigeminal cat. *Brain Res.*, **83**, 520–4.

Yaksh, T. L., Jessell, T. M., Gramse, R., Mudge, A. W., and Leeman, S. E. (1980). Intrathecal morphine inhibits substance P release from mammalian spinal cord *in vivo*. *Nature*, **286**, 155–7.

Yamada, M., Tokimasa, T., and Koketsu, K. (1982). Effects of histamine on acetylcholine release in bullfrog sympathetic ganglia. *Eur. J. Pharmacol.*, **82**, 15–20.

Yamaguchi, N., De Champlain, J., and Nadeau, R. A. (1977). Regulation of norepinephrine release from cardiac sympathetic fibers in the dog by presynaptic alpha and beta receptors. *Circ. Res.*, **41**, 108–17.

Yamamura, H. I., and Snyder, S. H. (1974a). Postsynaptic localization of muscarinic cholinergic receptor binding in rat hippocampus. *Brain Res.*, **78**, 320–6.

Yamamura, H. I., and Snyder, S. H. (1974b). Muscarinic cholinergic binding in rat brain. *Proc. Natl Acad. Sci., USA*, **71**, 1725–9.

Yang, H. Y. T., Fratta, W., Hong, J. S., DiGuilio, A. M., and Costa, E. (1978). Detection of two endorphin like peptides in nucleus caudatus. *Neuropharmacology*, **17**, 433–8.

Yau, W. M. (1978). Effect of substance P on intestinal muscle. *Gastroenterology*, **74**, 228–31.

Yau, W. M., and Youther, M. L. (1982). Direct evidence for a release of acetylcholine from the myenteric plexus of guinea-pig small intestine by substance P. *Eur. J. Pharmacol.*, **81**, 665–8.

Yoshida, H., Nagal, K., Ohashi, T., and Nakagawa, Y. (1969). K^+-dependent phosphatase activity observed in the presence of both adenosine triphosphate and Na^+; *Biochem. Biophys. Acta*, **171**, 178–85.

Young, W. S., and Kuhar, M. J. (1979). Noradrenergic α_1 and α_2 receptors—autoradiographic visualization. *Eur. J. Pharmacol*, **59**, 317–19.

Zetler, G. (1956). Substance P, ein Polypeptid aus Darm und Gehirn mit depressiven hyperalget-

256

ischen und Morphin-antagonistischen Wirkungen auf das Zentral-nerve system. *Naunyn-Schmiedeberg's Arch. Pharmacol.*, **228**, 438–513.

Zieglgänsberger, W., and Bayerl, H. (1976). The mechanism of inhibition of neuronal activity by opiates in the spinal cord of cat. *Brain Res.*, **115**, 111–28.

Zieglgänsberger, W., and Fry, J. (1978). Actions of opioids on single neurones, in *Developments in Opiate Research* (Ed. A Herz), Marcel Dekker, New York, pp. 193–239.

Zieglgänsberger, W., and Puil, E. (1973). Actions of glutamic acid on spinal neurones. *Exp. Brain Res.*, **17**, 35–49.

Zieglgänsberger, W., and Tulloch, I. F. (1979). The effects of methionine– and leucine–enkephalin on spinal neurones of the cat. *Brain Res.*, **167**, 53–64.

Zieglgänsberger, W., French, E. D., Siggins, G. R., and Bloom, F. E. (1979). Opioid peptides may excite hippocampal pyramidal neurons by inhibiting adjacent inhibitory interneurons. *Science*, **205**, 415–17.

Zieher, L. M., and De Robertis, E. (1963). Subcellular localization of 5-hydroxytryptamine in rat brain. *Biochem. Pharmac.*, **13**, 596–8.

Zieher, L. M., and Jaim-Etcheverry, G. (1971). Ultrastructural cytochemistry and pharmacology of 5-hydroxytrytamine on adrenergic nerve endings. II. Accumulation of 5-hydroxytryptamine in nerve vesicles containing norepinephrine in rat vas deferens. *J. Pharmac. exp. Ther.*, **178**, 30–41.

Zimmermann, B. G. (1978). Actions of angiotensin on adrenergic nerve endings. *Fedn Proc.*, **37**, 199–202.

Zimmerman, B. G., and Whitmore, L. (1967). Effect on angiotensin and phenoxybenzamine on release of norepinephrine in vessels during sympathetic nerve stimulation. *Int. J. Neuropharmacol.*, **6**, 27–31.

Zimmerman, B. G., and Kraft, E. (1979). Blockade by saralasin of adrenergic potentiation induced by renin–angiotensin system. *J. Pharmacol. exp. Ther.*, **210**, 101–5.

Zimmerman, H., and Whittaker, V. P. (1974). Effect of electrical stimulation on the yield and composition of synaptic vesicles from the cholinergic synapses of the electric organ of Torpedo: a combined biochemical, electrophysiological and morphological study. *J. Neurochem.*, **22**, 435–50.

Zukin, R. S., and Zukin, S. R. (1981). Multiple opiate receptors: emerging concepts. *Life Sci.*, **29**, 2681–91.

Zséli, J., Török, T., Vizi, E. S., and Knoll, J. (1979). Effect of prostaglandin E1 and indomethacin on responses of longitudinal muscle of guinea-pig ileum to cholecystokinin. *Eur. J. Pharmacol.*, **56**, 139–44.

Zsilla, G., Cheney, D. L., and Costa, E. (1976). Regional changes in the rate of turnover of acetylcholine in rat brain receiving diazepam or muscimol. *Naunyn-Schmiedeberg's Arch. Pharmacol.*, **294**, 251–5.

Index